LANDSCAPE PLANNING

Edited by **Murat Özyavuz**

Landscape Planning

Edited by Murat Özyavuz

CBS, Edition **2017**

Published by InTech

Janeza Trdine 9, 51000 Rijeka, Croatia

Notice

Publishing Process Manager Marina Jozipovic

Technical Editor Teodora Smiljanic

Cover Designer InTech Design Team

Additional hard copies can be obtained from orders@intechopen.com

Landscape Planning, Edited by Murat Özyavuz
p. cm.
ISBN 978-953-51-0654-8

Contents

Preface

The purpose of this book is to reveal of the landscape planning and design in recent years. For this purpose, chapters were selected on the topics of different landscape architecture study. Landscape architecture is the design of outdoor and public spaces to achieve environmental, socio-behavioral, and/or aesthetic outcomes. It involves the systematic investigation of existing social, ecological, and geological conditions and processes in the landscape, and the design of interventions that will produce the desired outcome. The scope of the profession includes: urban design; site planning; town or urban planning; environmental restoration; parks and recreation planning; visual resource management; green infrastructure planning and provision; and private estate and residence landscape master planning and design - all at varying scales of design, planning and management. Landscape planning is the key planning instrument for nature conservation and landscape management. Apart from the landscape plans at the local level, i.e. municipal level, there are landscape structure plans (Landschaftsrahmenplane) at the district or planning region level (Regierungsbezirk, Landkreis, Planungsregion) and the regional landscape programme (Landschaftsprogramm) for an entire regional state (Land). The local landscape plans (Artliche Landschaftsplane) are based on the specifications contained in the regional landscape programme and the landscape structure plans. At all levels landscape planning makes an important long-term contribution to the conservation of natural resources. It not only addresses the narrower areas of particularly valuable protected sites, but also devises strategies for full-coverage, sustainable conservation and the long-term development of nature and landscapes.

This book is for landscape architects and other planning professions. Theoretical foundations, theories, methods, and applications will be essential parts of this reference book. In addition, this book addresses several very different subjects of study; landscape management, biodiversity, landscape restoration, landscape design, and urban design related to theory, practice and the results will be covered. Due to the varied usage of the term Planning/Landscape Planning, the intended readership for this book is a broad audience including environmentalists, landscape architects, architects, environmentalists, botanists, urban and regional planners, government agencies, non-governmental organizations, agricultural organizations, students at all levels, research organizations, international organizations and all interested parties.

I would like to express my deep sense of gratitude and indebtedness to all the authors for their valuable contributions and also to the researchers who actually performed experiments and reported their findings. I must confess that it had been a rare privilege for me to be associated with InTech publishers. Thanks is the least word to offer to Ms. Marina Jozipovic, Ms. Sasa Leporic and Mr. Metin Ertufan, Intech Publishing Process Managers, yet I shall avail this opportunity to extend my sincere gratitude for their help and co-operation at various phases of book publication. Last but not least I express my sincere thanks and affection to my wife Ayten Özyavuz and my daughter Ayza Özyavuz for their patience and cheerfulness. I hope this book will be beneficial to the scientific world.

Assist. Prof. Dr. Murat Özyavuz
Namık Kemal University
Faculty of Agriculture
Department of Landscape Architecture
Tekirdağ, Turkey

Section 1

Landscape Planning

Protected Areas

Murat Özyavuz
*Namık Kemal University, Faculty of Agriculture,
Department of Landscape Architecture
Turkey*

1. Introduction

Protected areas are essential for biodiversity conservation. They are the cornerstones of virtually all national and international conservation strategies, set aside to maintain functioning natural ecosystems, to act as refuges for species and to maintain ecological processes that cannot survive in most intensely managed landscapes and seascapes. Protected areas act as benchmarks against which we understand human interactions with the natural world. Today they are often the only hope we have of stopping many threatened or endemic species from becoming extinct (Dudley, 2008).

The original intent of the IUCN Protected Area Management Categories system was to create a common understanding of protected areas, both within and between countries. This is set out in the introduction to the Guidelines by the then Chair of CNPPA (Commission on National Parks and Protected Areas, now known as the World Commission on Protected Areas), P.H.C. (Bing) Lucas who wrote: *"These guidelines have a special significance as they are intended for everyone involved in protected areas, providing a common language by which managers, planners, researchers, politicians and citizens groups in all countries can exchange information and views"* (The International Union For Conservation of Nature [IUCN], 1994).

IUCN defines a protected area as:

"An area of land and/or sea especially dedicated to the protection of biological diversity, and of natural and associated cultural resources, and managed through legal or other effective means" (IUCN 1994).

Protected areas can be categorized into six types, according to their management objectives (IUCN, 1994; 2003):

Category I

Protected area managed mainly for science or wilderness protection (I(a) Strict Nature Reserves, and I(b) Wilderness Areas).

An area of land and/or sea possessing some outstanding or representative ecosystems, geological or physiological features and/or species available primarily for research and/or environmental monitoring. A wilderness area is a large area of unmodified or slightly modified land and/or sea retaining its natural character and influence without permanent or significant habitation which is protected and managed so as to preserve its natural condition.

Category II

Protected area managed mainly for ecosystem protection and recreation (National Park).

A natural area of land and/or sea designated to (a) protect the ecological integrity of one or more ecosystems for present and future generations; (b) exclude exploitation or occupation inimical to the purposes of the area; and (c) provide foundation for spiritual, scientific, educational, recreational, and visitor opportunities all of which must be environmentally and culturally compatible.

Category III

Protected area managed mainly for conservation of specific natural features (Natural Monument).

An area containing one or more specific natural or natural/cultural feature which is of outstanding or unique value because of its inherent rarity, representative or aesthetic qualities or cultural significance.

Category IV

Protected area managed mainly for conservation through management intervention.

An area of land and/or sea subject to active intervention for management purposes so as to ensure the maintenance of habitats and/or to meet the requirements of specific species.

Category V

Protected area managed mainly for landscape/seascape conservation and recreation (Protected Landscape/Seascape).

An area with coast and sea, as appropriate, where the interaction of people and nature over time has produced an area with significant aesthetic, ecological and/or cultural value and often with high biological diversity. Safeguarding the integrity of this traditional interaction is vital to the protection, maintenance and evolution of such an area.

Category VI

Protected area managed mainly for the sustainable use of natural ecosystems (Managed Resource Protected Area)

An area containing predominantly unmodified natural systems managed to ensure long term protection and maintenance of biological diversity while providing at the same time a sustainable flow of natural products and services to meet community needs.

It is sometimes assumed that protected areas must be in conflict with the rights and traditions of indigenous and other traditional peoples on their terrestrial, coastal/marine, or freshwater domains. In reality, where indigenous peoples are interested in the conservation and traditional use of their lands, territories, waters, coastal seas and other resources, and their fundamental human rights are accorded, conflicts need not arise between those peoples' rights and interests, and protected area objectives. Moreover, formal protected areas can provide a means to recognize and guarantee the efforts of many communities of indigenous and other traditional peoples who have long protected certain areas, such as sacred groves and mountains, through their own cultures (IUCN, 2000).

Based on the advice in the protected areas management categories, on established WWF and IUCN policies on indigenous peoples and conservation, and on conclusions and recommendations of the IV World Congress on National Parks and Protected Areas, the two organizations, WWF and IUCN/WCPA, have adopted principles and guidelines concerning indigenous rights and knowledge systems, consultation processes, agreements between conservation institutions, decentralization, local participation, transparency, accountability, sharing benefits and international responsibility. The five principles are as follows (IUCN, 2000):

Principle 1.

Indigenous and other traditional peoples have long associations with nature and a deep understanding of it. Often they have made significant contributions to the maintenance of many of the earth's most fragile ecosystems, through their traditional sustainable resource use practices and culture-based respect for nature. Therefore, there should be no inherent conflict between the objectives of protected areas and the existence, within and around their borders, of indigenous and other traditional peoples.

Principle 2.

Agreements drawn up between conservation institutions, including protected area management agencies, and indigenous and other traditional peoples for the establishment and management of protected areas affecting their lands, territories, waters, coastal seas and other resources should be based on full respect for the rights of indigenous and other traditional peoples to traditional, sustainable use of their lands, territories, waters, coastal seas and other resources.

Principle 3.

The principles of decentralization, participation, transparency and accountability should be taken into account in all matters pertaining to the mutual interests of protected areas and indigenous and other traditional peoples.

Principle 4.

Indigenous and other traditional peoples should be able to share fully and equitably in the benefits associated with protected areas, with due recognition to the rights of other legitimate stakeholders.

Principle 5.

The rights of indigenous and other traditional peoples in connection with protected areas are often an international responsibility, since many of the lands, territories, waters, coastal seas and other resources which they own or otherwise occupy or use cross national boundaries, as indeed do many of the ecosystems in need of protection.

Financial Planning In Protected Areas

A financial plan is a tool which helps to determine the protected area's funding requirements, and to match income sources with those needs. Financial planning differs from a budget in that, in addition to identifying how much money is needed for different types of activities, it also identifies the most appropriate funding sources for short, medium, and long-term needs. (IUCN, 2001)

Seven steps are required to develop a financial plan:

1. define protected area goals and objectives;
2. identify the existing customer base;
3. list financial resources and demands on these resources;
4. identify new customers and relative levels of use versus contribution;
5. identify mechanisms to capture income from customers;
6. evaluate the feasibility of the proposed mechanisms; and
7. clearly state the financial plan.

Protected Area Economic Benefits

A protected area also provides its customers with a number of goods and services. These could include goods such as thatching grasses, wild berries and genetic materials, and services such as biodiversity conservation, crop pollination, water purification, game viewing and recreational opportunities. Such goods and services provide society with a stream of benefits from the existence of the protected area. The benefits can be divided into two categories: so-called 'use' (comprising direct and indirect values) and 'non-use' (comprising option, bequest and existence values) benefits (IUCN, 2001).

The structure of an ecosystem includes the species contained therein, their mass, their arrangement, and other relevant information. This is the ecosystem's standing stock—nature's free goods. The functions of an ecosystem, on the other hand, are characterized by the ways in which the components of the system interact. They provide nature's free services, maintaining clean air, pure water, a green earth, and a balance of creatures, enabling humans to obtain food, fiber, energy, and other material needs for survival. Evaluating the contribution of ecosystem functioning to human welfare is a complex task, involving human social values and political factors

Direct use values of protected areas derive from the actual use of the protected area for such activities as recreation, tourism, the harvesting of various natural or cultural resources, hunting and fishing, and educational services. Conversely, *indirect use value* sderive from the goods and services not directly provided by visits to protected areas. Notably these include ecological functions such as watershed protection, the provision of breeding or feeding habitat, climatic stabilization and nutrient recycling. Such indirect use values are often widespread and significant, but have been under-valued, if not totally ignored by past economic valuation systems. Indeed, most of the studies that have attempted to value these indirect goods and services have found that they have far greater value than the more easily measured direct values (Figure 1).

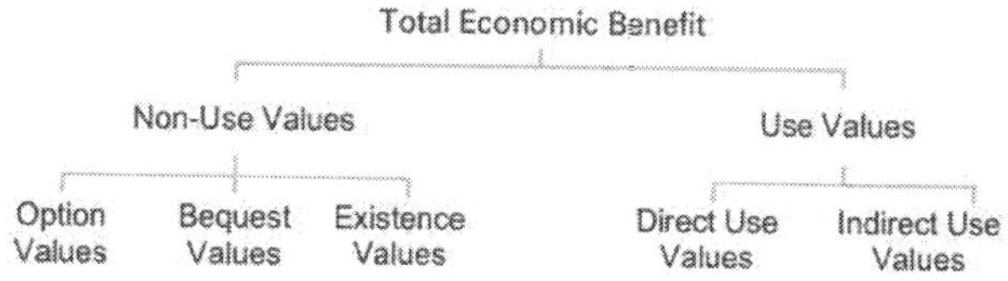

Fig. 1. Total economic benefit of protected areas (IUCN, 2001).

Option value refers to the potential for individuals or society to use the protected area in the future. For example, many people value a particular protected area even though they have never visited the park, but feel that at some future date they might like to do so.

Bequest value relates to the benefit of knowing that others (e.g. children or grandchildren) benefit or will benefit from the goods and services provided by the protected area. Finally, *existence value* derives from the benefit of knowing that the protected area exists and provides valuable goods and services. Even if they do not plan on ever visiting a particular protected area or protected area system, many people attach value to the more existence of such sites (e.g. for the indirect benefits they provide or as sources of local or national pride). (IUCN, 2001)

2. Protected areas in Turkey

In terms of biodiversity, Turkey is one of the richest countries in Europe and the Middle East, and ranks the ninth on the European Continent in this respect. There are a number of different ecological regions each with its own endemic species and natural ecosystems. The richness of biodiversity in Turkey is expressed in its 120 mammals, more than 400 bird species, 130 reptiles, and nearly 500 fish species. The diversity of the geographic formations of Turkey and its location at the intersection of two important Vavilovian gene centers (the Mediterranean and the Near Eastern) are the reasons for high endemism and genetic diversity (Ministry of Environment, 2002).

There have been various types of habitats formed in the earth since the beginning of the world and existence of the living beings. Human beings, animals, plants and microorganisms have been surviving in the ecosystems together for many years together with the non-living beings, like water, air, soil, rock and climatologically factors. However, due to technological developments starting from 1960's, there have been significant adverse impacts on the nature. Man can survive less dependent on the surrounding factors and has the ability to easily change the environmental factors with his technological power. The ecological balances have been greatly degraded due to increase in populations and rapidly developing technologies. In this regard, Turkey is relatively lucky when compared to the most of the countries in Europe and America. In Turkey, there are still number of ecosystems where natural balance has not been completely degraded and we still have a rich biodiversity throughout Turkey.

Turkey is home to 75% of the plant species that exist on the European continent, and one third of these species are endemic plants. The rich flora of Turkey includes more than 9,000 plant species and more than 500 bulbous plants. This flora, with a high endemism ratio, is also rich in medicinal and aromatic plants (Ministry of Environment, 2002). Most of the endemic plant species are found in the Taurus Mountains, the Nur Mountains and the Eastern Black Sea Coast (Ministry of Environment, 2001).

Located on the migration routes of many birds, Turkey is a key country for many bird species. 454 bird species have been sited. Several of its species are globally under threat (Ministry of Environment, 2002). Turkish wetlands are of crucial importance for many breeding species of birds.

There are 472 fish species in Turkey and 50 of these are at risk of extinction. Some 192 freshwater fish species belonging to 26 different families have been identified (Ministry of Environment,2002).

Approximately 3,000 plant and animal species have been identified in Turkey's seas (Ministry of Environment, 2001). There are about 20 species of mammals including the Mediterranean monk seal, whales and dolphins with mostly decreasing populations. The Turkish Straits and the Sea of Marmara form a special ecosystem (an ecotone) between the Mediterranean and the Black Sea. The Aegean Sea is especially important for the endangered Mediterranean monk seal (*Monachus monachus*), which is considered to be one of the 12 most endangered species in the world. Less than 50 specimens inhabit the coasts of Turkey (Ministry of Environment, 2001). The Aegean Sea and its islands contain numerous microhabitats (*Posidonia oceanica* and *Cystoseira species*) that play an important role in the sustainability of the ecosystem (Ministry of Environment, 2002).

Turkey has accepted the Action Plan (1989 and 1999) for the conservation of Mediterranean marine turtles within the framework of the Barcelona Convention. Several breeding habitats of marine turtles, including Dalyan, Fethiye, Patara, Goksu Delta, and Belek, were declared as Specially Protected Areas in 1988 and 1990. The Ministry of Environment established the Marine Turtles National Commission and the Marine Turtles Scientific Commission for the coordination of activities towards the protection of the two species. Turkey also accepted the action plan for the conservation of the Mediterranean monk seal, again developed in the framework of the Barcelona Convention (Ministry of Environment, 2002). In this context, Turkey has signed many international conventions and agreements.

In this context, Turkey has signed many international conventions. These conventions are;

International Conventions and Protocols on Nature Protection Ratified by Turkey

- Convention on Biodiversity Conservation (Rio Convention) (1997)
- Cartagena Protocol (2004)
- CITES (1996)
- Barcelona Convention (1988)
- Bucharest Convention (1994)
- Protection of Cultural and National Heritage (1983)
- Convention on Combating Erosion (1998)
- European Landscape Convention (2000)
- Bern Convention(1984)
- Ramsar Convention (1994)
- Kyoto Protocol (2009)

Depending on these conventions, by 2011, nearly 1800 sites had been identified by the Ministry of Forest and Water as warranted protection under the 1983 law (Table 1), by 2003, nearly 6 400 sites had been identified by the Ministry of Culture as warranted protection under the 1983 law (Table 2).

Conservation Status	Number	Related Law
National park	41	Law on National Parks
Nature conservation area	31	Law on National Parks
Natural monument	106	Law on National Parks
Nature park	41	Law on National Parks
Wild life reserve areas	79	Law on Terrestrial Hunting
Conservation forest	57	Law on Forest
Genetic conservation areas	214	Law on Forest
Seed stands	339	Law on Forest
Specially protected areas (SPAs)	14	Law on Environment
Natural sites	947	Law on Conservation of Cultural And Natural Heritage
Ramsar sites	13	Ramsar Convention By-law on Conservation of Wetlands
Biosphere Reserve	1	Law on National Parks -Law on Forest

Table 1. Protected areas which identified by the Ministry of Forest and Water.

Conservation Staus	Number
Archaeological Site	4,920
Natural Site	787
Urban Archaeological Site	182
Historical Site	121
Other Sites	371
Total Number	6,381

Table 2. Protected areas (especially cultural areas) which identified by the Ministry of Forest and Water.

National Parks

A national park refers to an plot of land set aside by a national government and usually designated as an area free of development. Often, national parks include pristine wilderness areas or other pieces of environmental heritage which the nation has deemed worthy of preservation. In the United States, national parks also include historic areas and monuments to scientific achievement.

Prepared by the IUCN classification of protected areas, national parks are in Categories 2. Definition of this category is below;

'Natural area of land and/or sea, designated to (a) protect the ecological integrity of one or more ecosystems for present and future generations, (b) exclude exploitation or occupation inimical to the purposes of designation of the area and (c) provide a foundation for spiritual, scientific, educational, recreational and visitor opportunities, all of which must be environmentally and culturally compatible.

Management Objectives of This Category

- To protect natural and scenic areas of national and international significance for spiritual, scientific, educational, recreational or tourist purposes
- To perpetuate, in as natural a state as possible, representative examples of physiographic regions, biotic communities, genetic resources, and species, to provide ecological stability and diversity;
- To manage visitor use for inspirational, educational, cultural and recreational purposes at a level which will maintain the area in a natural or near natural state
- To eliminate and thereafter prevent exploitation or occupation inimical to the purposes of designation;
- To maintain respect for the ecological, geomorphological, sacred o aesthetic attributes which warranted designation; and
- To take into account the needs of indigenous people, including subsistence resource use, in so far as these will not adversely affect the other objectives of management.

Guidance for Selection

- The area should contain a representative sample of major natural regions, features or scenery, where plant and animal species, habitats and geomorphological sites are of special spiritual, scientific, educational, recreational and touristic significance
- The area should be large enough to contain one or more entire ecosystems materially altered by current human occupation or exploitation.

National parks are natural areas that provide transcendental, adventure and educational experiences. One management goal, however, is to take into account the needs of indigenous people. In this way, parks serve multiple constituencies that have sometimes been at loggerheads (Weeks and Mehta, 2004).

National Parks Law in Turkey, scientific and aesthetic terms, national and international rare, natural and cultural resource values and conservation, recreation and tourism will have the values of nature.

The purpose of this Law is specified as the "identification of areas which possess values of national and international importance, as national park, nature park, nature monument, and nature protection area, and the protection, enhancement and management of these areas without degrading their values and characteristics" There are 43 national parks in Turkey (Figure 2)

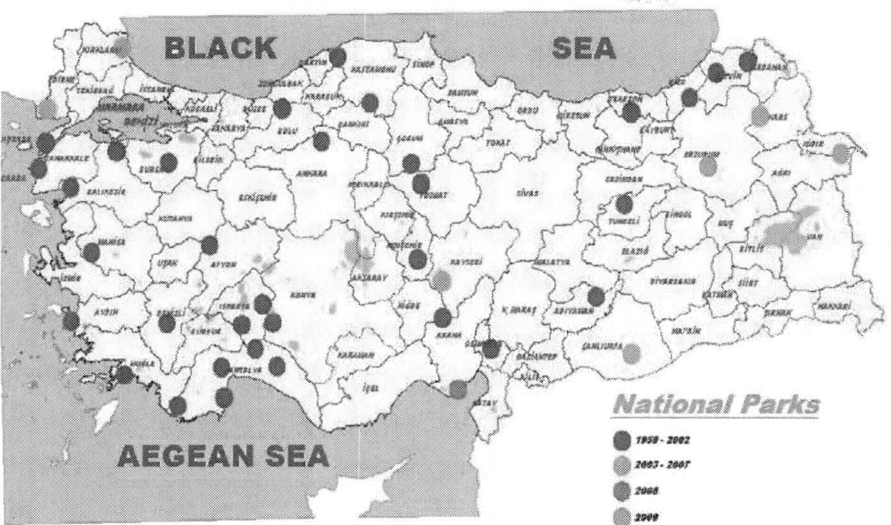

Fig. 2. National parks in Turkey.

The first national park in Turkey was established in 1958 (*The Yozgat Pine Grove National Park*) (Figure 3). Some of these parks, which were initially established for archaeological and historical purposes, are at the same time rich habitats where biological diversity is being protected.

Fig. 3. The Yozgat Pine Grove National Park, Turkey (www.milliparklar.gov.tr).

National Parks are defined as recreation and tourism areas which are rare in terms of scientific and scenic perspective in nature and are important for the conservation of the natural and cultural resource values (Yücel, 1995). These areas are in different regions of Turkey and were assigned as national parks at various dates and with various purposes; they are now under protective control and are kept open for public use (Güçlü and Karahan, 2008).

Some information and resource values of these national parks are given below (Table 1).

National Park	Area (ha)	Date	Resource Value
The Yozgat Pine Grove National Park	264	1958	Natural *Pinus sp.* (residual forest)
Karatepe Aslantaş National Park	7715	1958	Flora, visual landscape, historical value
Soğuksu National Park	1195	1959	Geological and geomorphologic value, thermal water
Bird Paradise National Park	24047	1959	Fauna, especially bird species
Uludağ National Park	12372	1961	Flora and fauna
Yedigöller (Sevenlakes) National Park	2019	1965	Lakes, flora, recreation
Dilek Peninsula National Park	27675	1966	Flora, fauna, wetlands
Spil Mountain National Park	6693,5	1968	Geological value, flora, historic and mythological value
Kızıldağ National Park	59400	1969	Geological value, flora
Termessos National Park	6702	1970	Ancient city, geological value, biodiversity
Kovada Lake National Park	6534	1970	Geological value, flora, karstic lake
Ilgaz Mountain National Park	1088	1976	Flora, winter sports, alpine flora
Munzur Valley National Park	42000	1971	Streams, flora, fauna, geomorphologic value
Olympos National Park	34425	1972	Archeological residual, flora
Gelibolu Peninsula Historical National Park	33000	1973	Historical war, geological and geomorphologic value
Köprülü Canyon National Park	36614	1973	Archeological residual, geological value
Başkomutan Historical National Park	40742	1981	Cultural and geological value
Göreme Historical National Park	9572	1986	Historical settlement, geomorphologic value
Altındere Valley National Park	4800	1987	Cultural value, landscape,
Boğazköy-Alacahöyük Historical National Park	2634	1988	Archeological residual

Nemrut Mountain National Park	13850	1988	Historical open air museum, landscape
Beyşehir Lake National Park	88750	1993	Historical residual, geomorphologic value, wetlands, fauna especially bird species
Kazdağları National Park	21463	1993	Flora, fauna, biodiversity,
Kaçkar Mountain National Park	51550	1994	Geological and geomorphologic value, flora and fauna
Hattila Valley National Park	16988	1994	Geological and geomorphologic value, flora and fauna
Altınbeşik Cavern National Park	1156	1994	Geological and geomorphologic value
Karagöl – Sahara National Park	3766	1994	Hydrographic structure, vegetation
Aladağlar National Park	54 524	1995	Landscape, waterfall
Honaz Mountain National Park	9616	1995	Geological and geomorphologic value, flora
Troya Historical National Park	13350	1996	Geomorphologic value, historical residual
Marmaris National Park	33350	1996	Geomorphologic value, flora and fauna
Saklıkent National Park	12390	1996	Flora, fauna, hydrological geomorphologic value
Küre Mountain National Park	37000	2000	Natural forest, biodiversity, geological and geomorphologic value
Sarıkamış-Allahuekber Mountain National Park	22980	2004	Historical value, fauna
Ağrı Mountain National Park	87 380	2004	Geomorphologic value
Gala Lake National Park	6090	2005	Wetland and forest ecosystem, bird species
Sultan Sazlığı National Park	24523	2006	Wetland ecosystem, bird species
Tek Tek Mountain National Park	19335	2007	Geomorphologic and historical value, fauna
İğneada Longos Forest National Park	3155	2007	Wetland and alluvial forest ecosystem, lagoon, flora, fauna
Yumurtalık Lagoon National Park	16 430	2008	Lagoon, swamp, sand dune
Nenehatun National Park	387	2009	Historical value

Table 3. National Parks in Turkey.

3. The case study, Iğneada Longos forest national park

The Igneada Longos Forests National Park, located on the Black Sea coast 15 km from the Turkish-Bulgarian border, is positioned between the northern latitudes 41′ 44′ 43′ and 41′ 58′ 27′ and the eastern longitudes 27′ 44′ 52′ and 28′ 3′ 17′.The Igneada area includes different kinds of ecosystems (sand dunes, wetlands, longos (flooded alluvial) forests, deciduous forests, and many streams) and a wide range of biodiversity; these characteristics make it one of the most important areas in Turkey (Ozyavuz, et al., 2006) (Table. Igneada and the surrounding environment have unique characteristics; these types (Igneada Longos Forests) of wild forest in other parts of Turkey and in Europe have been damaged due to anthropogenic effects (Figure 4).

Fig. 4. General view of this area.

Typically, flooded alluvial forests have high biological diversity, high productivity, and high habitat dynamism (Hughes et al., 2003). The surface area of these forests is around 3000 ha. Igneada alluvial longos forests are part of the Istranca forests; they are indeed "natural treasures" that have been formed by several ecosystems over thousands of years (Özyavuz and Yazgan, 2010).

Resource Value	Area (ha.)	Main Characteristic
Longos forest	1400	Alluvial flooded forest, rarity, sensitivity, flora, fauna
Wetlands and marshes	Wetlands (lagoon lake 52) (other 28) Marshes (315)	Geomorphological structure, flora, fauna,
Sand dunes	131	Geomorphological structure, flora (especially endemic plants)
Streams	-	Naturalness, flora
Deciduous forests	-	Naturalness, plant diversity, fauna

Table 4. Resource values of İğneda Longos forests National Park.

There are five lakes in the area. Lake Erikli Lagoon (43 ha) is adjacent to the northern part of I gneada subdistrict, which is not linked with the sea during the summer period. Lake Mert (266 ha) is located at the southern part of the subdistrict, where the stream reaches the Black Sea. Lake Saka, which is the smallest (5 ha), is at the southernmost part of the study area between the forest and sand. Lake Hamam (19 ha) and Lake Pedina (10 ha) are located in the inner part. The coastal dunes and the longos forests of Igneada constitute the most sensitive ecosystem in the study area. Most of the known endemic plants (*Silene sangaria, Crepis macropus, Centaurea kilaea*) in Igneada and its vicinity are found in the coastal dunes; other species found here, though not endemics, are of national and international concern (*Aurinia uechtritziana, Cakile maritima, Cionura erecta, Crambe maritima, Cyperus capitatus, Elymus elongatus subsp. elongatu, Eryngium maritimum, Euphorbia peplis, Eu. paralias, Jurinea kilae, Leymus racemosus, Otanthus maritimus, Pancratimum maritimum, Peucedanum obtusifolium, Stachys maritima*) (Figure 5-10) (Özyavuz and Yazgan, 2010).

Fig. 5. *Cianura erecta.*

Fig. 6. *Eryngium maritumum.*

Fig. 7. *Jurinea kilae.*

Fig. 8. *Leymus racemosus.*

Fig. 9. *Otanthus maritimus.*

Fig. 10. *Pancratimum maritimum.*

There are three longos forests in the area. The conserved natural longos forests in the study area are the Lake Mert longos (316 ha), Lake Erikli longos (456 ha), and Lake Saka longos (624 ha) forests (Figure 11-13) This type of ecosystem is unique and rare in Turkey and the world because these ecosystems are sensitive to environmental conditions. In general, deciduous mixed forest vegetation is found in the area outside of the longos forests, and in this area the forests have similar floristic composition to the longos forests. However, slopes are rather steep in the area where these forests are found, and therefore the water table is well below the surface. The different ecosystems in the area provide a diverse living environment for the fauna in the region. Nearly half (194) of the 454 bird species constituting the bird diversity of Turkey are seen in this area during the year (Özyavuz and Yazgan, 2010).

Fig. 11. Lake Mert.

Fig. 12. Lake Erikli.

Fig. 13. Lake Saka.

4. References

Dudley, N. (Editor) (2008). Guidelines for Applying Protected Area Management Categories, *International Union for Conservation of Nature and Natural Resources*, 86 pp., ISBN 978-2-8317-1086-0, Gland, Switzerland.

Güçlü, K., Karahan, F. (2004). A review: the history of conservation programs and development of the national parks concept in Turkey, Biodiversity and Conservation 13: 1373–1390, 2004.

IUCN (1994). Guidelines for Protected Area Management Categories, *IUCN Commission on National Parks and Protected Areas with the assistance of the World Conservation Monitoring Centre*, pp. 5-7, ISBN 2-8317-0201-1, Switzerland and Cambridge, UK.

IUCN (2000). Indigenous and Traditional Peoples and Protected Areas, IUCN, Gland, Switzerland and Cambridge, UK.

IUCN (2001). Guidelines for Financing Protected Areas in East Asia, World Commission on Protected Areas (WCPA), IUCN, Gland, Switzerland, and Cambridge, UK.

IUCN (2003). Guidelines for Management Planning of Protected Areas, IUCN, Gland, Switzerland and Cambridge, UK.

Ministry of Environment (2001). The National Strategy and Action Plan for Biodiversity in Turkey".

Ministry of Environment (2002). National Report on Sustainable Development 2002, Ministry of Republic of Turkey.

Özyavuz, M. and Yazgan, M. 2010. Planning of Igneada Longos (Flooded) Forests as a Biosphere Reserve, Journal of Coastal Research, (26)6:1104-1111.

Weeks, F. and Mehta, S. (2004). Managing People and Landscapes: IUCN's Protected Area Categories, J. Hum. Ecol, (16) 4: 253-263.
Yücel M. (1995). Protected Areas and Planning. Publication No. 104, C¸ ukurova University Press, Adana, Turkey, 255 pp. (in Turkish).

2

Land Use/Cover Classification Techniques Using Optical Remotely Sensed Data in Landscape Planning

Onur Şatır and Süha Berberoğlu
Cukurova University, Agriculture Faculty,
Department of Landscape Architecture,
Turkey

1. Introduction

The observed biophysical cover of the earth's surface, termed land-cover is composed of patterns that occur due to a variety of natural and human-derived processes. On the other hand Land-use is human activity on the land, influenced by economic, cultural, political, historical, and land-tenure factors. Remotely-sensed data (i.e., satellite or aerial imagery) can often be used to define land-use through observations of the land-cover (Brown, et al., 2000; Karl & Maurer, 2010). Up-to-date land-use information is of critical importance to planners, scientists, resource managers, and decision makers.

Optical remote sensing (RS) plays a vital role about defining LUC (land use/cover) and monitoring interactions between nature and human activities. Additionally, RS provides time, energy and cost saving. Today, optical RS data such as satellite sensor images and aerial photos are used widely to detect LUC dynamics. LUC mapping outcomes are used for global, regional, local mapping, change detection, landscape planning and driving landscape metrics.

RS image classification is a complex process and requires consideration of many factors. The major steps of image classification may include i) determination of a suitable classification system, ii) image preprocessing iii) selection of training samples, iv) selection of suitable classification approaches and post-classification processing, and v) accuracy assessment. Additionally, the user's need, scale of the study area, economic condition, and analyst's skills are important factors influencing the selection of remotely sensed data, the design of the classification procedure, and the quality of the classification results (Lu and Weng 2007).

LUC mapping has been used for various purposes in landscape planning and assessment such as, deriving landscape metrics (Southworth et al., 2010, Huang et al., 2007), landscape monitoring (Özyavuz et al., 2011, Berberoglu and Akin 2009), LUC change modeling (e.g., SLUETH (Clarke, 2008)), agricultural studies (agricultural policy environmental extender model (APEX) (Gassman et al., 2010); soil water assessment tool (SWAT) (Betrie et al. 2011)) and environmental processes (revised universal soil loss equation (RUSLE) (Renard et al., 1997)).

This chapter evaluates classification methods together with optical remote sensing data, and ancillary data integration to improve classification accuracy of LUC mapping.

2. LUC classification schemes

Standardization is one of the most discussed issues in LUC classification studies, and scientist and map developers were aware that using a common classification schemes might be more comparable and available. The first standardization works started in USA. Today there are several LUC schemes on the world according to region and scale. This chapter will discuss three largely used schemes; i) USGS (US geological survey) Anderson, ii) CORINE (Coordination of information on the environment) and iii) EUNIS (European Nature Information System) habitat schemes.

2.1 USGS Anderson classification schemes

This classification scheme was utilized within large number of models in the context of land physical dynamics and natural risk assessment. USGS classification scheme is based on James Anderson's system. This scheme is included nine main categories and four different levels (Anderson et al., 1976).

Level I is suitable for 1/250.000 – 1/150.000 scale imags like MODIS and Envisat MERIS. Level II is useful for higher spatial resolution satellite sensor images with a scale of 1:80,000. Level III is suitable for 1:20,000 to 1/80,000 scale images such as, Landsat 4-7 . Level IV is the most useful for images at scales larger than 1:20,000 (Ikonos, Kompsat, Rapid eye, Formosat, Geoeye, World view and aerial photos). Categories are designed to be adaptable to the local needs. Sample of Level I categories and forest land levels are showed in table 1.

LEVEL I	LEVEL II	LEVEL III	LEVEL IV
1 Urban or Built-up Land	41 Broadleaved Forest (generally deciduous)	**421 Upland conifers** 422 Lowland conifers	4211 White pine
2 Agricultural Land	**42 Coniferous Forest**		4212 Red pine
3 Rangeland	43 MixedConifer-		4213 Jack pine
4 Forest Land	Broadleaved Forest		4214 Scotch pine
5 Water			4215 White spruce
6 Wetland			4219 Other
7 Barren Land			
8 Tundra			
9 Perennial Ice or Snow			

Table 1. USGS classification scheme for level I of forest cover.

2.2 CORINE classification scheme

The European council found EEA (European environmental agency) in 1990 to search and discuss the environmental issues all around the Europe. LUC of Europe is one of the most important data to define environmental strategies. CORINE system aims at collecting comparable and consistent land cover data across Europe. This information system, offers the essential elements for the applications of nature conservation, urban planning and resource management. The European nomenclature distinguishes 44 different types of land cover. Individual countries can supplement these categories with a more detailed level if they desire so. CORINE Land Cover is bridging the gap between the local (micro) and the EU (macro) scales. CORINE Land Cover is a platform of communication not only for environmental information, but also for the policies that shape the environment (figure 1).

Fig. 1. The legend of CORINE LUC classes codes and colors (EEA 2012b).

2.3 EUNIS habitat classification scheme

The EUNIS habitat classification is a common reporting language on habitat types at European level, sponsored by the EEA. It originated from a combination of several habitat classifications - marine, terrestrial and freshwater. The terrestrial and freshwater classification builds upon previous initiatives, notably the CORINE biotopes classification (Devillers & Devillers-Terschuren 1991), the Palaearctic habitats classification (Devillers & Devillers-Terschuren 1996), of the EU Habitats Directive 92/43/EEC, the CORINE Land

Cover nomenclature (Bossard et al. 2000), and the Nordic habitat classification (Nordic Council of Ministers 1994). The marine part of the classification was originally based on the BioMar classification (Connor et al 1997), covering the North-East Atlantic. The EUNIS habitat classification introduced agreed criteria for the identification of each habitat unit, while providing a correspondence with these earlier classification systems.

The habitat classification forms an integral part of the EUNIS, developed and managed by the European Topic Centre for Nature Protection and Biodiversity (ETC/NPB in Paris) for the EEA and the European Environmental Information Observation Network (EIONET). The EUNIS web application (http://eunis.eea.europa.eu) (EEA 2012a) provides access to publicly available data in a consolidated database.

The information includes:

- Data on Species, Habitats and Sites compiled in the framework of NATURA2000 (EU Habitats and Birds Directives),
- Data collected from frameworks such as EIONET, data sources or material published by ETC/NPB (formerly the European Topic Centre for Nature Conservation).
- Information on Species, Habitats and Sites taken into account in relevant international conventions or from International Red Lists.
- Specific data collected in the framework of the EEA's reporting activities, which also constitute a core set of data to be updated periodically.

The resulting system of classification is still somewhat transitional. Down to level 3 (terrestrial and freshwater) and level 4 (marine), EUNIS habitats are now based on physiognomic and physical attributes, together with some floristic criteria. There are 10 main habitat categories in this scheme. Coastal habitats and main categories as an example were presented in this chapter (table 2). Detailed information can be found in revised EUNIS habitat classification report of Davies et al. (2004).

LEVEL I	LEVEL II	LEVEL III
A. Marine habitats	**B1 Coastal dunes and**	B1.1 Sand beach
B. Coastal habitats	**sandy shores**	driftlines
C. Inland surface waters	B2 Coastal shingle	B1.2 Sand beaches above
D. Mires, bogs and fens	B3 Rock cliffs, ledges	the driftline
E. Grasslands and lands dominated	and	B1.3 Shifting coastal
by forbs, mosses or lichens	shores, including the	dunes
F. Heathland, scrub and tundra	supralittoral	B1.4 Coastal stable dune
G. Woodland, forest and other		grassland (grey dunes)
wooded land		B1.5 Coastal dune heaths
H. Inland unvegetated and sparsely		B1.6 Coastal dune scrub
vegetated habitats		B1.7 Coastal dune woods
I. Regularly or recently cultivated		B1.8 Moist and wet dune
agricultural, horticultural and		slacks
domestic habitats		B1.9 Machair
J. Constructed, industrial and other		
artificial habitats		

Table 2. Main EUNIS habitat classes and sample levels of coastal habitats.

3. Remotely sensed data sources

Data characteristics are the most important issue to select appropriate available one for a LUC mapping. Both airborne and spaceborne data have various spatial, radiometric, spectral and temporal resolutions. Large numbers of studies have focused on characteristics of remotely sensed data (Barnsley 1999, Lefsky and Cohen 2003). Additionally, scan width (cover size in one scene), data availability (accessibility) and lunch date (data archive potential) are the other important factors (Table 3).

Sensor	Organization	Spatial resolution	Spectral resolution	Radiometric resolution	Temporal resolution	Imaging Swath	Lunch date
Geoeye	Geoeye ABD	Pan: 0.41m MS: 0.61m	4 VNIR bands (450-920nm)	11 bit	~3 days	15X15km	2008
Quickbird-2	Digital globe ABD	0.61m	4 VNIR bands (450-890)	11 bit	3.5 days	16.5X16.5km	2002
Ikonos	Geoeye ABD	Pan: 1m MS: 4m	4 VNIR bands (450-880nm)	11 bit	3.5 days	11X11km	1999
Rapideye	Rapideye AG Germany	5m	4 VNIR bands and red edge (440-850nm)	12 bit	5.5 days at nadir	77X77km	2008
World view (2) and (3)*	Digital globe ABD	Pan: 0.5m MS: 2m at nadir	4 VNIR standart bands & 4 VNIR unique bands (400-1040nm)	11 bit	1-3days	16.4X16.4km	2009 (2) 2014 (3)
Spot 5	SpotIMAGE France	Pan:2.5m VNIR:10m SWIR:20m	3 VNIR & 1 SWIR bands (490-1750)	8 bit	26 days	60X60km	2002
AVIRIS airborne hyperspectral	NASA ABD	17m	224 VNIR & SWIR bands (400-2500nm)	16 bit	airborne	11X11km	-
Alos (AVNIR2)	JAXA Japan	10m	4 VNIR bands	8 bit	46 days	70X70km	2006
ASTER	Japan&ABD	VNIR: 15m SWIR: 30m TIR: 90m	4 VNIR bands 6 SWIR bands 5 TIR bands (520-11650nm)	VNIR:8 bit SWIR: 8 bit TIR: 12 bit	16 days	60X60km	1999
Landsat 8*	NASA ABD	Pan: 15m MS: 30m	5 VNIR bands 2 SWIR bands 1 cirrus band (433-2300nm)	8 bit	16 days	185X185km	Dec. 2012
CHRIS proba hyperspectral	ESA EU	17-34m	18-62 VNIR bands (400-1100nm)	16 bit	16 days	15X15km	2002

Sensor	Organization	Spatial resolution	Spectral resolution	Radiometric resolution	Temporal resolution	Imaging Swath	Lunch date
Hyperion hyperspectral	NASA ABD	30m	220 VNIR & SWIR bands (400-2500nm)	16 bit	16 days	7.7X185km	2000
EnMAP*	DLR Germany	30m	244 VNIR & SWIR bands (420-2450)	14 bit	4 days	30X30km	2014-2015
MODIS	NASA ABD	250-1000m	36 VNIR & SWIR & TIR bands	12 bit	1 day	2330km	2000
MERIS	ESA EU	300-1200m	15 VNIR bands	16 bit	1 day	1150km	2002
AVHRR NOAA 15	NASA ABD	1090m	6 VNIR & TIR bands	10 bit	1 day	1446km	1978 1998

Table 3. The most used optical sensor specifications in LUC mapping. (*) planned missions.

4. LUC mapping techniques

Suitable remotely sensed data, classification systems, available classifier and number of training samples are prerequisites for a successful classification. Cingolani et al. (2004) identified three major problems when medium spatial resolution data are used for vegetation classifications: i) defining adequate hierarchical levels for mapping, ii) defining discrete land-cover units discernible by selected remote-sensing data, and iii) selecting representative training sites. In general, a classification system is designed based on the user's need, spatial resolution of selected remotely sensed data, compatibility with previous work, image-processing and classification algorithms, and time constraints. Such a system should be informative, exhaustive, and separable (Jensen 1996, Landgrebe 2003). In many cases, a hierarchical classification system is adopted to take different conditions into account (Lu and Weng 2007).

4.1 Image pre-processing

Image pre-processing includes geometric correction or image registration, atmospheric correction and radiometric calibration essentially. In addition, topographic correction and noise reduction may be applied if necessary. Optical images from current systems have already corrected geometrically (Landsat TM/ETM, MODIS) or can be corrected using freely available software or tools (e.g. BEAM for MERIS and CHRIS and MRT toolbox for MODIS).

Accurate geometric rectification or image registration of remotely sensed data is a prerequisite for a combination of different source data in a classification process. Many textbooks and articles have described this topic in detail (Jensen 1996, Toutin 2004). However, Geometric correction output should have the transformation rms. errors (RMSE) less than 1.0 pixel, indicating that the images are located with an accuracy of less than a pixel.

Atmospheric and radiometric corrections may not be necessary if a single image is used, but multitemporal or multisensor data are needed atmospheric and radiometric correction and calibration. A variety of methods, ranging from simple relative calibration such as, dark-object subtraction to calibration approaches based on complex models (e.g. MODTRAN, 6S,

ATCOR2), have been developed for radiometric and atmospheric normalization and correction (Chavez 1996, Heo and FitzHugh 2000, Hadjimitsis et al. 2004, Ozyavuz et al. 2011) (Figure 2).

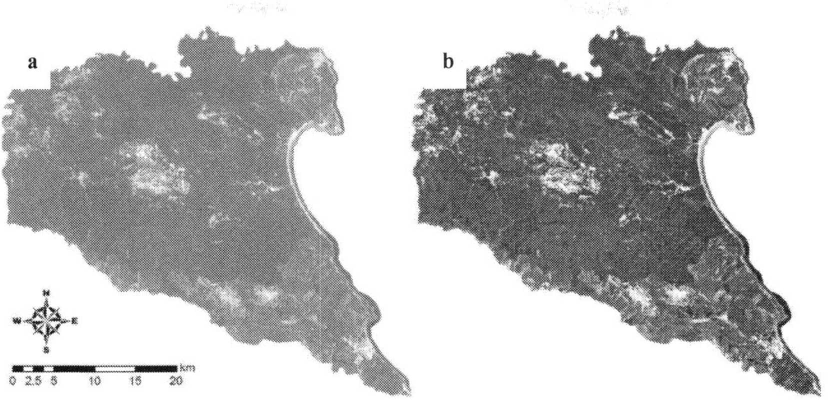

Fig. 2. (a) Non-corrected data, (b) Atmospherically corrected and haze removed Landsat TM data using ATCOR2 model (Özyavuz et al. 2011)

4.2 Classification techniques

There are two basic approaches to the classification process: supervised and unsupervised classification. With supervised classification, one provides a statistical description of the manner in which expected land cover classes should appear in the imagery, and then a procedure (known as a classifier) is used to evaluate the likelihood that each pixel belongs to one of these classes. With unsupervised classification, a very different approach is used. Here another type of classifier is used to uncover commonly occurring and distinctive reflectance patterns in the imagery, on the assumption that these represent major land cover classes. The analyst then determines the identity of each class by a combination of experience and ground truth (i.e., visiting the study area and observing the actual cover types) (Eastman 2003). Three essential parts are vital in a LUC mapping in classification stage; training, classifying and testing (accuracy assessment).

4.2.1 Classifiers

In this chapter four different classifiers and approaches were evaluated in the example of Landsat TM sub-scenes recorded over Eastern Mediterranean coastal part. Methods and performances were assessed based on accuracy, capability and applicability. This assessment covered traditional (minimum distance, maximum likelihood, linear discriminant analyses), machine learning (decision tree, artificial neural network, support vector machine), fuzzy (linear mixture modeling, fuzzy c-means, artificial neural network, regression tree) and object based classifiers for LUC mapping. The summary of the techniques and classifiers for various purposes were provided in table 4.

Criteria	Categories	Characteristics	Example of classifiers
Whether training samples are used or no	Supervised Classification approaches	Land cover classes are defined. Sufficient reference data are available and used as training samples. The signatures generated from the training samples are then used to train the classifier to classify the spectral data into a thematic map.	Maximum likelihood (MLC), minimum distance (MD), Artificial neural network (ANN), decision tree (DT) classifier.
	Unsupervised classification approaches	Clustering-based algorithms are used to partition the spectral image into a number of spectral classes based on the statistical information inherent in the image. No prior definitions of the classes are used. The analyst is responsible for labeling and merging the spectral classes into meaningful classes.	ISODATA, K-means clustering algorithm.
Whether parameters such as mean vector and covariance matrix are used or not	Parametric classifiers	Gaussian distribution is assumed. The parameters (e.g. mean vector and covariance matrix) are often generated from training samples. When landscape is complex, parametric classifiers often produce 'noisy' results. Another major drawback is that it is difficult to integrate ancillary data, spatial and contextual attributes, and non-statistical information into a classification procedure.	MLC and Linear discriminant analysis (LDA)
	Non-parametric classifiers	No assumption about the data is required. Non-parametric classifiers do not employ statistical parameters to calculate class separation and are especially suitable for incorporation of non-remote-sensing data into a classification procedure.	ANN, DT, Support vector machine (SVM), evidential reasoning, expert system.
Which kind of pixel information is used	Per-pixel classifiers	Traditional classifiers typically develop a signature by combining the spectra of all training-set pixels from a given feature. The resulting signature contains the contributions of all materials present in the training-set pixels, ignoring the mixed pixel problems.	MLC, MD, SVM, ANN, DT
	Subpixel classifiers	The spectral value of each pixel is assumed to be a linear or non-linear combination of defined pure materials (or endmembers), providing proportional membership of each pixel to each endmember.	Fuzzy-set classifiers, subpixel classifier, spectral mixture analysis.

Criteria	Categories	Characteristics	Example of classifiers
Which kind of pixel information is used	Object-oriented classifiers	Image segmentation merges pixels into objects and classification is conducted based on the objects, instead of an individual pixel. No GIS vector data are used.	eCognition.
	Per-field classifiers	GIS plays an important role in per-field classification, integrating raster and vector data in a classification. The vector data are often used to subdivide an image into parcels, and classification is based on the parcels, avoiding the spectral variation inherent in the same class.	GIS-based classification approaches.
Whether output is a definitive decision about land cover class or not	Hard classification	Making a definitive decision about the land cover class that each pixel is allocated to a single class. The area estimation by hard classification may produce large errors, especially from coarse spatial resolution data due to the mixed pixel problem.	MLC, MD, ANN, DT, SVM
	Soft (fuzzy) classification	Providing for each pixel a measure of the degree of similarity for every class. Soft classification provides more information and potentially a more accurate result, especially for coarse spatial resolution data classification.	Fuzzy-set classifiers, subpixel classifier, spectral mixture analysis.
Whether spatial information is used or not	Spectral classifiers	Pure spectral information is used in image classification. A 'noisy' classification result is often produced due to the high variation in the spatial distribution of the same class.	Maximum likelihood, minimum distance, artificial neural network.
	Contextual classifiers	The spatially neighbouring pixel information is used in image classification	Iterated conditional modes, point-to-point contextual correction, and frequency-based contextual classifier.
	Spectral-contextual classifiers	Spectral and spatial information is used in classification. Parametric or non-parametric classifiers are used to generate initial classification images and then contextual classifiers are implemented in the classified images.	ECHO, combination of para metric or non-parametric and contextual algorithms.

Table 4. A taxonomy of image classification methods (Lu and Weng 2007).

4.2.1.1 Model based classifiers (traditional)

Model based classifiers are run using basic statistical theories like mean, variance and standard deviation of the dataset. The most used ones at the literatures are supervised MLC, MD, LDA and unsupervised k-means.

The minimum distance classifier is used to classify unknown image data to classes which minimize the distance between the image data and the class in multi-feature space. The distance is defined as an index of similarity so that the minimum distance is identical to the maximum similarity. If a pixel closer than to mean of a signature pixels, it classifies as same as nearest one. In figure 3, the nearest signature mean to unclassified pixel is settlement, thus it will be assigned to settlement class according to MD classifier.

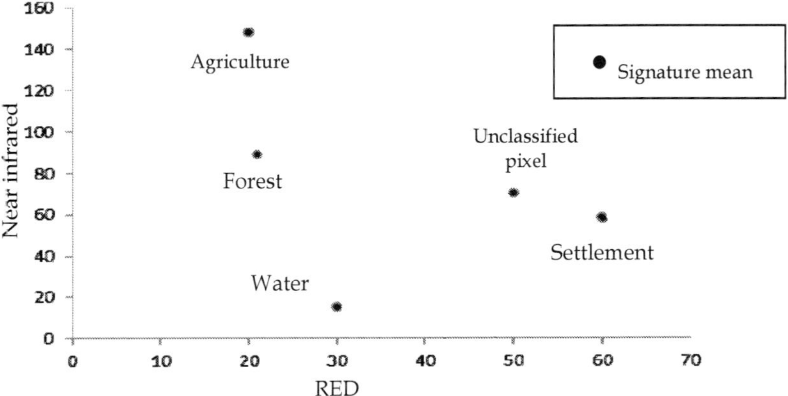

Fig. 3. MD classifier concept

The MLC procedure is based on Bayesian probability theory. Using the information from a set of training sites, MLC uses the mean and variance/covariance data of the signatures to estimate the posterior probability that a pixel belong to each class. MLC procedure is similar to MD with the standardized distance option. The difference is that MLC accounts for intercorrelation between bands. By incorporating information about the covariance between bands as well as their inherent variance, MLC procedures what can be conceptualized as an elliptical zone of characterization of signature. It calculates the posterior probability of belonging to each class, where the probability is highest at mean position of the class, and falls off in an elliptical pattern away from the mean.

The LDA classifier conducts linear discriminant analysis of the training site data to form a set of linear combination that expresses the degree of support for each class. The assigned class for each pixel is then that class which receives the highest support after evaluation of all functions. These functions have a form similar to that of a multivariate linear regression equation, where the independent variables are the image bands, and the dependent variable is the measure of support. In fact, the equations are calculated such that they maximize the variance between classes and minimize the variance within classes. So that class separation becomes easier.

In k-means unsupervised technique, K-means clustering technique is used to partition a n-dimensional imagery into K exclusive clusters. This method begins by initializing k centroids (means), then assigns each pixel to the cluster whose centroid is nearest, updates the cluster centroids, then repeats the process until the k centroids are fixed. This is a heuristic, greedy algorithm for minimizing SSE (Sum of Squared Errors), hence, it may not converge to a global optimum. Since its performance strongly depends on the initial estimation of the partition, a relatively large number of clusters are generally recommended to acquire as complete an initial pattern of centroids as possible (Richards & Jia, 1999).

All of the model based classifiers were compared each other using the same training data set in order to ensure the comparability of each technique. Landsat TM image recorded in August 2010 over the Eastern Mediterranean coastal zone of Turkey was used. Main LUC classes were coniferous tree, deciduous tree, permanent farmlands, temporary irrigated farmlands, temporary non-irrigated farmlands, bulrush, grassland, bareground, water bodies, settlement, sand dunes(figure 4).

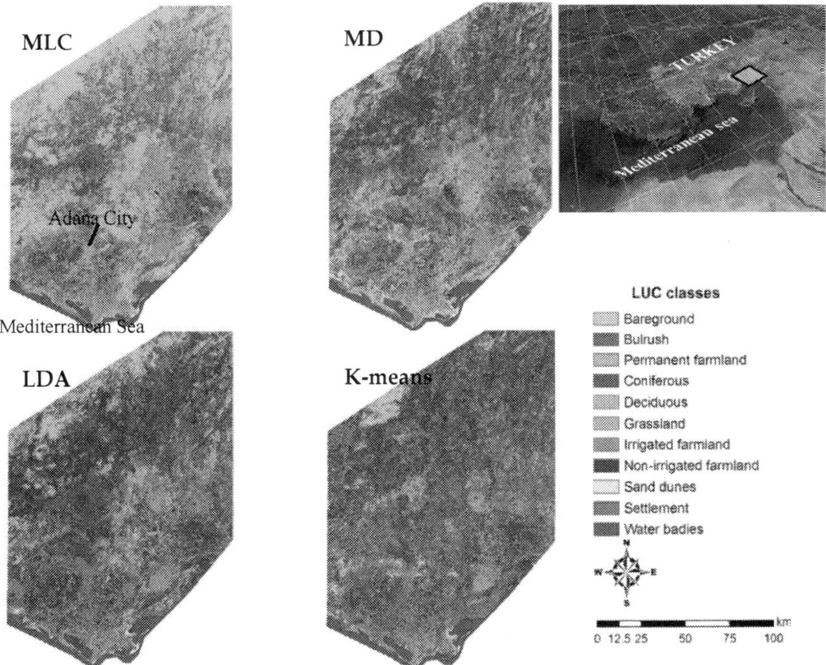

Fig. 4. Model based LUC classification results using strong training dataset and unsupervised K-means classification result in sample study area (in yellow).

4.2.1.2 Data dependent (machine learning classifiers)

Data dependent classifiers are based on non-parametric rules. Particularly, the machine learning classifiers use different approaches according to classifier type. In this chapter, largely used non-parametric classifiers were assessed such as ANN, DT and SVM.

The ANN is one of several artificial intelligence techniques that have been used for automated image classification as an alternative to conventional statistical approaches. Introductions to the use of ANNs in remote sensing are provided in (Kohonen 1988), (Bishop 1995) and (Atkinson and Tatnall 1997). The multilayer perceptron described by

(Rumelhart et al. 1986) is the most commonly encountered ANN model in remote sensing (because of its generalization capability). The accuracy of an ANN is affected primarily by five variables: (1) the size of the training set, (2) the network architecture, (3) the learning rate, (4) the learning momentum, and (5) the number of training cycles.

Size of the training set is the most important part in all LUC classifications. If training pixel counts are enough, accuracy of a LUC map would be better than less training pixels.

Network architecture of an ANN is similar to the small part of a neural network (NN) system of human brain. Essentially, there are 3 parts in a NN as input, hidden and output nodes (figure 5). Input nodes are the image bands (e.g. for Landsat TM 6 nodes except thermal band) in a LUC mapping using optical images. Hidden node count depends on the user or previous experiences. There are two way to detect optimal hidden node count; (i) user may check the literature deals with the similar or same area of study site to find the optimal hidden node counts, (ii) user may apply several possibilities itself to find optimal hidden node count checking the accuracy of each applications. According to literature, if a NN system uses the one hidden layer, it is two or three times more than the input nodes generally (Berberoglu et al. 2009). Output nodes counts are equal the class count. Each output nodes are produced a class probability.

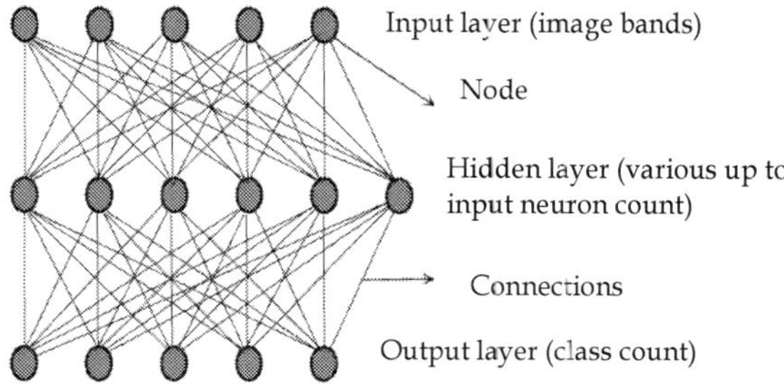

Fig. 5. NN architecture

The learning rate, determines the portion of the calculated weight change that will be used for weight adjustment. This acts like a low-pass filter, allowing the network to ignore small features in the error surface. Its value ranges between 0 and 1. The smaller the learning rate, the smaller the changes in the weights of the network at each cycle. The optimum value of the learning rate depends on the characteristics of the error surface. Lower learning rates are require more cycles than a larger learning rate.

Learning momentum is added to the learning rate to incorporate the previous changes in weight with the current direction of movement in the weight space. It is an additional correction to the learning rate to adjust the weights and ranges between 0.1 and 0.9.

Number of training cycles is defined according to training error of a NN system. When the training error became optimal, training cycles are sufficient.

In this chapter NN architecture was defined based on previous literature. Berberoglu 1999, designed a NN architecture for Eastern Mediterranean LUC mapping. Several NN architectures were tried in that literature and the highest performance was taken in four layer architecture. This NN was included 2 hidden layers. First hidden layer was included nodes two times more than input and the second hidden layer was contained nodes three times more than first hidden layer. Learning rate and learning momentum have defined according to training error (table 5).

Parameters	Values
Input layer	6 nodes
1. Hidden layer	12 nodes
2. Hidden layer	36 nodes
Output layer	11 nodes
Learning rate	0.001
Learning momentum	0.5
Number of cycles	44864

Table 5. ANN parameters and values

DT is a non-parametric image classification technique. A decision tree is composed of the root (starting point), active node or internode (rule node) and leaf (class). The root is starting point of the tree, active node creates leaves and the leaves are a group of pixels that either belong to same class or are assigned to a particular class (figure 6).

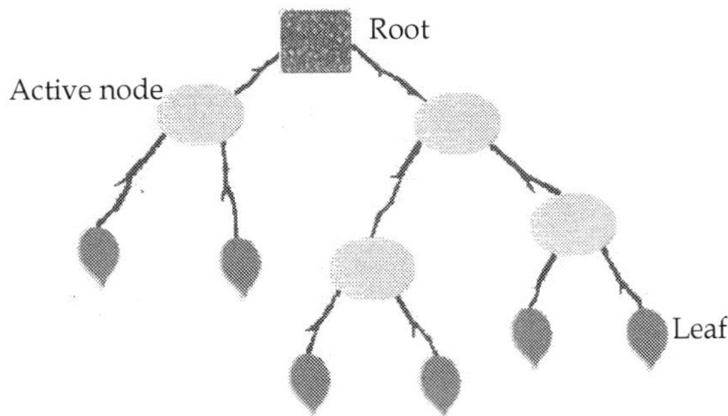

Fig. 6. Decision tree architecture

A Decision Tree is built from a training set, which consists of objects, each of which is completely described by a set of attributes and a class label. Attributes are a collection of properties containing all the information about one object. Unlike class, each attribute may have either ordered (integer or a real value) or unordered values (Boolean value) (Ghose et al. 2010).

Most of the DT algorithms generally use the recursive-partitioning algorithm, and its input requires a set of training examples, a splitting rule, and a stopping rule. Splitting rules are determined tree partitioning. Entropy, gini, twoing and gain ratio are the most used splitting rules at the literature (Quinlan 1993, Zambon et al. 2006, Ghose et al. 2010). The stopping rule determines if the training samples can split further. If a split is still possible, the samples in the training set are divided into subsets by performing a set of statistical test defined by the splitting rule. This procedure is recursively repeated for each subset until no more splitting is possible (Ghose et al 2010).

In this chapter, gain ratio, entropy and gini splitting algorithms have been used to find the most accurate one, and entropy accuracy was determined almost 3% more accurate than the gain ratio. Gini resulted the poorest performance for the study area. Stopping criteria and active nodes were determined according to fallowing rule;

If a subset of classes determined as pure, create a leaf and assign to interest class. If a subset having more than one class creates active nodes applying splitting algorithm, continue this processes until class leafs became purer.

The SVM represents a group of theoretically superior machine learning algorithms. SVM employs optimization algorithms to locate the optimal boundaries between classes. Statistically, the optimal boundaries should be generalized to unseen samples with least errors among all possible boundaries separating the classes, therefore minimizing the confusion between classes. In practice, the SVM has been applied to optical character recognition, handwritten digit recognition and text categorization (Vapnik 1995, Joachims 1998). SVM uses the pairwise classification strategy for multiclass classification. SVM can be used linear and non-linear form applying different kernel functions. In this chapter only sigmoidal non-linear kernel were used because, model based classifiers have already worked well if data histogram is linear. All data based models were run non-linearly, and sigmoidal application takes less time than other non-linear kernels. Different kernel functions like radial basis function, linear function or polynomial function may be applied. Even the accuracy of the SVM classifier may change when used the one kernel. For example, in polynomial kernel function, accuracy of SVM is various according to applied polynomial order (Huang et al. 2002).

All data dependent classifiers which were introduced in this chapter were evaluated in the Eastern Mediterranean environment (figure 7).

4.3 Accuracy assessments

A classification accuracy assessment generally includes three basic components: sampling design, response design, and estimation and analysis procedures (Stehman and Czaplewski 1998). Selection of a suitable sampling strategy is a critical step (Congalton 1991). The major components of a sampling strategy include sampling unit (pixels or polygons), sampling design, and sample size (Muller et al. 1998). Possible sampling designs include random, stratified random, systematic, double, and cluster sampling. A detailed description of sampling techniques can be found in previous literature such as Stehman and Czaplewski (1998) and Congalton and Green (1999).

The error matrix approach is the one most widely used in accuracy assessment (Foody 2002). In order to properly generate an error matrix, one must consider the following factors: (1) reference data collection, (2) classification scheme, (3) sampling scheme, (4) spatial autocorrelation, and (5) sample size and sample unit (Congalton and Plourde 2002). After generation of an error matrix, other important accuracy assessment elements, such as overall accuracy, user accuracy, producer accuracy (table 6), and kappa coefficient can be derived. Kappa is the difference between the observed accuracy and the chance agreement divided by one minus that chance agreement (Lillesand and Kiefer 1994).

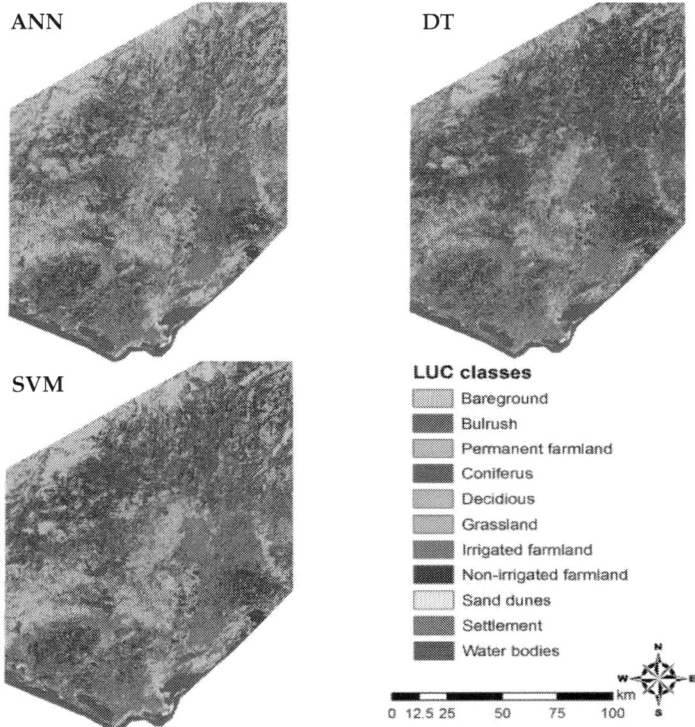

Fig. 7. Data dependent LUC classification results using strong training dataset

Error matrix	Agriculture	Forest	Water	Total classified pixels	User accuracy
Agriculture	32	7	1	40	32 / 40 = 80%
Forest	5	34	1	40	34 / 40 = 85%
Water	2	2	36	40	36 / 40 = 90%
Total ground truth pixels	39	43	38	120	
Producer accuracy	32 / 39 82%	34 / 43 79%	36 / 38 95%		
Overall accuracy				Correct pixels / Total pixel = 32+34+36 / 120 = **85%**	

Table 6. Error matrix and accuracy calculations

Overall classification accuracies and kappa coefficiencies of each classification using weak (6962 training pixels) and strong (16300 training pixels) training dataset were evaluated (table 7). In addition, each of the LUC user, producer and kappa accuracies were compared using strong training dataset to assess results in detail (table 8). No ancillary data integrated to the classifications, however, it was discussed in section 5.

	Classifier	Accuracies (%)			
		Weak		Strong	
		Overall	Kappa	Overall	Kappa
MB	**MLC**	69.5	65.6	85.6	83.8
	MD	67.6	63.3	72	68.5
	LDA	73.5	70	80.5	78
	K-means	Overall = 57.3		Kappa = 52.1	
DD	**ANN**	70.5	66.5	76.9	74.2
	DT	73	69.5	82.5	80
	SVM	74	70	79.2	76.7

Table 7. Overall and kappa accuracies of model based (MB) and data dependent (DD) classifiers using weak and strong training dataset

Overall classification accuracies indicated that MLC was the most accurate model based classifier when the strong training dataset was used. However, LDA with weak training dataset performed accurately because of its distance separation algorithm. On the other hand, unsupervised k-means classifier was the least accurate one due to the fact that no training pixels were used.

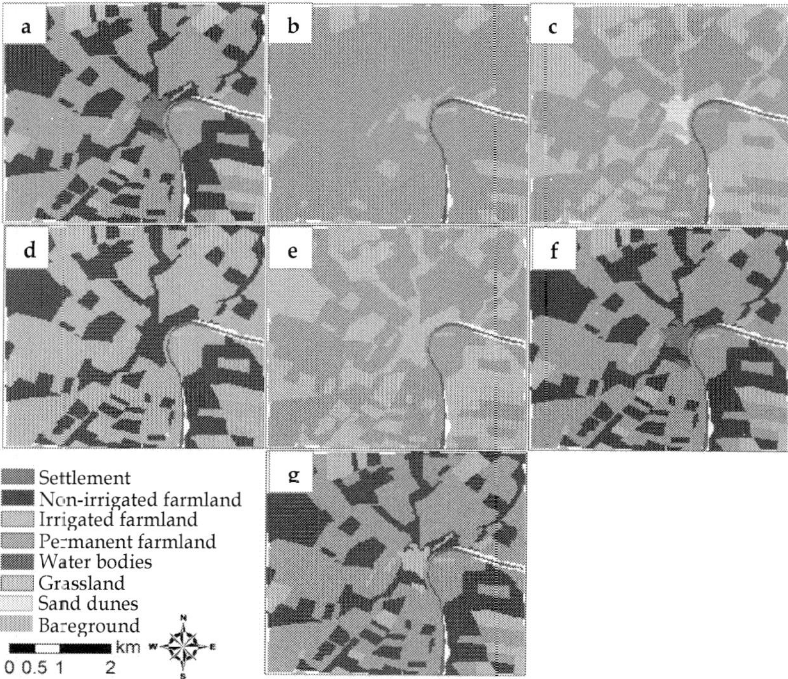

Fig. 8. Visual detail of a small subview; (a) Ground truth, (b) MLC, (c) MD, (d) LDA, (e) ANN, (f) DT and (g) SVM results

SVM has a reasonable performance than other data dependent classifiers using weak training dataset. However, the largest accuracy was resulted in DT classifier using strong dataset.

SVM classified forestlands, grassland and permanent farmlands more accurate than other classifiers. There was not significant difference in built up areas among classifiers. The most accurate sand dunes, bulrush and irrigated farmland class accuracies were resulted from DT classifier. DT, LDA, SVM showed reasonably well performance with both weak and strong training data sets (figure 8).

In general, data dependent classifiers performed well with weak training dataset. Especially SVM was successful in vegetative area separation. It is clear that if more detailed classification scheme required (e.g. forest tree species) using weak training dataset, SVM might be first option in terms of classification accuracy. On the other hand, application of SVM is time costly when using standard PC and laptops.

Three accuracy calculation methods were shown in table 8, however, major question is which one should be used? Large number of studies have utilized the kappa coefficiencies as an ideal approache for LUC classification.

A number of criteria were selected for the comparison of both model based and data dependent classifiers as (a) Overall accuracy, (b) classification speed, (c) input parameter handling, (d) hardness in application, (e) accuracy with different training sizes and accuracy difference between each class or classification stability (table 9).

Criteria	MLC	MD	LDA	k-means	ANN	DT	SVM
Overall accuracy	****	**	****	*	***	*****	****
Classification speed	*****	*****	****	****	**	***	*
Input parameter handling	*****	*****	*****	*****	***	***	**
Hardness in application	*****	*****	****	*****	***	**	*
Accuracy with different training sizes	****	*	****	No training	***	*****	*****
Classification stability	***	***	****	*	***	*****	***

Table 8. Comparing hard classifiers (***** stars and * star refer the most accurate and the poorest performances respectively).

4.3.1 Soft (fuzzy) classifiers

Defining "what is in a pixel?" numerically, very important for understanding the earth surface in remote sensing science. Increased spatial information may be valuable in a variety of situations. The forthcoming range of satellite spectrometers (e.g. MODIS, MERIS) provided detailed attribute information at relatively coarse spatial resolutions (e.g. 250m, 500m, 1km) (Aplin and Atkinson 2001).

Traditional hard per-pixel classification of remotely sensed images is limited by mixed pixels (Cracknell 1998). Soft classification overcomes this limitation by predicting the proportional membership of each pixel to each class. Mapping is generally achieved through the application of a conventional statistical classification, which allocates each image pixel to a land cover class. Such approaches are inappropriate for mixed pixels, which contain two or more land cover classes, and fuzzy classification approach is required (Foody 1996).

Classifiers and Accuracies (%)																					LUC
SVM			DT			ANN			K-means			LDA			MD			MLC			
K	P	U	K	P	U	K	P	U	K	P	U	K	P	U	K	P	U	K	P	U	
98	98	98	98	98	98	95	98	96	98	98	98	98	100	98	98	98	98	97	92	98	Water bodies
98	79	98	92	94	94	97	73	98	85	68	86	98	97	98	87	78	89	98	100	98	Coniferous tree
100	70	100	80	85	81	92	60	92	2	5	3	54	85	57	87	70	87	100	70	100	Deciduous tree
79	83	80	60	75	62	69	92	71	77	33	80	68	88	70	70	75	72	70	75	72	Grassland
86	89	87	97	82	97	85	84	86	100	42	100	100	58	100	85	84	86	89	92	90	Irrigated temporary
79	60	83	79	59	83	86	53	89	44	36	55	74	67	79	50	58	59	92	64	94	Non-irrigated temporary farmland
62	92	66	64	92	67	64	92	67	61	63	65	84	81	86	60	42	64	48	95	53	Bareground
55	100	57	75	100	76	46	100	48	45	75	48	50	100	52	39	100	41	93	94	94	Sand dunes
33	94	36	64	81	65	35	100	38	4	0	0	48	68	50	30	50	33	74	94	75	Bulrush
90	69	91	88	83	89	89	62	90	40	67	44	90	72	91	85	62	86	92	83	92	Settlement
91	50	92	71	73	83	72	64	74	22	65	26	68	64	70	67	82	69	100	91	100	Permanent farmland
79	80	81	79	84	88	75	80	77	53	50	55	76	80	77	69	73	71	87	86	88	Mean of accuracies

Table 9. (K) kappa, (P) producer, and (U) user accuracies of each LUC using hard classifiers in study area.

Fuzzy logic models constitute the modeling tools of soft computing. Fuzzy logic is a tool for embedding structured human knowledge into workable algorithms. There are two main types of sets. The 'crisp (or classic) sets' and the 'fuzzy sets'. For example, a crisp set can be defined by a membership function:

In crisp sets a function of this type is also called characteristic function. Fuzzy sets can be used to produce the rational and sensible clustering. For fuzzy sets there exists a degree of membership $\mu_s(X)$ that is mapped on [0, 1]. In the case of LUC map, every area simultaneously belongs to interest LUC clusters with a different degree of membership (Kandel, 1992).

$$\mu_s(X) = \begin{cases} 1 \text{ if } X \in S \\ \\ 0 \text{ if } X \notin S \end{cases} \qquad (1)$$

There are several soft classification techniques and these are variable according to training and testing dataset, scale of the study. In this frame, linear mixture modeling (LMM), Regression tree (RT), multi linear regression (MLR) and artificial neural network (ANN) soft classification techniques were evaluated in Eastern Mediterranean area called Upper Seyhan Plane (USP). Berberoglu et al. (2009), was focused on these four soft classification techniques to map percentage of tree cover using ENVISAT MERIS (full spatial resolution 300m) dataset and vegetation metrics. These metrics and more information about ancillary data integration were discussed in section 5.

For the accuracy assessment of a LUC or fuzzy map, we need to get high resolution ground truth data. Crisp data is adequate for the hard classifications, however assessment of a soft classification needs fuzzy ground truth like real forest cover in study scale quantitavely. High spatial resolution Ikonos (4m) satellite images of three selected plots were used to derive training and testing ground data. Ikonos images were classified as forest and non-forest classes and, results rescaled to MERIS spatial resolution. 80% of this tree cover dataset was used as training data and 20% were separated for accuracy assessment. Linear (LMM and MLR) and non-linear (ANN and RT) techniques were compared.

LMM is one of the most used fuzzy techniques in the literature (Berberoglu & Satir 2008) and based on the assumption that class mixing is performed in a linear manner and therefore adopts a least squares procedure to estimate the class proportions within each pixel. The idea is that a continuous scene can be modeled as the sum of the radiometric interactions between individual cover types weighted by their relative proportions (Graetz 1990). The form of mixture model is:

$$Vi = \sum_{j=1}^{n} f_j r_{ij} - e_i \qquad (2)$$

Where Vi is the value of a pixel in input i, f_j the fractional abundance of endmember j in input i, rii is the value of the highest endmember j in input i, ei is the residual error associated with input i and n is the number of endmembers. Equation (2) is constrained by the assumption that the sum of the input components in each grid should equate to 1.0 as defined by equation (3):

$$\sum_{i=I}^{n} f_j = 1 \tag{3}$$

LMM needs pure pixels for each class to define the endmembers. Class membership functions are obtained based on endmember spectral characteristics (figure 9).

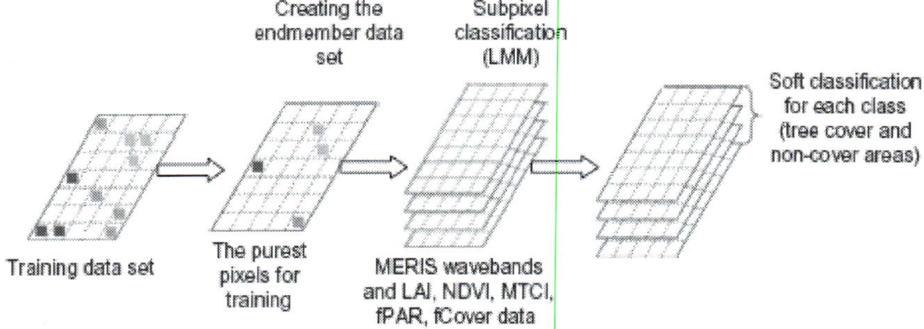

Fig. 9. Methodology for application of LMM

MLR refers to relating a response variable Y to a set of predictors xi in the form (e.g.Chatterjee and Price, 1991):

$$Y = b0 + b1.x1 + b2.x2 + ...+ bp.xp \tag{4}$$

Where the b0 is the constant value and b1 refers to coefficient of the first variable x1 (waveband). An advantage of linear regression is that it is easy to implement. MLR models are computationally efficient and can also predict confidence intervals for the obtained coefficients and the predicted data. Some of variable was eliminated using stepwise regression models.

The RT method has in recent years become a common alternative to conventional soft classification approaches, particularly with MODIS data (Hansen et al. 2005). The basic concept of a decision tree is to split a complex decision into several simpler decisions that can lead to a solution that is easier to interpret. When the target variable is discrete (e.g. class attribute in a land cover classification), the procedure is known as decision tree classification. By contrast, when the target variable is continuous, it is known as decision tree regression. In an RT, the target variable is a continuous numeric field such as percentage tree cover. Splitting algorithms were introduced in data dependent classifiers section. Splitting rules were contained only crisp equations. However, splitting rules were contained regression equation for each rule additionally in RT. In this study fallowing RT rules were applied to derive tree cover percentage (table 10).

	Condition	Target variable (percentage tree cover)
Rule 1	Band 3 > 4448 Band12<31651	66.2 – 0.1033 (band02) + 0.0509 (band03) + 0.0388 (band01) + 0.0014 (band07) – 0.0009 (band06) + 0.0012 (band04) – 0.0014 (band08) + 0.00012 (band12) – 0.0007 (band09) + 0.002 (band05)
Rule 2	Band 3>4448	–96.3 – 0.2479 (band02) + 0.1819 (band01) + 0.0672 (band07) + 0.0883 (band03) – 0.04 (band06) – 0.0459 (band08) – 0.0414 (band09) + 0.00472 (band12) + 0.095 (band05) – 0.00379 (band10) + 0.0101 (band04)
Rule 3	Band 3>4448 Band12>31651	–95.5 + 0.00571 (band10)

Table 10. Regression tree rules for tree cover percentage from MERIS data.

Correlation coefficiencies of each result with testing dataset from LMM, MLR, ANN and RT were 0.68, 0.69, 0.68 and 0.71 respectively. The most accurate result was obtained using RT technique.

These techniques are not only used to map two classes but also can be applied for more LUC class. In this frame, LMM and ANN fuzzy classification techniques were compared in almost same area as RT classification by Şatır (2006) (figure 10). Only forested areas were selected in Şatır's study. Training and testing dataset were derived from Landsat TM/ETM for each LUC.

Fig. 10. Study area boundary for LMM and ANN fuzzy classifiers comparison.

Turkish pine (*pinus brutia*), Crimean pine (*pinus nigra*), Lebanese cedar (*cedrus libani*), taurus fir (*abies cilicica*) and juniper (*juniperus excelsa*) were classified in detailed classification scheme. In addition, bareground, farmlands, forestlands and water bodies were classified in general classification scheme. LMM, ANN fuzzy classifications and ANN hard classification results were compared to see hard and soft classification accuracy difference and the best fuzzy classification technique (figure 11).

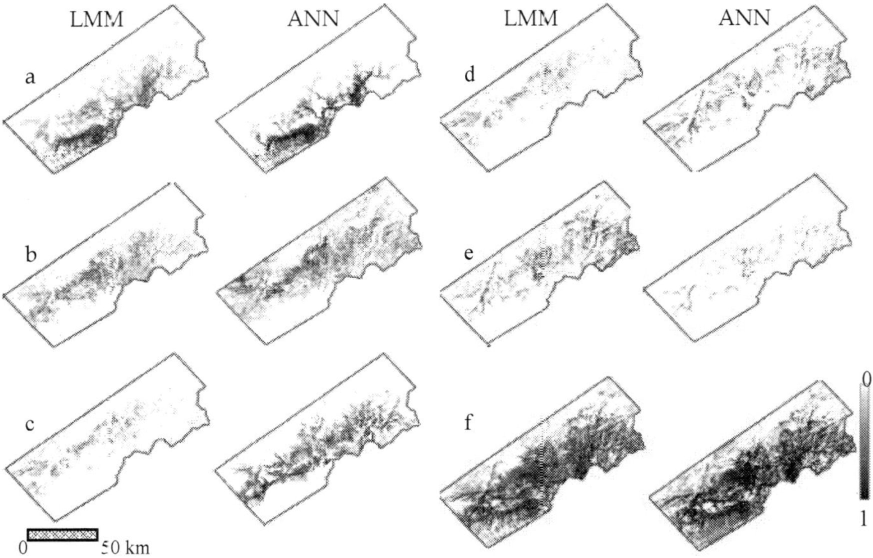

Fig. 11. Soft classification results (a: Turkish pine, b: Crimean pine, c: Juniper, d: Lebanese cedar, e: Taurus fir and f: General soft classification of forest).

LMM and ANN fuzzy classifications using medium spatial resolution data (300m) resulted reasonable classification outcomes if the training data set is large enough. On the other hand, in general both fuzzy classifications were more accurate than the hard classification results (table 11).

Fuzzy classifications are ideal for LUC mapping using coarse or medium spatial resolution data. However, fuzzy classification is not necessary in LUC mapping using very high spatial resolution data (e.g. 0.5m or 1m). High spatial resolution data have the characteristic that group of pixel shows the similar spectral characteristics. Object based classification techniques are suggested in this point.

	Accuracies (%)		
Detailed classifications	LMM	ANN (soft)	ANN (hard)
Bareground	78	72	60
Farmland	60	58	12
Water bodies	99	98	85
Turkish pine	85	89	56
Crimean pine	60	63	30
Lebanese cedar	35	40	8
Taurus fir	26	31	0
Juniper	34	44	7
Overall	60	62	33
General classifications	**LMM**	**ANN (soft)**	**ANN (hard)**
Bareground	79	72	70
Farmland	59	55	18
Forest	84	83	45
Water bodies	99	100	85
Overall	80	78	57

Table 11. Accuracy comparisons of fuzzy classification methods in different classification schemes.

4.3.2 Object based classification

Many complex land covers exhibit similar spectral characteristics making separation in feature space by simple per-pixel classifiers difficult, leading to inaccurate classification. Therefore, an object-based classification is a potential solution for the classification of such regions. The specific benefits are an increase in accuracy, a decrease in classification time and that it helps to eliminate within-field spectral mixing (Berberoglu et al., 2000). The object-based classification approach involved the integration of vector data and raster images within a geographical information system (GIS) and enabled the knowledge free extraction of image object primitives at different spatial resolutions, the so-called multiresolution segmentation. The segmentation operated as a heuristic optimization procedure which minimized the average heterogeneity of image objects at a given spatial resolution for the whole scene (Bian et al. 1992). The objective was to construct a hierarchical net of image objects, in which fine objects were sub-objects of coarser structures. Due to the hierarchical structure, the image data were simultaneously represented at different spatial resolutions. The defined local object-oriented context information was then used together with other (spectral, form, texture) features of the image objects for classification. At the next stage, supervised per-field classification was performed using the nearest neighbor algorithm utilizing field boundary data generated as a result of the segmentation procedure. Objects are segmented in the image and all objects are created object layer. Two or more object layer is called object hierarchy (figure 12).

Fig. 12. Image – object hierarchy

Basically, there are three steps in object based classification as segmentation, classification and per field integration. An image was divided segments dependent on pixel spectral similarities, structure of the image and surface texture characteristics. This progress is up to variables like scaling factor, smoothness vs. compactness and shape factors (figure 13).

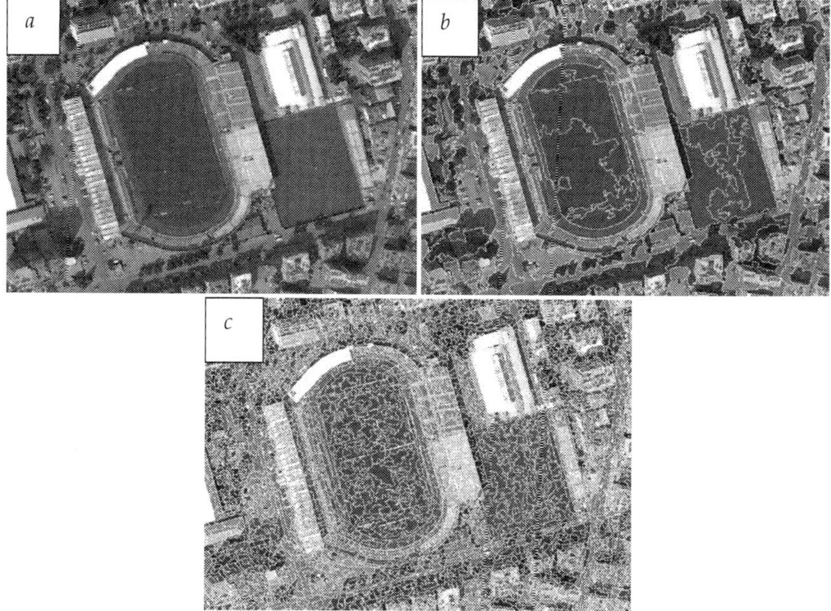

Fig. 13. *(a)* non-segmented image, *(b)* segmented image using scale factor 50, *(c)* segmented image using scale factor 10.

Each segments are contained a group of pixels and scaling factor is defined minimum pixel counts which have similar spectral characteristics in a segment. Compactness and smoothness are important for creating pixel groups. Shape factor is deal with boundary of a segment. Scale factor is variable according to the study scale and ideal scale can be found trying different scale factors. When the sensitive LUC analyze is necessary, compactness factor should be high and smoothness should be low (e.g. vegetation classification in CORINE level 3 and more). Shape factor is very important if shape of the LUC objects have a dominant characteristic (e.g. agricultural lands, roads and buildings).

Each LUC class can be defined using different dataset and rules according to characteristics of LUC. In this chapter, object and pixel based classifications were evaluated in a Mediterranean agricultural land called Lower Seyhan Plane (LSP) (figure 14).

Especially in agricultural land, object based classification is the most suitable technique. Most of the agricultural fields has regular shape and contains one dominant crop in a field in one time. In winter time dominant crop is wheat in the study area, summer period includes corn, soybean and cotton. Mapping the farmlands may be inappropriateusing only one optical image. Multitemporal object based classification approach was used to map LUC in LSP. Two Landsat TM images from March and April were classified together, and June, August images and some of physical variables like distance from cost line and distance from built up areas were added to create rules for each LUC. In this chapter only winter crop pattern discussed using LDA classifier, and rule dependent object based classification were compared each other to see accuracy difference in each LUC (figure 15).

Fig. 14. Lower Seyhan Plane (in yellow)

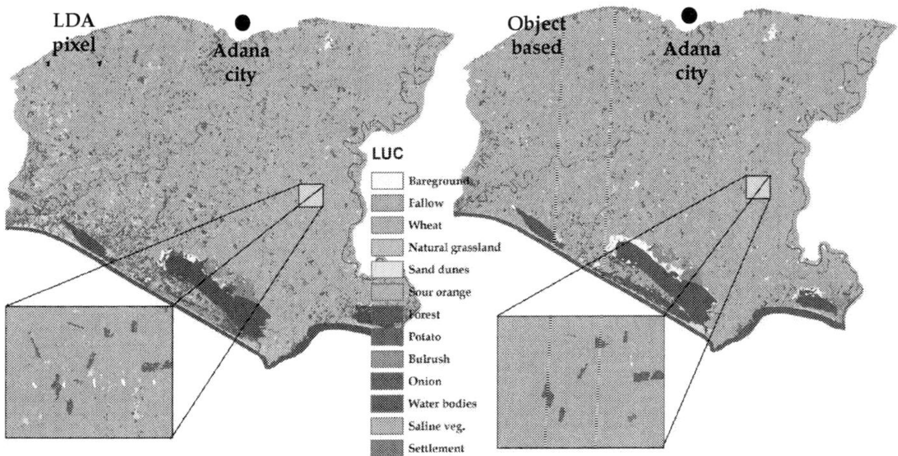

Fig. 15. LDA pixel based and object based classification results of LSP

Pixel based LDA classifier was failed in onion, sour orange, settlement, bulrush and sand dunes using March and April images. However, June and August images, distance from built up areas, distance from cost line were integrated in rule dependent object based classification and kappa coefficient was increased 28% in general. Sour orange, bulrush, sand dunes and settlement accuracy were raised impressively. One of the advantages in rule dependent classifiers was allowed to add new class during the classification. In this study, saline vegetation and natural grasslands were included to improve classification accuracy (table 12).

LUC	Object based classification	LDA	Difference
Wheat	%100	%100	-
Onion	%48	%15	%33
Potato	%84	%84	-
Citrus	%95	%23	%72
Water bodies	%100	%100	-
Fallow	%96	%96	-
Bareground	%85	%89	%4
Bulrush	%95	%34	%61
Forest	%100	%88	%12
Sand dunes	%100	%40	%60
Settlement	%95	%30	%65
Natural grasslands	%74	-	-
Saline vegetation	%95	-	-
Overall Kappa	%90	%62	%28

Table 12. Kappa accuracy of each LUC and difference between object and pixel based classifications

5. Ancillary data integration

Remotely sensed data may not be enough to map all LUC accurately alone. Ancillary data provide additional information on physical land dynamics, vegetation, climate, social geography and surface variability in LUC classification. When suitable ancillary dataset used, classification accuracy would be more accurate. In this chapter, only elevation (physical), texture (surface variability) and vegetation data (vegetation indices) were discussed in USP using DT and RT classifiers.

5.1 Physical data integration

Land physical dynamics such as elevation is vital physical input to LUC mapping. Digital elevation models (DEM) can be derived from stereo image pairs (e.g. ASTER) or radar (e.g. SRTM). Especially, vegetation formation and species vary according to elevation, aspect and climate. Using these ancillary data may improve accuracy of LUC maps (Coops et al. 2006, Şatır 2006). It is also possible to integrate soil characteristics into LUC mapping, because vegetation distribution and plant species are strongly dependent on soil depth, texture and moisture.

DEM was integrated to the DT and MLC classification in Eastern Mediterranean area discussed in section 4. Overall accuracy of the classification was increased approximately 4% and particularly bulrush, sand dunes and forestlands classified more accurately using DT. If topography vary in a study area, integrating the DEM may improve the LUC mapping accuracy. However, MLC classification overall accuracy was stable with and without DEM information. Most of the ancillary data increased the accuracy when using non-parametric techniques because parametric techniques like MLC uses the statistical equation to calculate distance of each LUC signature mean to the unknown pixel. However, DT creates rules based on the training data ranges, including elevation and spectral wavebands.

5.2 Surface texture data

Some of the variables can be produced using image wavebands such as surface texture and vegetation metrics. Surface textures are also used widely in LUC mapping. Many texture measures have been developed (Haralick et al. 1973, Kashyap et al. 1982, He and Wang 1990, Unser 1995, Emerson et al. 1999) and have been used for image classifications (Franklin and Peddle 1989, Narasimha Rao et al. 2002, Berberoglu et. al. 2000). Franklin and Peddle (1990) found that textures based on a grey-level co-occurrence matrix (GLCM) and spectral features of a SPOT HRV image improved the overall classification accuracy. Gong et al. (1992) compared GLCM, simple statistical transformations (SST), and texture spectrum (TS) approaches with SPOT HRV data, and found that some textures derived from GLCM and SST improved urban classification accuracy. Shaban and Dikshit (2001) investigated GLCM, grey-level difference histogram (GLDH), and sum and difference histogram (SADH) textures from SPOT spectral data in an Indian urban environment, and found that a combination of texture and spectral features improved the classification accuracy. The results based solely on spectral features increased about 9% to 17% with an addition of one or two texture measures. Furthermore, contrast, entropy, variance, and inverse difference moment provided larger accuracy and the most appropriate window size was 7X7 and 9X9. Multiscale texture measures should be incorporated with original spectral wavebands to improve classification accuracy (Shaban and Dikshit 2001, Podest and Saatchi 2002, Butusov 2003). Recently, the geostatistic-based texture measures were found to provide better classification accuracy than using the GLCM-based textures (Berberoglu et al. 2000). For a specific study, it is often difficult to identify a suitable texture because texture varies with the characteristics of the landscape under investigation and the image data used. Identification of suitable textures involves determination of texture measure, image band, the size of moving window, and other parameters (Chen et al. 2004). The difficulty in identifying suitable textures and the computation cost for calculating textures limit the extensive use of textures in image classification, especially in a large area (Lu and Weng 2007).

To test the texture data on classification accuracy, five different GLCM was derived such as, variance, contrast, dissimilarity, homogeneity, entropy. These measurements incorporated with Landsat spectral wavebands in Eastern Mediterranean region. Overall accuracy was unchanged, however accuracies of settlement and agricultural land classes were increased 4-5%. However, accuracy of bareground and sand dunes decreased using DT classifier.

5.3 Vegetation indices

Vegetation metrics are another ancillary data for more accurate LUC mapping. Deriving the metrics is dependent on the spectral resolution of an optical image. A vegetation index derived from combination of image wavebands. The most used vegetation indices are normalized difference vegetation index (NDVI), soil adjusted vegetation index (SAVI), normalized difference water index (NDWI), green vegetation index (GVI) and perpendicular vegetation index (PVI) at the literature. Vegetation indices indicate health condition (NDVI) and water content (NDWI) of the vegetation canopy. There are many textbooks and papers about calculation of vegetation indices. These indices provide extra information for LUC classification to discriminate subtle classes.

For instead, NDVI calculated using red and near infrared band (NIR) combination as shown in following equation (Rouse et al 1974);

$$NIR - RED / NIR + RED \qquad (5)$$

NDVI data was included to Landsat TM data to show the effect of a vegetation index on LUC mapping. Overall accuracy was unchanged significantly, but sand dunes, baregrounds, deciduous classes were classified more accurately.

Besides, there are some indices specifically designed for sensors. For example; Envisat MERIS data has own chlorophyll index called MERIS terrestrial chlorophyll index (MTCI). Additionally, vegetation metrics such as fraction of photosynthetically active radiation (fPAR), leaf area index (LAI) and fraction of green vegetation covering a unit area of horizontal soil (fCover) can be obtained using specific equations from MERIS data. Berberoglu et al. (2009) used this vegetation metrics to improve RT soft classification accuracy using MERIS data. When only MERIS wavebands used to determine the tree cover percentage, correlation coefficiency obtained as 0.58. Vegetation metrics and MERIS wavebands enhanced accuracy to 0.67.

6. Conclusions

This chapter has demonstrated various issues in LUC classification including, ability of optical remotely sensed data, different classifiers, training data size and ancillary data in the example of Eastern Mediterranean region. Parametric, non-parametric hard and soft LUC mapping techniques in local scale were assessed. Main findings of this chapter are:

Selection of a classification scheme and the optical data are vital for a reliable result in LUC mapping. Remotely sensed data must be defined according to the mapping scale and study purpose. LUC classification scheme and level should be defined based on optical data ability such as spatial and spectral resolution. Image pre-processing such as, geometric registration, atmospheric correction, geometric correction and radiometric calibration are essential parts in change detection studies.

Training data size, quality and mapping details are also important to select suitable classifier for LUC mapping. MLC, LDA, and DT techniques are useful for hard classification outputs. On the other hand, to derive a continuous map like cover percentage of each LUC or probability of each LUC needs soft classifiers such as RT and LMM. Training data size and quality affect the classification accuracy and classifier selection. Although model based classifiers has potential when strong training data set was used. In this case, data dependent classifiers can be chosen for better accurate LUC map. Linear techniques are suitable if mixture degree is small in a pixel. LMM is ideal if there are enough training data and pure pixel for each LUC. However, if training pixel size and pure pixels are weak, non-linear techniques like RT or ANN are suggested.

Hard classifiers were performed inaccurately with coarse spatial resolution images (e.g. MODIS, MERIS, NOAA, SPOTveg) because of mixed pixel problem. Fuzzy classifiers are reduced this problem and provided better accuracy than hard classification. Hard pixel based mapping techniques were successful using medium spatial resolution data (e.g. Landsat TM/ETM, Aster and Alos AVNIR) in regional and local scale, however, for the specific purposes like detailed crop pattern mapping or urban pattern mapping, object based classification approach was recommended for more reliable LUC mapping. Object based classification is appropriate when using very high spatial resolution data (e.g. rapid eye, Ikonos, Aerial photos, Geoeye). In segmentation stage of object based classification, pixels were merged to create each segment or object according to spectral, structural and textural similarities. This method is tolerated the pixel misclassification if there is a pixel noise in an area (Figure 15).

In this chapter ancillary data integration were also discussed using several data from satellite remote sensing sensors. Three types of ancillary data were integrated to the DT hard classifier. DEM resulted the largest improvement in overall classification accuracy among others. Surface texture and vegetation indices were improved the accuracy of specific land cover types. When all data used together, overall classification accuracy were reduced. Additionally, more ancillary data is not important to enhance classification accuracy. Success of the ancillary data varies based on classification target, study area characteristics and remotely sensed data.

7. References

Anderson, J. R.; Hardy, E. E.; Roach, J. T. & Witmer R. E. (1976). A Land Use and Land Cover Classification System For Use With Remote Sensor Data, *Geological Survey Professional Paper 964 A revision of the land use classification system as presented in U.S. Geological Survey Circular 671*, United States Government Printing Office, Washington.

Aplin, P. & Atkinson, P.M. (2001). Sub – Pixel Cover Mapping for Per – Field Classification. *International Journal of Remote Sensing*, 22 (14), pp. 2853 – 2858.

Atkinson, P.M. & Tatnall, A.R.L. (1997). Introduction: Neural Networks In Remote Sensing. *International Journal of Remote Sensing*, 18, pp. 699–709.

Barnsley, M.J. (1999). Digital Remote Sensing Data and Their Characteristics. In *Geographical Information Systems: Principles, techniques, applications, and management*, P. Longley, M. Goodchild, D.J. Maguire and D.W. Rhind (Eds), 2nd edn, pp. 451–466 John Wiley and Sons New York.

Berberoğlu, S. (1999). Optimising the Remote Sensing of Mediterranean Land Cover. Phd. Thesis, University of Southampton School of Geography.

Berberoglu, S.; Lloyd, C. D.; Atkinson, P. M. & Curran, P. J. (2000). The Integration of Spectral & Texture Information Using Neural Networks For Land Cover Mapping In The Mediterranean", *Computers and Geosciences* 26, pp. 385-396.

Berberoglu, S.; Satir, O. (2008). Fuzzy Classification of Mediterranean Type Forest Using ENVISAT-MERIS Data. *The International Archives of the Photogrammetry, Remote Sensing and Spatial Information Sciences*. ISPRS Beijing China. Vol.37, Part B8, pp. 1109 – 1114.

Berberog_u S. & Akin, A. (2009). Assessing Different Remote Sensing Techniques to Detect Land Use/Cover Changes In The Eastern Mediterranean. *International Journal of Applied Earth Observation and Geoinformation* 11, pp. 46–53.

Berberoglu, S.; Satir, O. & Atkinson, P. M. (2009). Mapping Percentage Tree Cover From Envisat MERIS Data Using Linear And Nonlinear Techniques. *International Journal of Remote Sensing*, 30:18, pp. 4747 – 4766.

Bian, L. & Walsh, S.J. (1992). Scale Dependencies of Vegetation and Topography In A Mountainous Environment Of Montana. *Professional Geographer*, 45, pp. 1–11.

Bishop, C.M. (1995). *Neural Networks for Pattern Recognition*. Oxford University Press, Oxford.

Betrie G. D.; Mohamed,; Y. A.; Van Griensven A. & Srinivasan, R. (2011). Sediment Management Modeling In The Blue Nile Basin Using SWAT Model. *Hydrol. Earth Syst. Sci.*, 15, pp. 807–818.

Brown, D. G.; Pijanowski,; B. C. & Duh, J. D. (2000). Modeling The Relationships Between Land Use and Land Cover on Private Lands in the Upper Midwest, USA. *Journal of Environmental Management*, 59, pp. 247-263.

Bossard, M., Feranec, J. & Otahel, J. (2000). *CORINE land cover technical guide - Addendum 2000*. European Environment Agency, Copenhagen.

Butusov, O.B. (2003). Textural Classification of Forest Types From Landsat 7 Imagery. *Mapping Sciences and Remote Sensing*. 40, pp. 91–104.

Cingolani, A.M.; Renison, D.; Zak, M.R. & Cabido, M.R. (2004). Mapping Vegetation in a Heterogeneous Mountain Rangeland Using Landsat Data: An Alternative Method to Define and Classify Land-Cover Units. *Remote Sensing of Environment*, 92, pp. 84–97.

Chatterjee, S.; Price, B. (1991). *Regression analysis by example*. Wiley NewYork.

Chavez, P.S. Jr, (1996). Image-Based Atmospheric Corrections—Revisited and Improved. *Photogrammetric Engineering and Remote Sensing*, 62, pp. 1025–1036.

Chen, D.; Stow, D.A. & Gong, P. (2004). Examining The Effect of Spatial Resolution and Texture Window Size On Classification Accuracy: An Urban Environment Case. *International Journal of Remote Sensing*, 25, pp. 2177–2192.

Clarke, K.C. (2008). A Decade of Cellular Urban Modeling with SLEUTH: Unresolved Issues and Problems, Ch. 3 in *Planning Support Systems for Cities and Regions*, Brail, R. K.,(Ed.), pp. 47-60, Lincoln Institute of Land Policy, Cambridge, MA.

Congalton, R.G. (1991). A Review Of Assessing The Accuracy of Classification of Remotely Sensed Data. *Remote Sensing of Environment*, 37, pp. 35–46.

Congalton, R.G. & Green, K. (1999). Assessing the Accuracy of Remotely Sensed Data: *Principles and practices*, Lewis Publishers Boca Raton, London, New York.

Congalton, R.G. and Plourde, L. (2002). Manual of Geospatial Science and Technology, In: *Quality assurance and accuracy assessment of information derived from remotely sensed data* J. Bossler (Ed.), pp. 349–361, Taylor & Francis, London.

Connor, D.W.; Brazier, D.P.; Hill, T.O. & Northen, K.O. (1997). *Marine Nature Conservation Review*: marine biotope classification for Britain and Ireland. Vol. 1, Littoral biotopes; Vol. 2, Sublittoral biotopes. Joint Nature Conservation Committee, Peterborough.

Coops, N. C.; Wulder, M. A.; White, J. C. (2006). Integrating Remotely Sensed and Ancillary Data Sources To Characterize A Mountain Pine Beetle Infestation, Mountain Pine Beetle, *Initiative Working Paper 2006-07*, Canadian Forest Service, Canada.

Cracknell, A.P. (1998). Synergy in Remote Sensing: What's in A Pixel? *International Journal of Remote Sensing*, 19, pp. 2025 - 2047.

Davies, C.E. & Moss, D. (2004). EUNIS Habitat Classification. Marine Habitat Types: Proposals for Revised Criteria, July 2004. *Report to the European Topic Centre on Nature Protection and Biodiversity*, European Environment Agency. July 2004.

Devillers, P.; Devillers-Terschuren, J. & Ledant, J.P. (1991). *Habitats of the European Community, CORINE biotopes manual. Vol. 2*. Office for Official Publications of the European Communities, Luxembourg.

Devillers, P. & Devillers-Terschuren, J. (1996). A classification of Palearctic habitats. Council of Europe, Strasbourg, *Nature and environment*, No 78.

Eastman, R.J. (2003). Classification of Remotely Sensed Imagery, In: *IDRISI Klimanjaro Guide to GIS and Image Processing*, 10.02.2012, Available from http://www.gis.unbc.ca/help/software/idrisi/kilimanjaro_manual.pdf

Emerson, C.W.; Lam, N.S. & Quattrochi, D.A. (1999). Multi-Scale Fractal Analysis of Image Texture And Pattern. *Photogrammetric Engineering and Remote Sensing*, 65, pp. 51–61.

European Environment Agency (EEA) (2012a). EUNIS Biodiversity Database, 03.01.2012, available from http://eunis.eea.europa.eu/.

European Environment Agency (EEA) (2012b). CORINE Land Cover Classes legend, 16.01.2012, available from http://www.eea.europa.eu/data-and-maps/figures/corine-land-cover-2000-launch-event posters/legend.png/?searchterm=None.

Foody, G.M. (1996). Approaches for the production and evaluation of fuzzy land cover classifications from remotely sensed data. *International Journal of Remote Sensing*, 17, pp. 1317 – 1340.

Foody, G.M. (2002). Status of Land Cover Classification Accuracy Assessment. *Remote Sensing of Environment*, 80, pp. 185–201.

Franklin, S.E. & Peddle, D.R. (1989). Spectral Texture For Improved Class Discrimination In Complex Terrain. *International Journal of Remote Sensing*, 10, pp. 1437–1443.

Franklin, S.E. & Peddle, D.R. (1990). Classification of SPOT HRV Imagery and Texture Features. *International Journal of Remote Sensing*, 11, pp. 551–556.

Gassman, P. W.; Williams, J. R.; Wang, X.; Saleh, A.; Osei, E.; Hauck, L. M.; Izaurralde, R. C.; Flowers, J. D. (2010). The agricultural policy/environmental extender (apex) model: an emerging tool for landscape and watershed environmental analyses. *American Society of Agricultural and Biological Engineers*. 53, pp. 711 – 740.

Ghose, M. K.; Pradha R.; Ghose, S. S. (2010). Decision Tree Classification Of Remotely Sensed Satellite Data Using Spectral Separability Matrix. *International Journal of Advanced Computer Science and Applications*, Vol. 1, No.5, pp. 93-101.

Gong, P.; Marceau, D.J. & Howarth, P.J. (1992). A Comparison Of Spatial Feature Extraction Algorithms For Land-Use Classification With SPOT HRV Data. *Remote Sensing of Environment*, 40, pp. 137–151.

Graetz, R.D. (1990). Remote Sensing of Terrestrial Structure: An Ecologist's Pragmatic View. In: *Remote Sensing of Biosphere Functioning*. R.J. Hobbs & H.A. Mooney (Eds), *Ecological Studies*, Vol. 79, pp. 5–30 New York: Springer Verlag.

Hadjimitsis, D.G.; Clayton, C.R.I. & Hope, V.S. (2004). An Assessment of the Effectiveness of Atmospheric Correction Algorithms Through The Remote Sensing of Some Reservoirs. *International Journal of Remote Sensing*, 25, pp. 3651–3674.

Hansen, M.C.; Townshend, J.R.G.; Defries, R.S. & Carroll, M. (2005). Estimation of Tree Cover Using MODIS Data At Global, Continental And Regional/Local Scales. *International Journal of Remote Sensing*, 26, pp. 4359–4380.

Haralick, R.M.; Shanmugam, K. & Dinstein, I. (1973). Textural Features For Image Classification. *IEEE Transactions on Systems, Man and Cybernetics*. 3, pp. 610–620.

He, D.C. & Wang, L. (1990). Texture Unit, Textural Spectrum and Texture Analysis. *IEEE Transactions on Geoscience and Remote Sensing*, 28, pp. 509–512.

Heo, J. & Fitzhugh, T.W. (2000). A Standardized Radiometric Normalization Method For Change Detection Using Remotely Sensed Imagery. *Photogrammetric Engineering and Remote Sensing*, 66, pp. 173–182.

Huang, C.; Davis, L. S. & Townshend J.R.G. (2002). An Assessment of Support Vector Machines For Land Cover Classification, *International Journal of Remote Sensing*, Vol.23, No. 4, pp. 725-749.

Huang, C.; Geiger E. L. & Kupfer, J. A. (2006). Sensitivity of Landscape Metrics to Classification Scheme, *International Journal of Remote Sensing*, 27:14, pp. 2927-2948.

Jensen, J.R. (1996). *Introduction to Digital Image Processing: A remote sensing perspective*, 2nd edn. Prentice Hall, Piscataway, NJ.

Joachims, T. (1998). Making Large Scale SVM Learning Practical. In Advances in *Kernel Methods – Support Vector Learning*. B. Scholkopf, C. Burges and A. Smola (eds). MIT Press New York.

Kohonen, T. (1988). An Introduction To Neural Computing. *Neural Networks*, 1, pp. 3–6.

Kandel, A. (1992). *Fuzzy Expert Systems*. CRC Press, Florida.

Karl, J. W. & Maurer, B. A. (2010). Multivariate Correlations Between İmagery and Field Measurements Across Scales: Comparing Pixel Aggregation and Image Segmentation. *Landscape Ecology*, 25, pp. 591-605.

Kashyap, R.L.; Chellappa, R. & Khotanzad, A. (1982). Texture Classification Using Features Derived From Random Field Models. *Pattern Recognition Letters*, 1, pp. 43–50.

Landgrebe, D.A. (2003). *Signal Theory Methods in Multispectral Remote Sensing Hoboken*, John Wiley and Sons NJ.

Lefsky, M.A. & Cohen, W.B. (2003). Selection of Remotely Sensed Data. In *Remote Sensing of Forest Environments: Concepts and case studies*, M.A. Wulder and S.E. Franklin (Eds), pp. 13–46 Kluwer Academic Publishers Boston.

Lillesand, T.M. & Kiefer, R.W. (1994). *Remote Sensing and Image Interpretation*, New York: Oxford University Press.

Lu, D. & Weng, Q. (2007). A Survey of Image Classification Methods and Techniques for Improving Classification Performance, *International Journal of Remote Sensing*, 28:5, pp. 823-870.

Muller, S.V.; Walker, D.A.; Nelson, F.E.; Auerbach, N.A.; Bockheim, J.G.; Guyer, S. & Sherba, D. (1998). Accuracy Assessment of A Land-Cover Map of The Kuparuk River Basin, Alaska: Considerations For Remote Regions. *Photogrammetric Engineering and Remote Sensing*, 64, pp. 619–628.

Narasimha Rao, P.V.; Sesha Sai, M.V.R.; Sreenivas, K.; Krishna Rao, M.V.; Rao, B.R.M.; Dwivedi, R.S. & Venkataratnam, L. (2002). Textural Analysis Of IRS-1D Panchromatic Data For Land Cover Classification. *International Journal of Remote Sensing*, 23, pp. 3327–3345.

Nordic Council of Ministers (1994). *Vegetation types of the Nordic Countries*. Nordic Council of Ministers, Copenhagen.

Özyavuz, M.; Şatır, O & Bilgili, B. C. (2011). A Change Vector Analysis Technique to Monitor Land Use/Land Cover In Yıldız Mountains, Turkey. *Fresenius Environmental Bulletin*, 20:5, pp. 1190 – 1199.

Podest, E. & Saatchi, S. (2002). Application of Multiscale Texture in Classifying JERS-1 Radar Data over Tropical Vegetation. *International Journal of Remote Sensing*. 23, pp. 1487–1506.

Renard, K.G.; Foster, G.R.; Weesies, G.A.; McCool, D.K. & Yoder, D.C. (1997). Predicting Soil Erosion by Water: A Guide to Conservation Planning With the Revised Universal Soil Loss Equation (RUSLE), *Agricultural Handbook*, 703. U.S. Government Printing Office, Washington, D.C.

Richards, J.A. & Jia, X. (1999). *Remote Sensing Digital Image Analysis*, Springer New York.

Rumelhart, D.E. Hinton, G.E. & Williams, R.J. (1986). Learning Internal Representations By Error Propagation. Parallel Distributed Processing: Explorations in the Microstructure of Cognition, Volume 1: Foundations, D. E. Rumelhart and J. L. McClelland (Eds). pp. 318-362. Cambridge, MA: The MIT Press.

Şatır, O. (2006). Land Cover Classification Using Appropriate Fuzzy Image Classification Techniques: Aladag Case Study, MSc Thesis, Cukurova University Institute of Basic and Applied Sciences, Landscape Architecture Dept. Adana, Turkey.

Shaban, M.A. & Dikshit, O. (2001). Improvement of Classification in Urban Areas By The Use of Textural Features: The Case Study of Lucknow City, Uttar Pradesh. *International Journal of Remote Sensing*. 22, pp. 565-593.

Southworth, J.; Nagendra, H. & Tucker, C. (2002). Fragmentation of A Landscape: Incorporating Landscape Metrics Into Satellite Analyses of Land-Cover Change, *Landscape Research*, 27:3, pp. 253-269.

Stehman, S.V. & Czaplewski, R.L. (1998). Design and Analysis For Thematic Map Accuracy Assessment: Fundamental Principles. *Remcte Sensing of Environment*, 64, pp. 331-344.

Toutin, T. (2004). Geometric Processing of Remote Sensing Images: Models, Algorithms and Methods. *International Journal of Remote Sensing*, 25, pp. 1893-1924.

Unser, M. (1995). Texture Classification and Segmentation Using Wavelet Frames. *IEEE Transactions on Image Processing*, 4, pp. 1549-1560.

Quinlan, J.R. (1993). *Programs for Machine Learning*. Morgan Kaufmann, San Mateo, CA.

Vapnik, V.N. (1995). *The Nature of Statistical Learning Theory*. New York: Springer-Verlag.

Zambon, M.; Lawrence R.; Bunn, A. & Powell S. (2006). Effect Of Alternative Splitting Rules On Image Processing Using Classification Tree Analysis. *Photogrammetric Engineering and Remote Sensing*, 72(1): 25-30.

3

GIS in Landscape Planning

Matthias Pietsch
Anhalt University of Applied Science
Germany

1. Introduction

Landscape planning supports sustainable development by creating planning prerequisites that will enable future generations to live in an ecological intact environment (Bfn, 2002). It breeds to a full-coverage strategy with the aim of maintaining landscape and nature as well as facilitating municipal and industrial development (von Haaren, 2004; BfN 2002). Contrary to the design approach (McHarg, 1969) it has been developed to an institutionalized planning system based on analytical processes (Schwarz- v. Raumer & Stokman, 2011). Objectives will be derived from scientifically based analysis and normative democratically legitimized goals (Riedel & Lange, 2001; Jessel & Tobias, 2002; von Haaren, 2004 a.o.).

Existing Geographic Information Systems (GIS) offer the needed capabilities concerning the whole planning cycle (Harms et al., 1993; Blaschke, 1997; von Haaren, 2004; Pietsch & Buhmann, 1999; Lang & Blaschke, 2007; a.o.). Data capturing for inventory purpose, scientific-based analysis, defining objectives, scenarios and alternative futures and planning measures can be carried out by using GIS (Schwarz-v. Raumer, 2011; Ervin, 2010; Steinitz, 2010; Flaxman, 2010). For the implementation and sometimes necessary updates environmental information systems can be developed for specific purpose (Zölitz-Möller, 1999; Lang & Blaschke, 2007). Nowadays required models (e.g. process, evaluation, decision) can be defined and interchanged for different scopes (Schaller & Mattos, 2010). The technical evolution of hard- and software enable planners and designers to improve participation processes and decision-making using visualization and WebGIS-technologies (Warren-Kretzschmar & Tiedtke, 2005; Paar, 2006; Lange, 1994; Wissen, 2009; Bishop et al., 2010; Buhmann & Pietsch, 2008a and b; Pietsch & Spitzer, 2011; Richter, 2009; Lipp, 2007). Transforming the existing planning process to a process-oriented one with new ways of interaction technical enhancements are necessary as well as a new planning and design style (Ervin, 2011). Therefore teaching methods must be changed to a more process- and workflow oriented thinking (Steinitz, 2010; Ervin, 2011) using the advantages of the different software tools like GIS, CAD, visualization and Building Information Models (BIM) (Flaxman, 2010). GeoDesign as a "new" term had been discussed the last years. While some planners contribute that they are doing this for years (Schwarz-v. Raumer & Stokman, 2011) requirements had been defined for a more collaborative and process-oriented planning (Francica, 2012; Ervin, 2011; Steinitz; 2010; Flaxman, 2010; Dangermond, 2009, 2010).

In this paper the different terms will be explained and the special German definition of landscape planning will be described. Based on this definition the use of GIS in the different working steps will be described and useful methods (e.g. habitat suitability analysis) will be explained. The needs for standardization and existing information management will be discussed and future improvements realizing the GeoDesign framework will be shown.

2. Definition

A lot of different terms are used in the context of spatial planning in the literature. Often spatial planning, physical planning, conservation planning, environmental planning, landscape planning, landscape design and a lot of other terms are used (Steinitz, 2010; Flaxman, 2010; Opdam et al., 2002; von Haaren, 2004; Jessel & Tobias, 2002; Kaule, 1991; Gontier, 2007; Weiers et al., 2004). But the common sense is the analyzing, evaluation, modeling, designing or the planning of measures (Opdam et al., 2002; Harms et al., 1993; Flaxman, 2010; Gontier et al., 2010) to achieve a more sustainable land management (McHarg, 1969; BfN, 2002). They deal with different scales (spatial, temporal) and pursue different goals (e.g. select conservation sites, design alternative futures, reduce environmental impact; improve urban development in a sustainable way). Therefore it's necessary to describe some of these terms that will be used in this article especially because some of them are specific German instruments. In all planning tasks you have to deal with spatial datasets (e.g. presence of species, habitats, soil types, land use, water bodies) and create in the planning process new datasets based on this information. Existing evaluation models will be used or new ones will be created using different scientific methods. New visions or alternative futures will be designed and have to be discussed with the public and explained to decision-makers using appropriate media and techniques. And at the end the results must be implemented (e.g. urban development plans, conservation agencies). Planning is not science, which means there are often different ways for the same direction. Scientific models and methods (e.g. landscape ecology) must be used to get the best results but in the end the decision is made by politicians in discussion with the public. Therefore it's necessary to work with as much as possible transparency (Steinitz, 2010; von Haaren, 2004) to produce convincing results with great acceptance (von Haaren, 2004). GIS tools and methods offer capabilities that are helpful in the whole planning process (Blaschke, 1997; Pietsch & Buhmann, 1999; Lang & Blaschke, 2007; a.o.).

2.1 Landscape design

Landscapes have been designed for thousands of years by human beings with impact from local to global scale (e.g. climate change, degradation). On the one hand intensive land use had and has a great environmental impact (e.g. fragmentation, biodiversity loss, soil erosion, water contamination) on the other hand specific land use types occurred that are now representative for specific regions (e.g. farmland in the Alps) and that are habitats for endangered species (plants and animals).

While societal and environmental issues are changing six basic questions must be asked in any situation of design (Steinitz, 1990, 2010) (see Fig. 1). Taking this framework into account the technology we are using might help to develop the necessary models and methodologies to manage landscapes in the complex world and to help the decision-makers in a confident way to design sustainable landscapes (Steinitz, 2010).

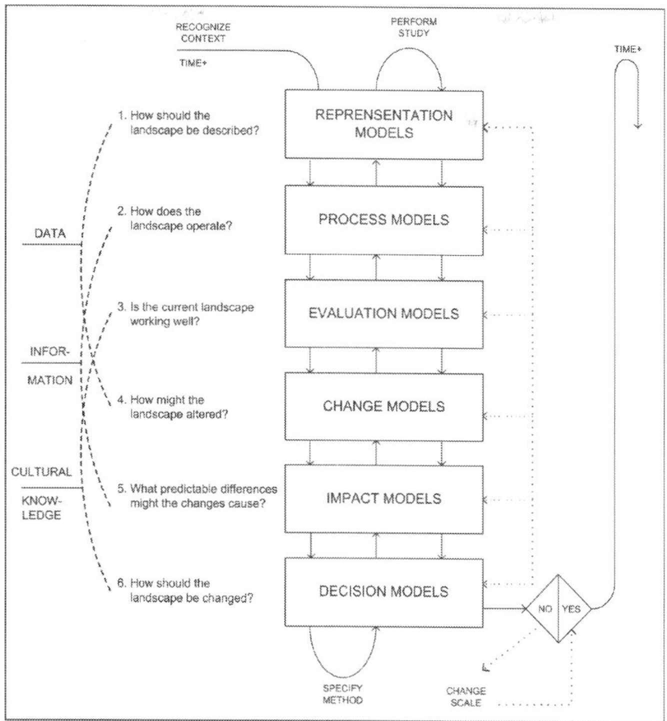

Fig. 1. „Steinitz Framework" (Steinitz, 1990, 2010)

It is mentioned that there is not one answer which are the best models and what are the spatial-analytic needs for designers or planners (Ervin, 2011). It depends on scale and complexity. The more detailed the planning task is the simpler the models may be and the experience of the planner or designer may be sufficient. For a large scale, the complexity is increasing and appropriate methods must be decided. Two basic strategies exist. The first one is to design the future state and then ask: "By what scenario might it be achieved" (Steinitz, 2010). The second one is to design a scenario and then ask: "In what future might it result?" (Steinitz, 2010).

Steinitz (2010) defined seven basic ways of designing that can be adapted using GIS technologies (at least in part) but that are not dependent upon them. They are accepted and often used in a specific way or in various combinations. They are:

- anticipatory
- participatory
- sequential
- combinatorial
- constraining
- optimizing or
- agent-based.

But there is a great relationship between the factors scale, decision models, process models and the way of designing (Flaxman, 2010, Steinitz, 2010). It is a difference between doing a design at small scale or even larger and larger. Therefore increasingly research is needed focused around the following six themes:

- content-problem seen over varied scales and locations
- decision model and its implementation (e.g. public participation)
- comparative studies of landscape processes (pattern and functions) and its models
- design methods and its applications (e.g. decision support systems, agent-based modeling)
- representation of the results (e.g. visualization techniques, animations, simulations)
- new technologies and its applications (e.g. augmented-reality, mobile devices, GIS on demand, WebGIS) (Steinitz, 2010).

2.2 The "German" situation

In Germany environmental planning and landscape planning can be distinguished (von Haaren, 2004; Klöppel et al., 2004; Jessel & Tobias, 2002; Riedel & Lange, 2001). They are the fundamentals for sustainable planning and the basics for decision-makers to take landscape functions into consideration (BfN, 2002).

2.2.1 Environmental planning

There are a lot of policies influencing the landscape. These are e.g. the European Landscape Convention, the Environmental Impact Assessment Act, the Strategic Environmental Impact Assessment Act or the Habitat Directive (Köppel et al., 2004; Umweltbundesamt, 2008). This leads to the recognition of the value of landscapes in law and the requirement of public participation processes (Steinitz, 2010; von Haaren, 2004; a.o.).

In contrast to Environmental impact assessment (EIA) on the project level, Strategic Environmental Impact Assessment (SEA) covers a wider range of activities, a wider area or sometimes a longer time span and identifies and evaluates the environmental consequences of proposed policies, plans or programs to ensure that the consequences are addressed at the earliest possible stage of the decision-making process (European Commission Environment DG, 2001; Gontier, 2007; Köppel et. al., 2004; Gontier et al., 2010). For this reason the focus of implementing sustainability aspects into the decision-making process is more effective and more sustainable-driven than being reactive on the EIA-level.

There exist three combinations of SEA and planning process (see Fig. 2).

The objectives of the intervention and environmental impact assessment provisions are to ensure that individual projects, plans or programs are performed with as little impact on the environment and nature as possible. This does not entail a separate procedure in its own right but rather takes place as part of the regular planning, permit and approval procedures e.g. of a road planning project or the urban development planning process (Heins & Pietsch, 2010; Köppel et al., 2004). The intervention provisions require registration and assessment of the effects that the planned project or plan can be expected to have on the functional capacity of the ecosystem and appearance of landscapes. In addition, the EIA and SEA is also concerned with the effects that such measures can have on human health and on cultural heritage (Federal Ministry for the Environment, Nature Conservation and Nuclear Safety 1998; Riedel & Lange, 2001; Busse et al., 2005; Köppel et al., 2004; Georgi & Peters, 2003; Georgi, 2003; Umweltbundesamt, 2008; Höhn, 2008; Jessel et al., 2009; Hendler, 2009; Wiegleb, 2009). Therefore the fundamental components of an EIA and SEA although there are variations around the world are:

- Screening
- Scoping
- Assessment and evaluation of impacts and development of alternatives
- Reporting
- Review
- Decision-making
- Monitoring, compliance, enforcement and environmental auditing (Commission for Environmental Assessment, 2006; von Haaren, 2004; Riedel & Lange, 2001; Köppel et al., 2004).

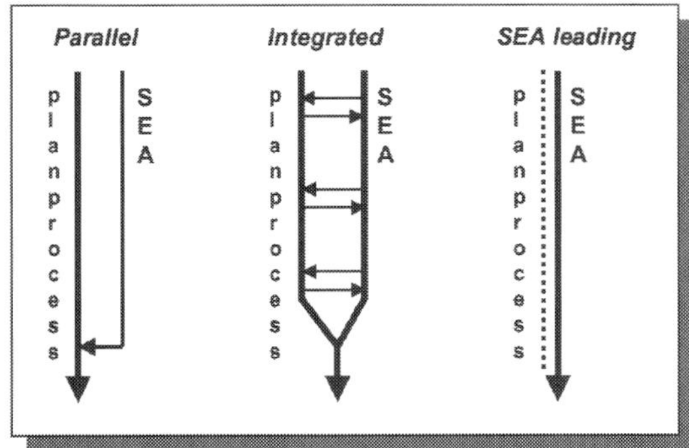

Fig. 2. Combinations of SEA and planning process (Commission for Environmental assessment, 2006)

In Germany mitigation and landscape envelope planning have a special role in the whole spectrum of environmental planning and nature conservancy. It is an instrument that is always tied in with the formal legal planning process such as e.g. road planning, zoning of areas for wind energy or solar panels (Riedel & Lange, 2001; Schultze & Buhmann, 2008; Köppel et al. 2004). Landscape envelope and mitigation planning is that instrument that optimizes the ecological balance of the overall project and to describe the compensation measures (BMV 1998). In the logical sequence of work steps Landscape envelope and mitigation planning uses the results of an EIA and is the legally binding part of the planning (Köppel et al., 2004; Schultze & Buhmann, 2008).

Based on the Habitats Directive and the Birds Directive sites had been selected according to EU standards to create the NATURA 2000 network. If any plan or project either itself or in combination with other plans or projects might have significance impact on a given site an impact assessment is required (Köppel et al., 2004; European Commission Environment DG, 2002).

The key issue to identify the significance impact is an assessment that covers:

-   habitats listed in Annex I of the Habitats Directive, including their characteristic species.
-   species listed in Annex I of the Habitats Directive and bird species listed in Annex I and Article 4 (2) of the Birds Directive, including their habitats and locations.
-   biotic and abiotic locational factors, spatial and functional relationships, structures, and site-specific functions and features of importance to the above mentioned habitats and species (Bundesamt für Naturschutz, 2012).

The impact assessment should lead to alternative solutions, compensatory measures and to preserve the overall coherence of the NATURA 2000 network (European Commission, 2007). For specific information about the procedure, the exceptional circumstances when a plan or project might still be allowed to go ahead in spite of a negative assessment see European Commission Environment DG (2002).

2.2.2 Landscape planning

In Germany, landscape planning, based on the Federal Nature Conservation Act, is the planning instrument for nature conservation and landscape management as opposed to other planning instruments and administrative procedures (The Federal Ministry for the Environment, Nature Conservation and Nuclear Safety 1998; Bfn 2002). It makes important contributions to the conservation of natural resources at all levels (local, district or entire regional state) and for full-coverage, sustainable conservation and long-term development of nature and landscapes in the built and non-built environment (see Fig. 3).

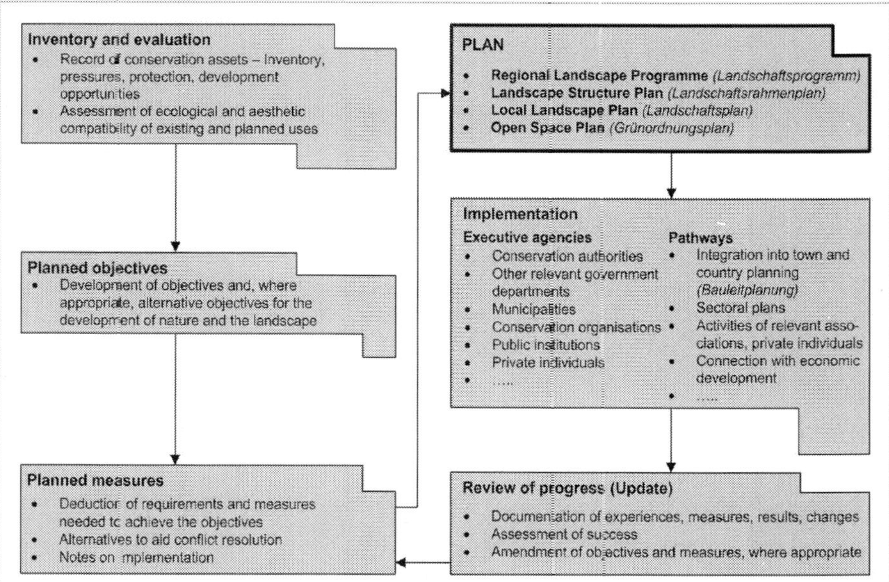

Fig. 3. Tasks and contents of landscape planning in logical sequence of work steps (BfN 2002)

In the different steps of the planning process a lot of information are needed to analyze and evaluate the current state of nature and the landscape, the functional capacity of the natural environment, the scenic qualities of the landscape, the development potential and the existing and foreseeable problems and conflicts with other uses (e.g. agriculture, traffic, housing) (Riedel & Lange, 2001; Jessel & Tobias, 2002; BfN, 2002; von Haaren, 2004;. a.o.). Based on that information a guiding vision and a set of objectives and measures have to be developed together with all stakeholders. Scenarios and visualizations are helpful to explain the details and the timescale for implementation (Lange, 1994; Ervin & Hasbrouck, 2001; Lange, 2002; Appleton & Lovett, 2003; Bishop & Lange, 2005; Paar, 2006; Warren-Kretzschmar & Tiedtke, 2005; Sheppard et al. 2008; Schroth, 2010; Schwarz-v. Raumer, 2011). At the end of the planning process the implementation and review (sometimes updates) are the most important tasks (BfN 2002).

2.3 Conclusions

Environmental and landscape planning in Germany differs in scope and aims. Landscape planning makes important contributions at all levels and full-coverage while environmental planning deals with specific scales (spatial and temporal) and territories (Köppel et al., 2004; Lambrecht et al., 2007; Riedel & Lange, 2001; von Haaren, 2004). The results of the landscape planning process can be used in the evaluation process of an EIA or SEA to analyze the negative effects caused by a project, plan or program. Otherwise the results of an SEA might be used in the following EIA or results of an EIA must be used in the following envelope and mitigation planning (Lambrecht et al., 2007; Schultze & Buhmann, 2008; Heins & Pietsch, 2010). That implies the necessity to realize an efficient information management and minimum standards to provide data and information in the right way at all work steps in the planning process (Pietsch & Heins, 2008; Heins & Pietsch, 2010; Schauerte-Lücke, 2008). Taking all frameworks in consideration information about the current state of nature and landscape, landscape functions, negative impacts or foreseeable conflicts must be evaluated and measures must be planned. Therefore different spatial information using scientific methods must be analyzed using e.g. GIS. Because the requirements are similar the following remarks will be limited to the landscape planning process while they may be assigned to the other planning tasks.

3. Capabilities of using GIS in landscape planning

Information technologies especially GIS can help to improve the landscape planning process and to capture the results in existing information systems for future use or as part of environmental information or decision support systems (Arnold et al., 2005; Pietsch & Buhmann, 1999; Blaschke, 1997; Lang & Blaschke, 2007; Gontier et al., 2007) (see Fig. 4).

GIS can be used in the different working steps of the landscape planning process. At the beginning data capturing in all planning tasks is necessary. This can be done by fieldwork, using existing thematic datasets (e.g. via Web Services) or by converting from existing databases or other monitoring systems (e.g. remote sensing). Checking the data quality is one of the most important tasks in the first step, to appreciate the necessity of data conversion, field work or usability of the existing material. After analyzing the existing situation for the defined scope the evaluation of potential of development, environmental functions, ecosystem services, scenic qualities, conflicts, previous and future impacts must be evaluated. Methods and tools analyzing datasets in different formats (raster and vector) or scales like spatial (resolution, grain, 2D, 3D) and temporal (historic, present, future conditions) are available. Some of them can be used for different purposes (e.g. evaluation, scenario and planning objectives) or they are only comfortable for a specific level or topic. Afterwards in landscape planning (e.g. local level) a guided vision or alternative futures should be planned and discussed (BfN, 2002; von Haaren, 2004). GIS can help to improve the participation process using visualization and multimedia techniques and Web GIS functionality. After the discussion process measurements should be planned and implemented in different ways. Monitoring of the landscape transformation must be done to check if it is really doing what it is expected to do (Opdam et al., 2002). Information management and a basic standardization are necessary to make sure that the life cycle works and in all work steps the right information in the right quality is available (Heins & Pietsch, 2010; Schauerte-Lücke, 2008; Opdam et al., 2002).

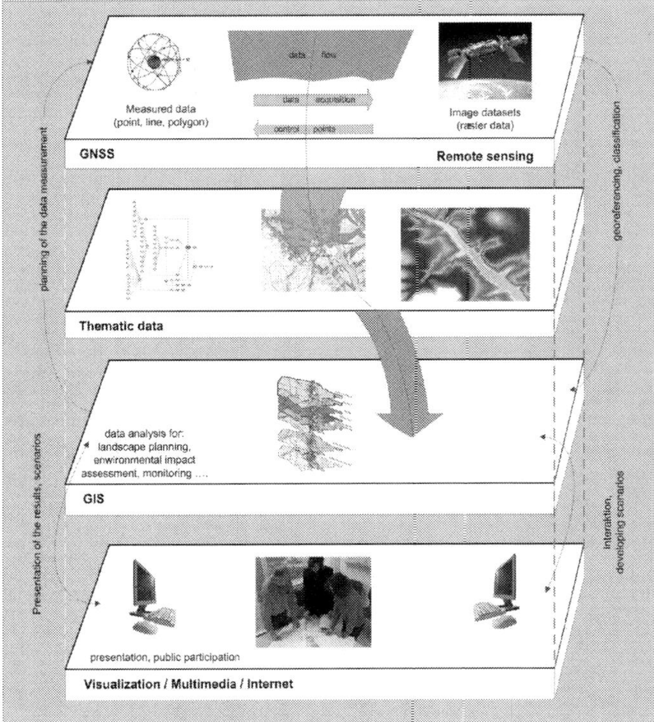

Fig. 4. Interaction of modern technologies in landscape planning (von Haaren, 2004)

3.1 Data capturing

Nowadays mobile devices are used for data capturing using Global Navigation Satellite Systems (GNSS) (e.g. GPS, GLONASS) in a standardized and formalized way (Dangermond, 2009; Brandt, 2007) to reduce effort in data conversion to implement and use the results in the

                                                                63

planning process. They are used e.g. to create tree cadastre (Pietsch, 2007; Brandt, 2007; GALK-DST, 2006), to collect species presence datasets (Dangermond, 2009), to reduce time and costs capturing land use types or habitats and for monitoring (e.g. checking mitigation measures). Depending on the hardware capacity and performance and the receiver accuracy improvements in data collections are possible. Using UMTS or other online services datasets might be send to a server (e.g. in the office) on the fly without necessary active copying or basic datasets like aerial images, top maps, thematic layers (e.g. streets) or the datasets that have to be checked can be received via Web Services (e.g. WMS, WFS) to be used online in the field. Digital cameras with a GNNS module facilitate documenting the investigation area. Images with coordinates are stored and some cameras and applications are able to analyze the viewing direction automatically. Using techniques like that enable the planner to create automatically documentations based on the existing images to present them e.g. online via Google Earth or to use them in the planning process (e.g. visualization, sketches, participation).

3.2 Analyzing

In the landscape planning process landscape functions like regulation, carrier, production and information functions must be analyzed (Groot, 1992; Pietsch & Buhmann, 1999; Jessel & Tobias, 2002; von Haaren, 2004; Lang & Blaschke, 2007). For nature conservation the regulation function is the most relevant (Weiers et al., 2004). Therefore landscape ecology defined as a problem-oriented science can provide methods for the different planning steps. But to optimize the knowledge-transfer between landscape ecology and spatial planning landscape ecology must co-evolve (Opdam et al., 2002). "In decision-making on future landscapes, landscape planners, landscape managers and politicians are involved in a cycling process" (Opdam et al., 2002) (see Fig. 5)

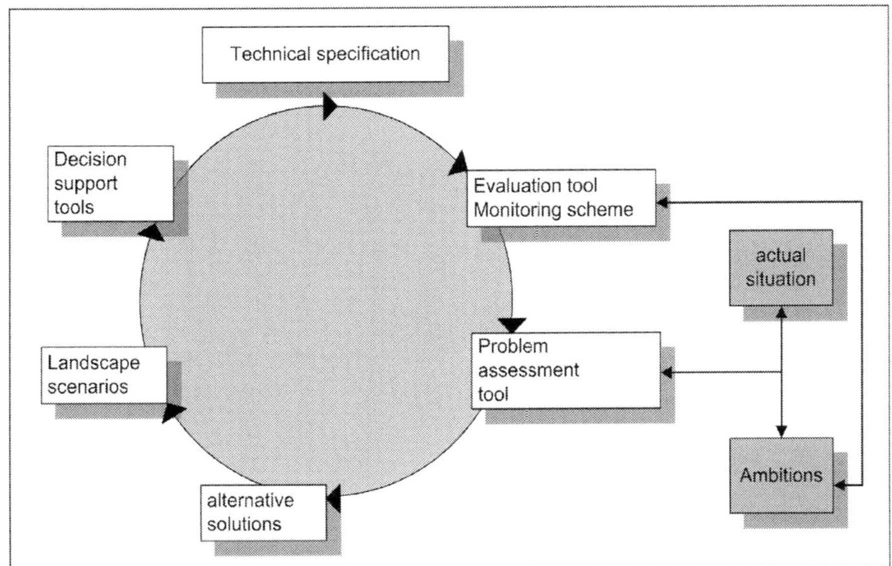

Fig. 5. Planning cycle (adapted from Harms et al., 1993)

That means that different models and methods are needed to integrate science in the planning process (Blaschke, 1997; Lang & Blaschke, 2007; Schwarz-v. Raumer & Stokman, 2011). Some examples will be given in the following chapters.

3.2.1 Multi-criteria evaluation (MCE)

Analyzing landscape functions (e.g. soil erosion) different information (e.g. land use, gradient, rainfall) must be taken in consideration. Using scientific-based methods the potential, risk or existing conflicts can be calculated. Depending on the selected methodology and the available information / datasets multi-criteria evaluation (MCE) is a very powerful tool. Therefore a reduction of complex environmental factors into a cohesive spatial concept is necessary. Overlay-functions (raster or vector-based) in combination with evaluation or impact models can be used to calculate e.g. suitable areas for farming or settlement or to perform impact analyses (see Fig. 6.)

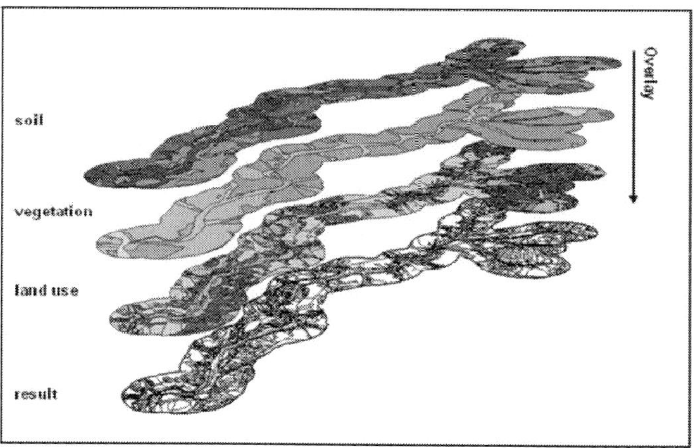

Fig. 6. Overlay (example)

3.2.2 GIS-based habitat models

In conservation biology and conservation planning there is a great diversity of GIS-based species distribution, habitat or population models (Blaschke, 1997; Blaschke, 2003; Taeger, 2010; Guisan & Zimmermann, 2000; Gontier, 2007; Gontier et al., 2010; Pietsch et al., 2007; Amler et al., 1999). Habitat suitability models based on empirical data versus models based on expert knowledge can be distinguished. On the other hand the level of detail (e.g. individuals, populations, species occurrences or species communities) is another way to describe the different models (Gontier et al. 2010).

There has been a lot of discussion about the possibilities to implement habitat suitability analysis (HSI) in environmental and landscape planning (Kleyer et al., 1999/2000; Schröder, 2000; Blaschke, 1999 and 2003; Rudner et al., 2003). They are established in environmental- and bio-science but because of the data requirements and the time- and cost-consuming modeling used only in a few planning examples (Jooß, 2003 and 2005; Pietsch et al., 2007; Gontier, 2007; Rudner et al., 2004; Schröder, 2000).

GIS-based models based on expert knowledge normally use presence datasets of specific species. Using the knowledge about the habitat preferences it's possible to analyze the suitability. Actual land use maps or other thematic information about habitats and specific structures or qualities (e.g. hydrological situation, soils, water quality) are used to evaluate the actual situation (Blaschke, 1997; Jooß, 2003; Taeger 2010). In contrast to ecological models using statistical methods models based on expert knowledge have a great potential to be used in landscape planning (see table 1). They are not as precise as ecological models but easier to interprete and applicable in larger areas.

	ecological model	"planning" model
method	- statistical approach -	- expert knowledge - prediction model
datasets	- actual presence data (based on field work) - (knowledge about habitat preferences)	- knowledge about habitat preferences
criteria / habitat information	- very detailed - for a small area	- more general - based on existing planning information - depends on the expert knowledge
quality/validity	- precise models, but only usable for a small area - scientific models	- fuzzy models, but for larger areas
validation	Easy to validate	Validation based on control samples
usability	- precise model for one specific species - maximum quality	- much more general, but useful for planning tasks - cumulative models for several species - scoping for species research

Table 1. Types of habitat models (adapted from Jooß, 2003)

They can be used for several species using existing species information. Based on the prediction model future conditions and different scenarios can be simulated to evaluate the impact of land use changes in the planning process (Gontier et al., 2010; Taeger, 2010; Pietsch et al., 2007; Blaschke, 1997). The visualization of future habitat suitability is possible. GIS offers the capability to create models (Fig. 7) based on existing datasets (species, land use, previous impact, structures, qualities) to analyze the actual and future suitability.

It's possible to create evaluation models, to create scenarios to improve the situation for one specific or several species, to develop measures to reduce negative impacts or create new habitats (Hunger, 2002; Hennig & Bögel, 2004) and they are useful to evaluate the negative impact of a plan, project or program in the context of an SEA or EIA (Gontier, 2007; Pietsch et al, 2007; Lang & Blaschke, 2007; Blaschke, 1999). In combination with connectivity analysis (see chapter 3.2.3) physical and functional links in ecological networks can be examined.

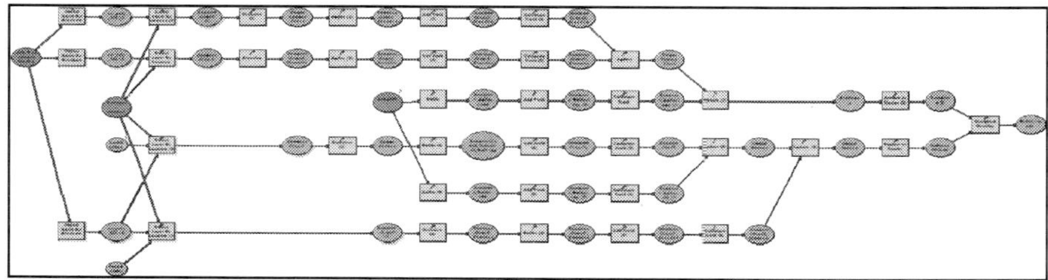

Fig. 7. Example for a habitat suitable model (Schmidt, 2007)

3.2.3 Connectivity analysis

Land use change and the physical and functional disconnection of ecological networks represent one of the driving forces of biodiversity loss (Zetterberg et al., 2010; Bundesamt für Naturschutz, 2004; Spangenberg, 2007; Reck et al., 2010). Beside a lot of different methodologies (see Fig. 8) network analysis and graph theory provide powerful tools and methods for analyzing ecological networks (Pietsch & Krämer, 2009; Urban et al., 2009; Zetterberg et al., 2010). There are three different types of connectivity analysis that can be classified according to the increasing data requirements and detail (Calabrese & Fagan, 2004).

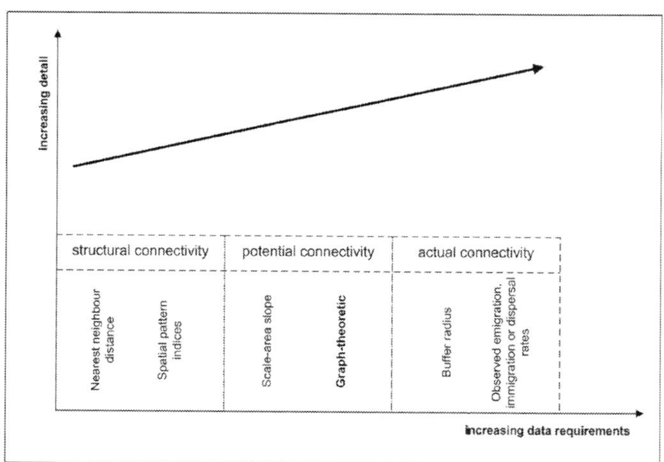

Fig. 8. Different types of quantitative connectivity analysis (adapted from Calabrese & Fagan, 2004; Wolfrum, 2006)

Graph-theory can be used as a method with very little data requirements, easy to use and not as sensitive as other methods against changes in scale (Urban et al., 2009; Bunn et al., 2000; Calabrese & Fagan, 2004; Urban & Keitt, 2001; Saura & Pascual-Hortal, 2007; Pascual-Hortal & Saura, 2006). Several graph-theoretic metrics related to classical network analysis problems had been developed and tested and ecologically interpreted (Bunn et al., 2000; Urban & Keitt, 2001; Wolfrum, 2006).

In graph-theory a graph is represented by nodes (e.g. habitats) and links (dispersal). A link connects node 1 and node 2 (see Fig. 9) (Tittmann, 2003; Urban & Keitt, 2001; Saura & Pascual-Hortal, 2007; Wolfrum, 2006). If the distance between two nodes is longer than the specific dispersal rate the link is missing, if the distance is in the dispersal range there is an existing link (Pietsch & Krämer, 2009; Zetterberg et al., 2010).

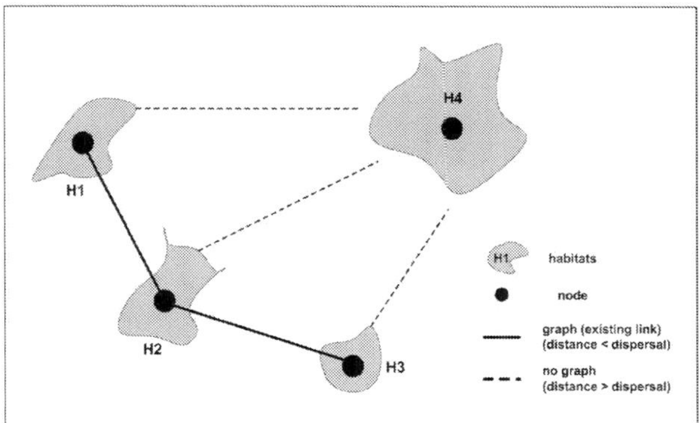

Fig. 9. Scheme of nodes and landscape graph representing habitats and connections (Pietsch & Krämer, 2009)

The graph-theory models can be distinguished in binary and probability models (Pascual-Hortal & Saura, 2006; Saura & Pascual –Hortal, 2007; Bunn et al., 2000; Urban & Keitt, 2001). Using binary models it's only possible to analyze if there is a link or not, while using probability models it's possible to analyze the existing situation (if there are links or not) and to evaluate each specific patch (habitat) (see Fig. 10) (Bunn et al., 2000; Urban & Keitt, 2001; Zetterberg et al., 2010). The distance between the nodes can be represented as edge-to-edge interpatch distance, as Euclidian distance or as least-cost path (Tischendorf & Fahrig, 2000; Ray et al., 2002; Adriaensen et al., 2003; Nikolakaki, 2004; Theobald, 2006, Zetterberg et al., 2010).

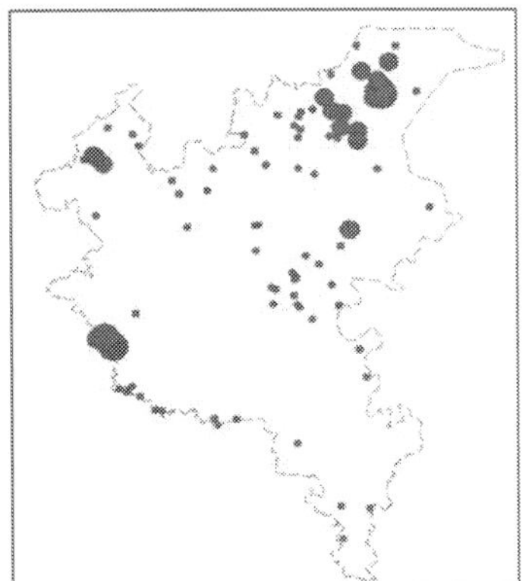
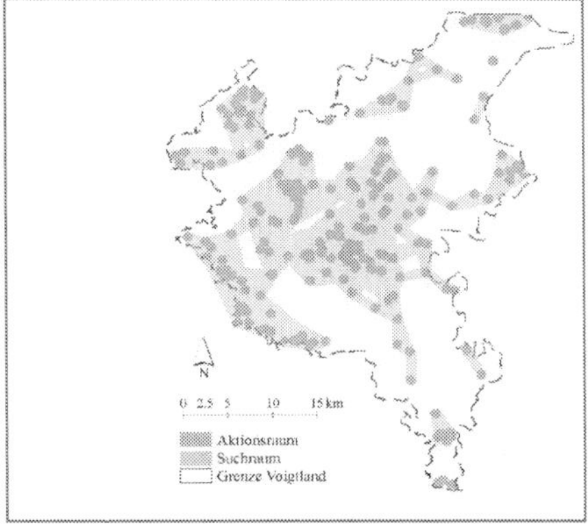

Fig. 10. Evaluation of specific habitats of Zootoca vivipara (example) (the bigger the more valuable) (left); patches and connectivity zones (right)

Using these techniques critical parts of a network can be identified e.g. through patch removal (Keitt et al., 1997; Zetterberg et al., 2010; Pietsch & Krämer, 2009), the different patches can be ranked according to their importance (Urban & Keitt, 2001) and natural and man-made barriers and breaks can be found (Zetterberg et al., 2010). They can be used as evaluation tools in the planning process or to analyze and visualize different possible scenarios for the participation process or to define areas that are most important for specific measures. In combination with cost-distance modeling (Adriaensen et al., 2003; Theobald, 2006; Zetterberg et al., 2010) and improved knowledge about species preferences and dispersal (Pietsch & Krämer, 2009) the tools are helpful to reduce negative ecological impact and find appropriate solutions in the landscape planning process.

3.3 Participation

The results of the landscape planning process are planned objectives or planned measures to be implemented into town and country planning, sectoral plans or executed by executive agencies (e.g. public institutions, conservation authorities, private individuals) (BfN, 2002; Riedel & Lange, 2001). Therefore landscape planning must be extended from an expert planning to a process-oriented planning where the participation process is one of the most important topics (Steinitz, 2010; von Haaren, 2004; Wissen, 2009). Based on the communication model of Norbert Wiener (Steinitz, 2010) the process has three elements: the message, the medium and the meaning. In landscape planning that means that the planner has a vision (a plan), the landscape is the medium and the viewer (public, stakeholder etc.) gains an impression of the changed landscape. In existing planning processes often the communication starts by the designer and ends by the viewer. But there must be a two-way alternate communication between the designer and the viewer to improve the results and the acceptance (Wissen 2009; Steinitz 2010; a.o.) (Fig. 11).

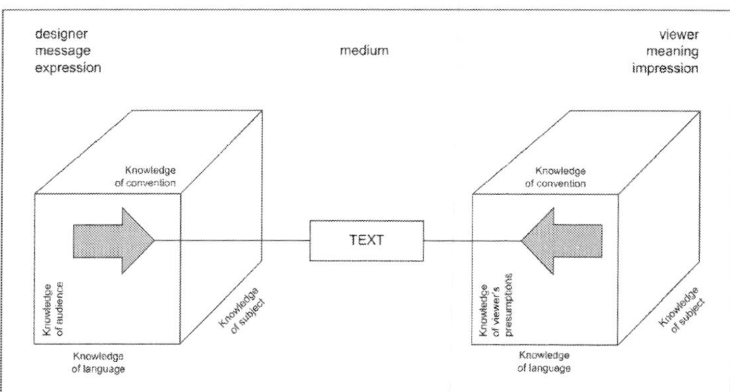

Fig. 11. Nobert Wiener's Communication Model (Steinitz, 2010)

Communication and information are the basic elements of participation (Warren-Kretzschmar & Tiedtke, 2005; Wissen, 2009). The advantages of computer-generated visualizations (plans, photomontages, 2D visualizations, 3D visualizations, real-time visualizations) in decision-making processes have been recognized for a long period (Lange 1994; Al-Kodmany 1999; Warren-Kretzschmar & Tiedtke 2005; Wissen 2009 a.o.). There are a lot of different media or visualization techniques that can be used for citizen participation (see table 2.). For non-experts it's often difficult to understand the planning ideas. On the other side it's necessary for the planner to express and communicate his thoughts in order to promote more sensitive landscape managing (Buhmann et al. 2010).

	Dynamic navigation	Interactivity (of image)	Photorealistic	GIS-supported	Internet
Interactive maps/ Aerial photos	+	-	-/+	++/-	++
Panorama photos	+	-	++	-	+
Photomontage	-	+	++	-	+
Sketches	-	++	-	-	-
Rendering of 3D-Model	-	+	+	++	+
VRML	++	-	+	+	++
Real-time	++	+	++	++	-

Legend: - unsuitable, + suitable, ++ very suitable

Table 2. Overview of visualization methods and their attributes (see Warren-Kretzschmar & Tiedtke, 2005)

But in all cases the questions remain:

- "Which characteristics of the visualizations are crucial for the support of citizen participation in the planning process?
- Which of the visualization methods are best suited for the different landscape planning tasks?
- How can visualization be successfully employed in citizen participation activities, both online and offline and which organizational aspects are important?" (Warren-Kretzschmar & Tiedtke, 2005).

Using appropriate techniques and media it's possible to explain complex environmental issues to layman. The combination of modeling techniques and GIS permit to open a "window to the future" to show scenarios, 3D- and 4D-simulations in different level of details and realism (Sheppard et al., 2008; Bishop & Lange, 2005; Paar & Malte, 2007; Säck-da Silva, 2007; Schroth, 2010; Pietsch & Spitzer, 2011).

A possibility to analyze relevant observation points for detailed visualizations are GIS based viewshed analysis. Digital Elevation Models (DEM) in combination with actual landform based on topography maps, orthophotos or thematic land use maps can be analyzed to select important vistas and areas from which a specific project might be visible or which area might be affected realizing a specific project (e.g. in the context of impact assessment for wind turbines) (see Fig. 12). After calculating the results they can be checked in the field and detailed visualizations of the before and after situation can be developed (Buhmann & Pietsch 2008a).

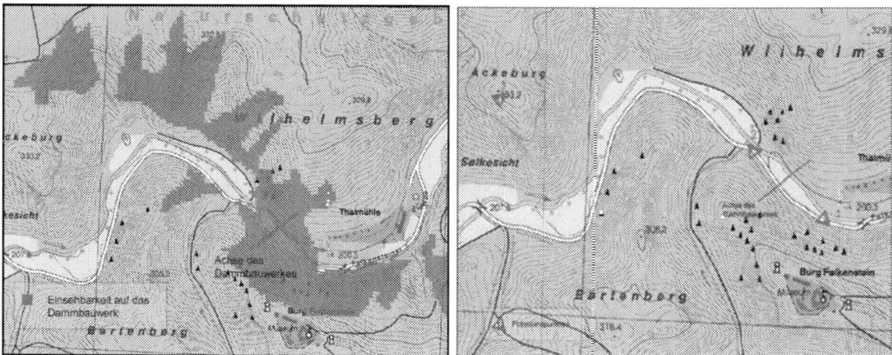

Fig. 12. Viewshed analysis (areas in green from which proposed dam in red is visible) (left);
seleceted observation points for detailed visualization (right)
(Buhmann & Pietsch, 2008a)

Creating a 3D model of the investigation area enables to calculate scenarios and simulations
through the integration of GIS data and generated 3D models (e.g. buildings, plants). Visual
impact assessment or aesthetic analysis (Kretzler, 2003; Ozimek & Ozimek, 2008; Bishop et
al., 2010) are possible as well as calculating affected settlements by different levels of
flooding and evaluating different measurements to reduce the impact (see Fig. 13)
(Buhmann & Pietsch, 2008a). Using these techniques different landscape ecological impacts
can be spatially defined, evaluated and visualized.

Fig. 13. Simulation of affected areas (blue) and not affected areas (white) during a flood 1994
(right); simulation with planned dams (right) (Buhmann & Pietsch, 2008a)

In detailed scale seasonal changes can be presented and discussed and visual impact
analysis based on temporal changes can be evaluated especially in sensitive (e.g. areas with
a high touristic potential) areas (see Fig. 14).

For public participation processes planning results can be presented interactively in real-time for offline or online purpose (Paar, 2006; Paar & Malte, 2007; Buhmann & Pietsch 2008a and b; Warren-Kretzschmar & Tiedtke, 2005; Wissen, 2009; Kretzler, 2002; a.o.). Visualization using different media and specific level of detail is a useful methodology to explain the results of the different working steps as well as to explain complex ecological issues in a way everybody (experts, public, stakeholder, politicians) is able to understand. In the context of Wieners communication model visualization techniques are a possibility to improve the meaning and understanding of the planers vision (Wissen, 2009).

Fig. 14. Simulation of a dam (left: winter, right: summer) (created by Lenné3D GmbH) (Buhmann & Pietsch, 2008b)

To create a two-way alternate communication between planer/designer and viewer Web GIS-technologies can be used (Warren-Kretzschmar & Tiedtke, 2005; Lipp, 2007; Richter, 2009). Through Web GIS-technologies it's possible to present spatial information via the internet, combine them with other media (Dangermond, 2009) and offer GIS capabilities (e.g. zoom, pan, spatial analysis, upload and download, network-analysis, editing) to enhance the communication process. Thematic maps using WebMap- or WebFeature-Services can be presented via Internet (Richter 2009; Krause 2011). In the landscape planning context the thematic maps and the belonging text explaining the landscape functions, conflicts, the guiding vision and objectives and measures can be published. Users can be enabled to navigate through the whole documents and give feedback using drawing or editing tools to locate the response and the possibility for textual information (see Fig. 15). All these information can be stored in a database to analyze the results of the public participation process, to redesign the plan and if necessary to reply to each of the user.

In combination with visualization techniques the communication between planner/designer and viewer can be improved. In addition to town meetings or specific workshops much more people can be involved using these techniques. Especially for landscape planning projects on regional level or for the entire state it might be helpful to improve the participation process simultaneously reducing planning period and costs.

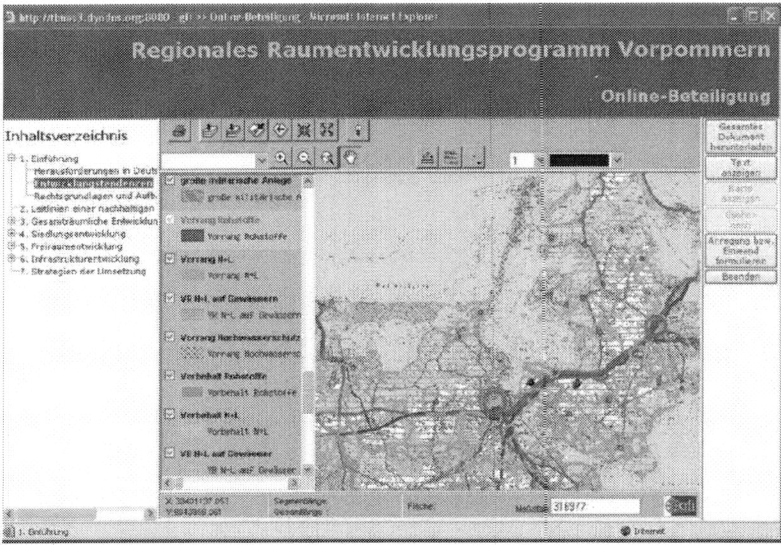

Fig. 15. Screenshot public participation server (Richter, 2009)

3.4 Objectives / vision / measures

Based on the inventory and evaluation process objectives and if necessary alternative objectives must be developed (Bfn 2002; von Haaren, 2004). A methodology to meet these requirements is to define for each landscape function (e.g. species/habitats/biotopes) two categories. All patches in the first category are most suitable for the defined function. All other land use types have no negative effect (e.g. habitats of endangered species). The second category involves patches with high relevance that has to be weighted with other functions (e.g. parks with relevance for endangered species and recreation). Defining all these categories and selecting the different patches might be very complex. GIS may help selecting the specific areas and create a summary of all demands. Based on these information using weighted overlay algorithms the vision and objectives can be defined (see Fig. 16).

Creating maps step-by-step in consideration of the defined criteria the discussion with decision-makers, public, stakeholders and experts can be improved. In town meeting or participation using internet technologies it's easier to understand the analysis and the requirements for each category and landscape function. During an interactive presentation different scenarios might be tested and the results can be visualized. Transparency can be increased and general agreement achieved.

Based on the defined objectives specific ad spatial concrete measures must be planned (Riedel & Lange, 2001; von Haaren, 2004; BfN, 2002). The results should be implemented in standardized environmental information systems to ensure the possibility for implementation, update and monitoring as well as publishing using Web services for other planning tasks or following working steps.

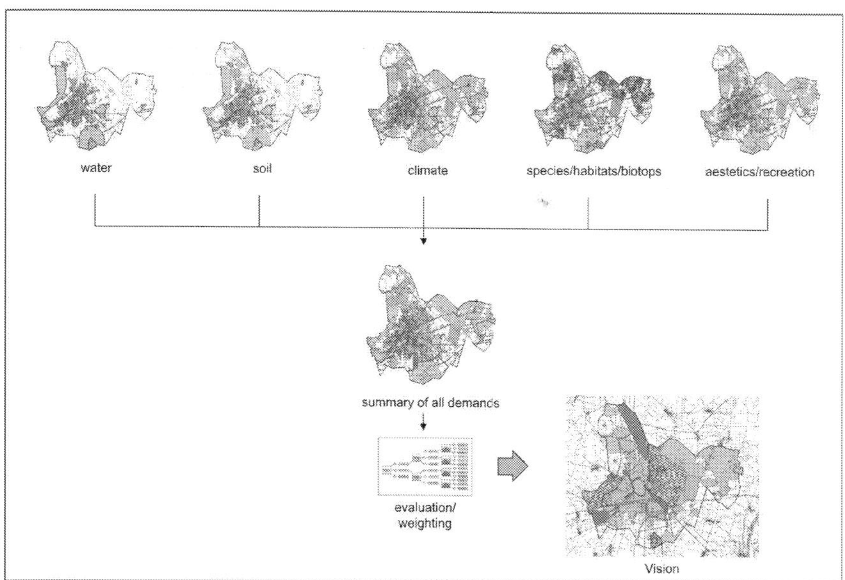

Fig. 16. Generating a vision based on weighted overlay of different plan concepts

4. Information management

In the past there had been a lot of problems exchanging information in horizontal and vertical ways between different planners and different landscape planning procedures (Krämer, 2008; Dembinsky, 2008; Arnold et al., 2005; Pietsch et al. 2010). In the context of environmental planning the whole planning process can be described as a life cycle of information (see Fig. 17).

To improve data exchange standardized, conceptual data models had been created e.g. for various areas of roads and transport (Hettwer, 2008) or regional, municipal land management and landscape planning in Germany (Benner et al. 2008; Benner & Krause, 2007). The purpose is to ensure a consistent object representation and a unified data exchange of graphic and geometric data (Ernstling & Portele, 1996; Hettwer, 2008; Pietsch et al., 2010; a.o.). The defined data models allow software developer to create specific application for landscape planning purposes and develop interface for data exchange.

For the representation guidelines and standard maps for different purposes had been developed to achieve a unified design in creating maps (Schultze & Buhmann, 2008). Taking the communication model of Norbert Wiener (Steinitz, 2010) in consideration defining and using data models lead to standardized communication without loss of information and meaning and improves data quality (Pietsch & Heins, 2009; Heins & Pietsch, 2010; Hettwer, 2008). Otherwise producing standardized datasets allows the implementation and development of Web GIS-applications for public participation or in monitoring / environmental information systems. Validation checks may be implemented to ensure data quality and to guarantee integrity. This allows to choose and develop scientific (process, evaluation, change, impact, decision) models for the planning process (Flaxman, 2010). Therefore existing data models must be extended using actual technical (e.g. OGC) and functional (e.g. guidelines, standard maps) standards. This might cause to a homogeneous terminology for planners and designers and a consistent presentation of results in the decision-making process. First steps had been done and some examples exist, but there is a lot of research to do to become these things reality.

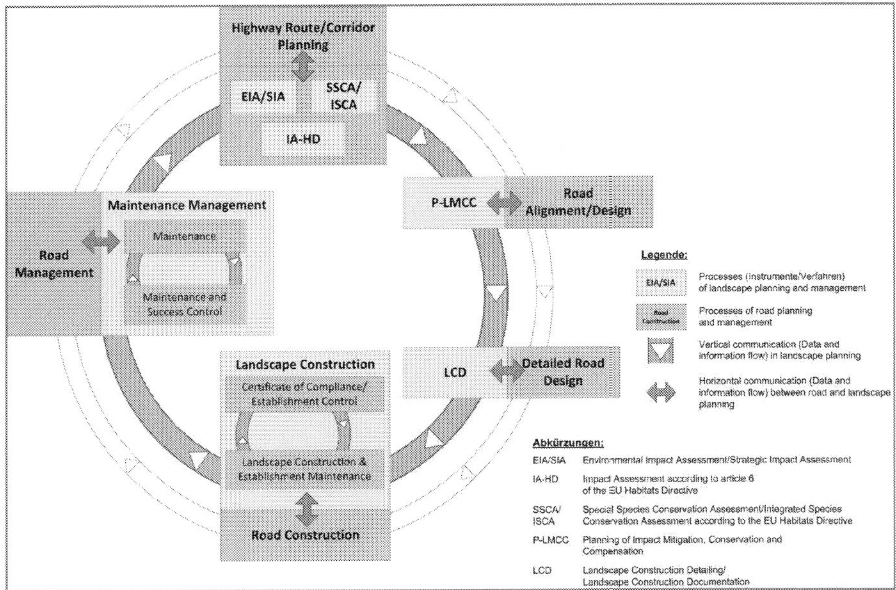

Fig. 17. Cycle of information in the context of landscape and road planning (Pietsch et al., 2010)

5. Geodesign - A new approach?

Since ESRI started the GeoDesign Summit in 2010 the term started his triumphal procession. But what is GeoDesign? According to MICHAEL FLAXMAN (2010) "GeoDesign is a design and planning method which tightly couples the creation of a design proposal with impact simulations informed by geographic context". The idea is that the planner or designer receives at every working step real-time guidance using contextual geographic information. The design can be evaluated relative to the local conditions and continuous feedback on multiple aspects will be provided through the whole planning process instead of post-hoc evaluation (Flaxman, 2010). GeoDesign may improve the design and planning process combining the potentials of CAD, GIS, Building Information Models (BIM) and visualization tools (Dangermond, 2009 and 2010; Flaxman, 2010; Ervin, 2011) and improve interaction and collaboration in the planning process (Tomlin, 2011; Francica, 2012). In contrast to the specific GIS or CAD workflow a hypothetical one for the GeoDesign workflow will look like Fig. 18.

Design Instantiation

> Pick a site or area of study
> Pick suitable feature representations, based on standard or custom data models
> Adjust visual portrayal (symbology) as desired
> Select suitable evaluation models based on availability, project needs

Integrated Design/Sketch Evaluation

> Sketch features (semantically rich and georeferenced by default)
> Sketch evaluation tools give feedback without blocking drawing
> Running selected models on design iterations is default and automatic

Full Impact Evaluation

> Same technical structure as sketch models (simply take longer to run)
> Models run as background tasks (typically as web geoprocessing services)
> Models results streamed back to design client incrementally as computed
> Evaluation models recognize design context in addition to input design data
> Appropriate analysis context can vary by model

Fig. 18. GeoDesign Process-Flow (Flaxman, 2010)

But actual the full process remains hypothetically while aspects are already available in existing software tools (Dangermond, 2009, 2010; Flaxman, 2010). The concepts had been embedded in Decision Support Systems (Brail et al., 2008) or GIS-based planning tools (Flaxman, 2010). GeoDesign is not a new concept. It's a refinement and restatement of ideas that had been discussed in the past multiple times (Flaxman, 2010; Ervin, 2011; Schwarz-v. Raumer & Stokman, 2011). But thinking about context-sensitive impact evaluation leads to an evolving concept. While multi-criteria analysis are not new (Schwarz-v. Raumer & Stokman, 2011; von Haaren, 2004; Jessel & Tobias, 2002) using them in real-time is a very complex issue and only a few GIS systems are able to do so (Flaxman, 2010). Sharing and deploying a variety of models and indicators using web services will radically reduce software installation and configuration time. The enhancement of web services to "geodesign evaluation services" (Flaxman, 2010) using open and interoperable formats will enlarge the development of tools and software systems. Standardized data models like CityGML (Flaxman, 2010) or XPlanung (Pietsch et al., 2010; Benner & Krause, 2007; Benner et al., 2008) in Germany are necessary as semantic representations of design domains but have to be expanded to evaluate the compliance of a plan for sustainable planning (Flaxman, 2010). The necessary elements that a hybrid GeoDesign System (GDS) requires are described by ERVIN (2011). He mentions sixteen essential components knowing that additional to the technical evolution some shifts in working styles are necessary. However the inevitable complications remain the GeoDesign concept remains enormous potential to improve design and planning processes if new ways of interaction towards a process-driven planning and project implementation will be achieved (Tomlin, 2011; Stockman & von Haaren, 2010; Schwarz-v. Raumer & Stokman, 2011).

6. Outlook

The rapid technical evolution in combination with internet technologies (e.g. Web 2.0) offers a chance for more collaboration and participation. New hardware like smart phones or tablet pc's with integrated GNSS facilitate the development of location based services or location sensitive services. They can be used for collecting data by the public (e.g. species data) or to publish information in the context of public participation processes. Using augmented reality technologies alternative futures can be presented in the real spatial context to improve decision-making processes. While GIS moved from workstations to desktop pc Web technologies and mobile devices are arising. GIS on demand using cloud technologies will be the next step.

Taking technical evolution in consideration standardization and a qualified information management will get more and more relevant. Moving the planning cycle from a step-by-step framework to a more process-oriented one standardized data models are needed. A unified terminology as a base for developing scientific models is necessary as well. Research in new design methods and the integration of science in the decision-making process is needed as well as the discussion about required changes in teaching students.

7. References

Adriaensen, F., Chardon, J.P., De Blust, G., Swinnen, E., Villalba, S., Gulinck, H., Matthysen, E. (2003): The application of least-cost modelling as a functional landscape model. in: Landscape Urban Planning, 64 (4), S. 233-247

Al-Kodmany, K. (1999): Combining Artistry and Technology in Participatory Community Planning, in: Berkeley Planning Journal, 13, p. 28-36

Amler, K., Bahl, A., Henle, K., Kaule, G., Poschlod, P. & Settele, J. (Hrsg.) (1999): Populationsbiologie in der Naturschutzpraxis, Ulmer Verlag, Stuttgart, 336 S.

Appleton, K., Lovett, A. (2003): GIS-based visualization of rural landscapes: defining 'sufficient' realism for environmental decision-making, in: Landscape and Urban Planning, 65, p. 216-224

Arnold, V., Lipp, T., Pietsch, M., Schaal, P. (2005): Effektivierung der kommunalen Landschaftsplanung durch den Einsatz Geographischer Informationssysteme, in: Naturschutz und Landschaftsplanung, Heft 11/2005, S. 349

Benner, J., Krause, K.-U. (2007): Das GDI-DE Modellprojekt XPlanung – Erste Erfahrungen mit der Umsetzung des XPlanGML-Standards, in: Schrenk, M., Popovich, V., Benedikt, J. (Eds.): Real Corp 007 Proceedings, S. 379-388

Benner, J., Köppen, A., Kleinschmidt, B., Krause, K.-U., Neubert, J., Wickel, M. (2008): XPlanung – Neue Standards in der Bauleit- und Landschaftsplanung, in: Buhmann/Pietsch/Heins (Eds.): Digital Design in Landscape Architecture 2008, Wichmann Verlag, Heidelberg, p. 240-248

BfN (The Federal Agency for Nature Conservation) (2002): Landscape planning for sustainable municipal development, Leipzig, 24 S.

Bishop, I., Lange, E. (2005): Visualization in Landscape and Environmental Planning. Technology and Applications. Taylor & Francis, London/New York, 296 S., 2005

Bishop, I., Smith, E., Williams, K., Ford, R. (2010): Scenario Based Evaluation of Landscape Futures – Tools for Development, Communication and Assessment, in: Buhmann/Pietsch/Kretzler (Eds.): Digital Landscape Architecture 2010, Wichmann Verlag, VDE Verlag GmbH, Berlin and Offenbach, p. 296-312

Blaschke, T. (1997): Landschaftsanalyse und –bewertung mit GIS, Forschungen zur Deutschen Landeskunde, Heft 243, Trier, 320 S.

Blaschke, T. (1999): Habitatanalyse und Modellierung mit Desktop-GIS, in: Blaschke, T.: Umweltmonitoring und Umweltmodellierung, Wichmann Verlag, p. 259-274

Blaschke, T. (2003): Habitatmodellierung im Naturschutz: Unterschiedlich komplexe Modelle und deren Zusammenführung. In: Dormann/Blaschke/Lausch/Schröder und Söndgerath: Habitatmodelle: Methodik, Anwendung, Nutzen – Tagungsband zum Workshop 8.-10. Oktober 2003, Ufz-Berichte 9/2004, S. 135-140

BMV (Bundesministerium für Verkehr) (1998): Musterkarten für die einheitliche Gestaltung landschaftspflegerischer Begleitpläne im Straßenbau, Verlags-Kartographie, Alsfeld

Botequilha, L., Ahern, J. (2002): Applying landscape ecological concepts and metrics in sustainable landscape planning, in: Landscape and Urban Planning, 59, p. 65-93

Brandt, M. (2007): Nutzen und Einsatzschwerpunkte von Freiflächeninformationssystemen, in: Die Wohnungswirtschaft, Sonderheft Freiflächenmanagement, 4/2007, S. 32-33

Brail, R. (Ed.): Planning support systems for cities and regions, Cambridge, 312 p.

Buhmann, E., Pietsch, M. (2008a): Interactive Visualization of Flooding Impact of Selke River, Harz. in: Buhmann/ Pietsch/Heins (Eds.): Digital Design in Landscape Architecture, Herbert Wichmann Verlag, S. 152-162

Buhmann, E., Pietsch, M. (2008b): Interaktion mit digitalen Landschaften, Szenarien und Modellen. in: Garten+Landschaft Heft 03/2008, S. 36-39

Buhmann, E., Pietsch, M., Langenhaun, D., Santosh, C., Paar, P. (2010): Visualizing Boathouses of Dwejra Bay in Gozo, Malta for Access Through WebGIS Applications, in: Buhmann/Pietsch/Kretzler (Eds.): Digital Landscape Architecture 2010, Wichmann Verlag, VDE Verlag GmbH, Berlin and Offenbach, p. 406-414

Bundesamt für Naturschutz (Hrsg.) (2004): Empfehlungen zur Umsetzung des §3 BNatSchG „Biotopverbund", Reihe Naturschutz und Biologische Vielfalt, Heft 2, Bonn-Bad Godesberg

Bunn, AG., Urban, DL, Keitt, TH. (2000): Landscape connectivity: a conservation application of graph theory, in: Journal Environmental Management, 59, S. 265-278

Busse, J., Dirnberger, F., Pröbstl, U. & Schmid, W. (2005): Die neue Umweltprüfung in der Bauleitplanung, 1. Auflage, Verlagsgruppe Hüthig Jehle Rehm GmbH, Heidelberg/München/Landsberg/Berlin, 316 S.

Calabrese, J. M., Fagan, W. (2004): A comparison-shopper's guide to connectivity metrics, in: Front Ecol Environ, 2 (19), S. 529-536

Commission for Environmental Assessment (Eds.) (2006): Biodiversity in EIA & SEA – Voluntary Guidelines on Biodiversity-Inclusive Impact Assessment, Utrecht

Dangermond, J. (2009): GIS: Design and Evolving Technology, ArcNews, ESRI, Fall 2009

Dangermond, J. (2010): GeoDesign and GIS – Designing our Futures, in: Buhmann/Pietsch/Kretzler (Eds.): Digital Landscape Architecture 2010, Wichmann Verlag, VDE Verlag GmbH, Offenbach and Berlin, p. 502-514

Dembinsky, K. (2008): Die Kommunikation in gemeinsamen PLanungsprozessen der Straßen- und Landschaftsplanung aus Sicht eines Straßenplaners, in: Buhmann/ Pietsch/Heins (Eds.): Digital Design in Landscape Architecture 2008, Wichmann Verlag, Heidelberg, p. 281-286

Erstling, R. & Portele, C. (1996): Standardisierung graphischer Daten im Straßen- und Verkehrswesen. Teil 1 - Studie. - Forschung Straßenbau und Straßenverkehrstechnik 724, Bundesministerium für Verkehr, Abteilung Straßenbau, Bonn-Bad Godesberg.

Ervin, S., Hasbrouck, H. (2001): Landscape Modeling: Digital Techniques for Landscape Visualization, McGraw-Hill, 289 S.

Ervin, S. (2011): A System for Geodesign in: Buhmann/Ervin/Tomlin/Pietsch (Eds.): Teaching Landscape Architecture – Prelimenary Proceedings, Bernburg, S. 145-154

European Commission Environment DG (2002): Assessment of plans and projects significantly affecting Natura 2000 sites, Luxembourg

Flaxman, M. (2010): Fundamentals of Geodesign, in Buhmann/Pietsch/Kretzler (Eds.): Digital Landscape Architecture 2010, Digital Landscape Architecture 2010, Wichmann Verlag, VDE Verlag GmbH, Berlin and Offenbach, p. 28-41

Francica, J. (2012): GeoDesign Is About Collaboration and Process, DirectionsMagazine, http://www.directionsmag.com/articles/geodesign-is-about-collaboration-and-process/225065

Galk-DST (Hrsg.): Grünflächen-Management, Beckmann Verlag

Georgi, B. (2003): Die Umweltfolgenprüfung als Instrument zur Umsetzung der Biodiversitätskonvention, UVP-report 17 (3+4), S. 148-150

Georgi, B., Peters, W. (2003): National Expert Workshop: Further development of the draft guidelines for incorporating biodiversity-related issues into environmental impact assessment legislation and/or process and strategic environmental assessment, Berlin 2003

Gontier, M. (2007): Scale issue in the assessment of ecological impacts using a GIS-based habitat model – A case study for the Stockholm region, in: Environmental Imapct Assessment Review, 27, p. 440-459

Gontier, M., Mörtberg, U., Balfors, B. (2010): Comparing GIS-based habitat models for application in EIA and SEA, in: Environmental Impact Assessment Review, 30, p. 8-18

Groot de, R.S. (1992): Function of nature. Evaluation of Nature in Environmental Planning, Management and Decision Making, Wolters-Noordhoff

Guisan, A., Zimmermann, N.E. (2000): Predictive habitat distribution models in ecology, in: Ecol. Moel., 135, p. 147-186

Harms, B., Knaapen, J. P. and Rademakers, J. G. (1993): Landscape planning for nature restoration: comparing regional scenarios, in: Landscape Ecology of a Stressed Environment, p. 197–218.

Hendler, R. (2009): Biodiversität in der Umweltprüfung – Rechtliche Sicht, in: Umweltbundesamt (Hrsg.): Umwelt im Wandel – Herausforderungen für die Umweltprüfungen (SUP/UVP), Berichte 1/09, Berlin, S. 17-24

Hennig, S. Bögel, R. (2004): Multivariate Statistik als Grundlage zur Habitatmodellierung – Am Beispiel der Habitatnutzung der Gams (Rudicapra rudicapra) im Nationalpark Berchtesgaden, in: Z. Naturschutz und Landschaftsplanung, 36, p. 363-370

Heins, M., Pietsch, M. (2010): Fachtechnische Standards für die vorhabenbezogene Landschaftsplanung, Edition Hochschule Anhalt, Selbstverlag, S. 125

Hettwer, J. (2008): Objektkatalog für das Straßen- und Verkehrswesen – OKSTRA, in: Buhmann/ Pietsch/Heins (Eds.): Digital Design in Landscape Architecture 2008, Wichmann Verlag, Heidelberg, p. 234-239

Höhn, K. (2008): Monitoring in der Bauleitplanung, VDM Verlag Dr. Müller, Saarbrücken, 146 S.

Hunger, H. (2002): Anwendungsorientiertes Habitatmodell für die Helm-Azurjungfer aus amtlichen GIS-Grundlagendaten, in: Natur und Landschaft, 77, p. 261-265

Jessel, B., Tobias, K. (2002): Ökologisch orientierte Planung, Verlag Eugen Ulmer, Stuttgart, 470 S.

Jessel, B., Böttcher, M. & Wilke, T. (2009): Biodiversität in der Umweltprüfung, in: Umweltbundesamt (Hrsg.): Umwelt im Wandel – Herausforderungen für die Umweltprüfungen (SUP/UVP), Berichte 1/09, Berlin, S. 1-16

Jooß, R. (2003):Ermittlung von Habitatpotenzialen für Zielartenkollektive der Fauna – Expertensysteme und empirische Ansätze im Landschaftsmaßstab. In: Dormann/Blaschke/Lausch/Schröder und Söndgerath: Habitatmodelle: Methodik, Anwendung, Nutzen – Tagungsband zum Workshop 8.-10. Oktober 2003, Ufz-Berichte 9/2004, S. 151-166

Jooß, R. (2005): Planungsorientierter Einsatz von Habitatmodellen im Landschaftsmaßstab-Schutzverantwortung für Zielarten der Fauna. In Korn, H./Feit, U. (Bearb.) Treffpunkt Biologische Vielfalt V – wissenschaftliche Expertentagung an der Naturschutzakademie Insel Vilm 23.-27. August 2004, Hrsg. BfN

Kaule, G. (1991): Arten- und Biotopschutz, 2. Auflage, Eugen Ulmer Verlag, Stuttgart, S.519

Keitt, T.H., Urban, D.L., Milne, B.T. (1997): Detecting critical scales in fragmented landscapes, in: Conserv. Ecol., 1 (1), p. 15-16

Kleyer, M., Kratz, R., Lutze, B., Schröder, B. (1999/2000): Habitatmodelle für Tierarten: Entwicklung, Methoden und Perspektiven für die Anwendung, in: Z. Ökologie und Naturschutz 8: p. 177-194

Köppel, J., Peters, W., Wende, W. (2004): Eingriffsregelung – Umweltverträglichkeitsprüfung, FFH-Verträglichkeitsprüfung, Eugen Ulmer Verlag, Stuttgart, 367 S.

Krämer, B. (2008): Die Kommunikation in gemeinsamen Planungsprozessen der Straßen- und Landschaftsplanung aus Sicht eines Landschaftsplaners, in: Buhmann/ Pietsch/Heins (Eds.): Digital Design in Landscape Architecture 2008, Wichmann Verlag, Heidelberg, p. 287-301

Krause, K.-U. (2011): Stand der Einführung von XPlanung in Norddeutschland, in: Schrenk/Popovich/Zeile (Eds.): Proceedings Real Corp 2011, p. 1399-1403

Kretzler, E. (2002): Computer Visualization of Environmental Impacts, in: Buhmann/Nothhelfer/Pietsch (Eds.): Trend in GIS and Virtualization in Environmental Planning and Design, Wichmann Verlag, Heidleberg, p. 58-67

Kretzler, E. (2003): Improving Landscape Architecture Design Using Real Time Engins, – in: Buhmann/Ervin (Eds.), Trends in Landscape Modeling, Herbert Wichmann Verlag, Heidelberg, p. 95-101

Lambrecht, H., Peters, W., Köppel, J., Beckmann, M., Weingarten, E., Wende, W. (Bearb.) (2007): Bestimmung des Verhältnisses von Eingriffsregelung, FFH-VP, UVP und SUP im Vorhabensbereich, BfN-Skripten 216, Bonn-Bad-Godesberg

Lang, S., Blaschke, T. (2007): Landschaftsanalyse mit GIS, UTB Verlag, 404 S.

Lange, E. (1994): Integration of Computerized Visual Simulation and Visual Assessment in Environmental Planning, in: Landscape and Urban Planning, 30 p. 99-112

Lange, E. (2002): Visualization in Landscape Architecture and Planning, in: Buhmann/Nothhelfer/Pietsch (Eds.): Trends in GIS and Virtualization in Environmental Planning and Design, Wichmann Verlag, Heidelberg, p. 8-18

Lipp, T. (2007): Der Landschaftsplan Güstrow online, in: Naturschutz und Landschaftsplanung, 39 (1), p. 25-29

McHarg, I.L. (1969): Design with Nature, The Natural History Press, New York, USA

Nikolakaki, P. (2004): A GIS site-selection process for habitat creation: estimating connectivity of habitat patches. in: Landscape Urban Planning, 68, S. 77-94

Opdam, P., Foppen, R., Vos, C. (2002): Bridging the gap between ecology and spatial planning in landscape ecology, in: Landscape Ecology, 16, p. 767-779

Ozimek, A., Ozimek, P. (2008): Computer-Aided Method of Visual Absorption Capacity Estimation, in: Buhmann/Pietsch/Heins (Eds.): Digital Design in Landscape Architecture 2008, Wichmann Verlag, Heidelberg, p. 105-114

Paar,P. (2006): Landscape visualizations: Applications and requirements of 3D visualization software for environmental planning, in: Computers, Environment and Urban Systems, 30, p. 815-839

Paar, P., Malte, C. (2007): Earth, Landscape, Biotope, Plant. Interactive Visualization with Biosphere3D, in: Schrenk/Popovich/Benedikt (Eds.): REAL CORP 007 Proceedings, Vienna, 207-214

Pascual-Hortal, L., Saura, S. (2006): Comparison and development of new graph-based landscape connectivity indices: towards the priorization of habitat patches and corridors for conservation. In: Landscape Ecology, 21 (7), S. 959-967

Pascual-Hortal, L., Saura, S. (2007): Impact of spatial scale on the identification of critical habitat patches for the maintenance of landscape connectivity. in: Landscape and Urban Planning, 83/2007, S. 176-186

Penko Seidel, N., Cof, A., Breskvar Zaucer, L., Marusic, I., (2010): The Problem of Large Protected Areas in the Process of Planning: A Case Study in the Municipality Ig, Slovenia, in: Buhmann/Pietsch/Kretzler (Eds.): Digital Landscape Architecture 2010, Wichmann Verlag, VDE Verlag GmbH, Berlin and Offenbach, p. 154-162

Pietsch, M., Buhmann, E. (1999): Auf dem Weg zur GIS-gestützten Landschaftsplanung – Die Hürden in der Praxis am Beispiel des Landschaftsplans der Verwaltungsgemeinschaft Sandersleben, in: Strobl/Blaschke/Griesebner (Hrsg.): Beiträge zur AGIT 1999, Herbert Wichmann Verlag, Heidelberg, S. 392-398

Pietsch, M. (2007): Softwarelösungen im Freiflächenmanagement, in: Die Wohnungswirtschaft, Sonderheft Freiflächenmanagement, 4/2007, S. 30-31

Pietsch, M., Schmidt, D., Richter, K. (2007): Ansatz zur Integration von Habitateignungsmodellen in die FFH-Managementplanung am Beispiel der

Flussperlmuschel. in: Strobl/Blaschke/Griesebner: Angewandte Geoinformatik 2007 – Beiträge zum 19. AGIT-Symposium Salzburg, Wichmann Verlag, S. 571-576

Pietsch, M., Heins, M. (2008): Qualifizierung und Optimierung der Landschaftsplanung durch Standardisierung und Informationsmanagement am Beispiel des OKSTRA®, in: Buhmann/Pietsch/Heins (Eds.): Digital Design in Landscape Architecture 2008, Herbert Wichmann Verlag, Heidelberg, p. 321-336

Pietsch, M., Krämer, M. (2009): Analyse der Verbundsituation von Habitaten und – strukturen unter Verwendung graphentheoretischer Ansätze am Beispiel dreier Zielarten, in Strobl/Blaschke/Griesebner (Hrsg.): Angewandte Geoinformatik 2009, Herbert Wichmann Verlag, Heidelberg, S. 564-573

Pietsch, M., Heins, M., Buhmann, E., Schultze, C. (2010): Object-based, process-oriented and conceptual Landscape models for standardizing landscape planning procedures in the context of road planning projects, in: Buhmann, Pietsch, Kretzler (Eds.): Digital Landscpae Architecture 2010, Wichmann Verlag, VDE Verlag, Berlin and Offenbach p. 62-68

Pietsch, M., Spitzer, S. (2011): Landschaftsbildvisualisierung im Hochwasserrisikomanagement unter Verwendung von open source Software, in: DWA (Hrsg.): GIS und GDI in der Wasserwirtschaft – Die Hochwasserrisikomanagementrichtlinie im Fokus von GIS und GDI, Selbstverlag, S. 1-16

Ray, N., Lehmann, A., Joly, P., (2002): Modelling spatial distribution of amphibian populations: a GIS approach based on habitat matrix permeability. in: Biodivers. Conserv. 11 (12), S. 2143-2165

Reck, H., Hänel, K., Huckauf, A. (2010): Nationwide Priorities for Re-Linking Ecosystems: Overcoming Road-Related Barriers – Summary, Leipzig, 23 p.

Richter, A. (2009): Öffentlichkeitbeteiligung durch den Einsatz von Web-GIS-Lösungen – Praxisbeispiel "Beteiligung-Online" der Landes- und Regionalplanung Mecklenburg-Vorpommerns, Vortrag im Rahmen des 1. Geofachtags Sachsen-Anhalt am 18.02.2009 in Bernburg, http://www.netzwerk-gis.de/files/richter_oeffentlichkeitsbeteiligung_2.pdf

Riedel, W., Lange., H. (Hrsg.) (2001): Landschaftsplanung, Spektrum Akademischer Verlag, Heidelberg – Berlin, 364 S.

Riedl, U., Taeger, S. (2009): GIS-basierte Habitatmodelle für das Biosphärenreservat Spreewald – Ergebnisse eines Forschungsvorhabens, in: Natur und Landschaft, 84 (4), p. 145-152

Rudner, M., Schadek., U., Damken, C. (2003): Habitatmodelle und ihre mögliche Integration in die Planungspraxis-ein Diskussionsbeitrag. In: Dormann/Blaschke/Lausch/Schröder und Söndgerath: Habitatmodelle: Methodik, Anwendung, Nutzen – Tagungsband zum Workshop 8.-10. Oktober 2003, Ufz-Berichte 9/2004, p. 167- 171

Saura, S., Pascual-Hortal, L. (2007): A new habitat availability index to integrate connectivity in landscape conservation planning: Comparison with existing indices and application to a case study. in: Landscape and Urban Planning, 83/2007, S. 91-103

Säck-da Silva, S. (2007): Visualisierung und Interaktivität: wie GIS die Öffentlichkeitsbeteiligung unterstützt, in: Haustein/Demel (Hrsg.): GIS in der Stadt- und Landschaftsplanung, Heft 164, Universität Kassel, S. 87-96

Schauerte-Lücke, N. (2008): Informations- und Datenmanagement in gemeinsamen Planungsprozessen der Straßen- und Landschaftsplanung, in: Buhmann/Pietsch/Heins (Eds.): Digital Design in Landscape Architecture 2008, Herbert Wichmann Verlag, Heidelberg, p. 302-309

Schwarz-v. Raumer (2011): GeoDesign - Approximations of a catchphrase in: Buhmann/Ervin/Tomlin/Pietsch (Eds.): Teaching Landscape Architecture - Prelimenary Proceedings, Bernburg, p. 106-115

Sheppard, S., Shaw, A., Falnders, D., Burch, S. (2008): Can Visualisation Save the Worlds? - Lessons for Landscape Architects from Visualizing Local Climate Change, in: Buhmann/Pietsch/Heins (Eds.): Digital Design in Landscape Architecture 2008, Herbert Wichmann Verlag, Heidelberg, p. 2-21

Spangenberg, J.H. (2007): Biodiversity pressure and the driving forces behind, in: Ecological Economics, p. 146-158

Steinitz, C. (2010): Landscape Architecture into the 21st Century - Methods for Digital Techniques, in: Buhmann/Pietsch/Kretzler (Eds.): Digital Landscape Architecture 2010, Wichmann Verlag, VDE Verlag GmbH, Berlin and Offenbach, p. 2-26

Schaller, J., Mattos, C. (2010): ArcGIS ModelBuilder Applications for Landscape Development Planning in the Region of Munich, Bavaria, in: Buhmann/Pietsch/Kretzler (Eds.): Digital Landscape Architecture 2010, Wichmann Verlag, VDE Verlag GmbH, Berlin and Offenbach, p. 42-52

Schmidt, D. (2007): Ein wissensbasiertes Habitateignungsmodell für die Flussperlmuschel (*Margaritifera margaritifera L.*) im Rahmen der FFH-Managementplanung, Diplomarbeit an der Hochschule Anhalt (unveröffentlicht)

Schröder, B. (2000): Habitatmodelle für ein modernes Naturschutzmanagement in Gnauck, A.: Theorie und Modellierung von Ökosystemen - Workshop Kölpingsee 2000, Aachen

Schroth, O. (2010): Tools for the understanding of spatio-temporal climate scenarios in local planning: Kimberley (BC) case study, University of British Columbia, Vancouver

Schultze, C., Buhmann, E. (2008): Developing the OKSTRA Standard for the Needs of Landscape Planning in Context of Implementation for Mitigation and Landscape Envelope Planning of Road Projects, in: Buhmann/Pietsch/Heins (Eds.): Digital Design in Landscape Architecture 2008, Wichmann Verlag, Heidelberg, p. 310-320

Schwarz-v. Raumer (2011): GeoDesign - Approximations of a catchphrase in: Buhmann/Ervin/Tomlin/Pietsch (Eds.): Teaching Landscape Architecture - Prelimenary Proceedings, Bernburg, S. 106-115

Stockman, A., von Haaren, C. (2010): Integrating Science and Creativity for Landscape Planning and Design of Urban Areas, in: Weiland/Richter (Hrsg.): Urban Ecology - a global Framework, Oxford, Blackwell Publishing

Taeger, S. (2010): GIS and Hydrological Models as Tools for Habitat Modelling in Alluvial Plains, in: Buhmann/Pietsch/Kretzler (Eds.): Digital Landscape Architecture 2010, Wichmann Verlag, VDE Verlag GmbH, Berlin and Offenbach, p. 370-378

The Federal Ministry for the Environment, Nature Conservation and Nuclear Safety (Eds.) (1998): Landscape Planning, Bonn

Theobald, D.M. (2006): Exploring the functional connectivity of landscapes using landscape networks. in: Crooks, K.R., Sanjayan, M. (Eds.): Connectivity Conservation, Cambridge University Press, New York, S. 416-443

Tischendorf, L., Fahrig, L. (2000): On the usage and measurement of landscape connectivity. in: Oikos 90 (1), S. 7-19

Tittmann, P. (2003): Graphentheorie: eine anwendungsorientierte Einführung, München

Tomlin, C.D. (2011): Speaking of Geodesign in: Buhmann/Ervin/Tomlin/Pietsch (Eds.): Teaching Landscape Architecture – Prelimenary Proceedings, Bernburg, S. 136-144

Umweltbundesamt (Hrsg.) (2008): Leitfaden zur Strategischen Umweltprüfung (Langfassung) – Forschungsvorhaben 206 13 100, Berlin, 52 S.

Urban, D., Keitt, T. (2001): Landscape connectivity: A graph –theoretic perspective. in: Ecology 82 (5), 1205-1218

Urban, D.L., Minor, E.S., Treml, E.A., Schick, R.S. (2009): Graph models of habitat mosaics, Ecol. Lett., 12 (3), p. 143-157

von Haaren, C. (Hrsg.) (2004): Landschaftsplanung, Eugen Ulmer Verlag, Stuttgart, 528 S.

Warren-Kretzschmar, B., Tiedtke, S. (2005): What Role Does Visualization Play in Communication with Citizens? – A Field Study from the Interactive Landscape Plan, in Buhmann/Paar/Bishop/Lange (Eds.), Trends in Real-Time Landscape Visualization and Participation, Herbert Wichmann Verlag, p. 156-167

Weiers, S., Bock, M., Wissen, M., Rossner, G. (2004): Mapping and indicator aproaches for the assessment of habitats at different scales using remote sensing and GIS methods, in: Landscape and Urban Planning, 67, p. 43-65

Wiegleb, G. (2009): Methodik der Erfassung und Bewertung von Biodiversitätsschäden aus ökologischer Sicht, in: Knopp, L., Wiegleb, G. (Hrsg.): Der Biodiversitätsschaden des Umweltschadensgesetzes, Schriftenreihe Natur und Recht, Band 11; Springer Verlag, Dordrecht Heidelberg London New York, S. 29-148

Wissen, U. (2009): Virtuelle Landschaften zur partizipativen Planung – Optimierung von 3D Landschaftsbildvisualisierungen zur Informationsvermittlung, IRL-Bericht 5, vdf Hochschulverlag an der ETH Zürich, 242 p.

Wolfrum, S. (2006): Theorie und Methodik der Quantifizierung von Biotopverbundfunktion – Landschaftsgraphen als Alternative zu Landschaftsstrukturmaßen bei der Erfassung eines funktionalen Zusammenhangs in der Landschaft. in: Kleinschmit, B., Walz, U. (Hrsg): Landschaftsstrukturmaße in der Umweltplanung, Schriftenreihe der Fakultät für Architektur Umwelt Gesellschaft, Band 19, Berlin, S. 73-83

Zetterberg, A., Mörtberg, U., Balfors, B. (2010): Making graph theory operational for landscape ecological assessments, planning and design, in: Landscape and Urban Planning, 95, S. 181-191

Zölitz-Möller, R. (1999): Umweltinformationssystem, Modelle und GIS für Planung und Verwaltung? – Kritische Thesen zum aktuellen Stand der Dinge, in: Blaschke, T. (Hrsg.): Umweltmonitoring und Umweltmodellierung, Herbert Wichmann Verlag, Heidelberg, p. 183-190

4

An Approach to Landscape Planning in Borders

Gloria Aponte
Pontifical Bolivarian University
Colombia

1. Introduction

The so-called *urban-rural borders* represent a territorial phenomenon that presents itself as different kinds of landscape, according to the social dynamics of each settlement. Some of those are representative of their historical sprout or boom time, and others of their location. Urban-rural borders represent nowadays a very outstanding development in major cities particularly in developing countries.

This chapter randomly revises first, as a broad context, the very carefully treated and built borders of walled ancient towns, as representative of the self-centred urban attitude, where landscape is seen as an external reality distant from everyday interests. And second, the growth without borders or, better, without control, originating from the beginning of industry, that manifests itself as an invasive and underhand force that devours natural landscape by slowly ruminating and digesting it.

Following, as the core of the reflection, the fact that in the second half of the 20th century and beginning of the 21st when it becomes a centre of attention as the border could mean a crucial place to stop destruction of resources essential for life, is addressed. In the developing world the situation is not just severe because of its rapid rendering, expansion and consequent deterioration of landscape, but it is aggravated by social unbalance and complex socio-political situations.

Landscape studies in urban-rural fringe have not been abundant. Nevertheless, some representatives from very different corners of the planet can be quoted: Qvistr \u00f6m and Saltzman (2003, 2006, 2007) from Sweden; Wang, Gu and Li (2207) from China; The Landscape Partnership Ltd. (2007) from the United Kingdom, and Pellegrino (2003) from Brazil. In Colombia some academics have talked about borders, mainly recently, but not precisely about "landscape in borders". For example: Toro, Velasco and Niño (2005), Velasco, Díaz, López (2010).

As a local application, an academic approach towards the solution to this threatening problem is shown, in a very special and intricate situation: the urban-rural border on steep slope. This is exemplified in the urban fringe of Medellín, settled in the Aburrá river valley. The topographical difficulty in this region is overlapped by a quite difficult social situation derived from rural forced displacement that makes the population, and consequently the settlement, grow not only from inside to outside but also by groups coming from distant places attracted by the urban imagery, but stopped at the periphery.

This is a very dynamic and complex landscape that deserves on the one hand a deep analysis and, on the other, creative solutions to cope with preservation of natural resources, satisfaction of social needs and development of cultural identity. A related research focused on the structural role of streams in the landscape of urban-rural borders on steep slopes has been carried out in the Landscape Design Master Programme at Universidad Pontificia Bolivariana."

The research team wondered: *How could we structure the landscape on borders?*

For the specific case of this research, the question was concretized as: How to value the structuring role of the water streams in the fringe landscape on the steep slopes of Medellín?

The purpose has been to produce a set of landscape guidelines to be presented to local authorities, with the aim that those be applied when planning, developing or reorganizing urban-rural borders in the conditions previously mentioned.

The method of the research consisted of disaggregation and aggregation. That is to say that the landscape universe was analysed from the diverse points of view that allowed a reasonable panorama of the situation. It meant to focus on the following landscape components: natural, social, morphologic, normative, and spatial/perceptual, for a clear and balanced approach. Although the research team is not properly interdisciplinary, each member took the responsibility of one component. The process was enriched with the advice of four landscape professionals, visiting lecturers from abroad, who came mainly to share their knowledge with the Master´s students.

2. Borders in the past

As a broad time context following there is a selection of urban development milestones, commenting on them their particularities in relation to the landscape first represented as a menace from outside and later as an injured party of urban growth.

2.1 Walled towns

When ancient settlements, through the specialization of jobs, grew into villages and then into towns, different reasons drove their inhabitants to identify and separate themselves from the surrounding fields. As a physical consequence, a strong and conspicuous feature emerged in the landscape: the defensive wall. Undoubtedly this feature was seen, recognized and perceived by people who approached them or worked in agricultural fields outside the towns, but probably it was not interpreted as part of the landscape, because that concept did not exist in the vocabulary or imagination of older civilizations.

Of course, by the time that walled cities flourished, landscape was not a planner's worry, or even a simple purpose. The walls, promoted mainly by military and political causes, as well as every huge human construction, had a strong effect on people's perception of landscape, although this was an unconscious perception. Nature was "there", or outside the town, and life, property and safety were "here" inside the town.

> *"Sumerian cities...from the III millennium B. C.... were surrounded by a wall and a moat that defended them and separated –for the first time- the natural open environment from the close city environment (L., 1977)"*

The wall signified another difference as well. The dominant people lived inside while the subjugated people lived and/or worked outside. That idea persisted worldwide, and still persists in many places, for a very long time up to the moment when "suburbia" started to mean economic power and high status outside. That is to say that there are at least two ways to inhabit the non-urban territory that surrounds the cities core: one not being able to reach their standards and the other passing those standards. Both ways are observed, in a strong contrast in many cities today.

In some cases, such as Arbela (or Erbil) in Mesopotamia (Figure 1), this division was totally defined by walls, while in Babylon and many other ancient cities the walls that conformed the border were combined with natural or managed watercourses. An outstanding and surprising case is that of Carcassonne, in France, surrounded by a double wall, with one quite close to the other. Some of the walled cities remained firmly throughout the centuries while some others underwent several changes and re-constructions. One such a case is that of the city of Athens, quite didactically expressed by Benévolo in his five volumes work Design of cities (1977).

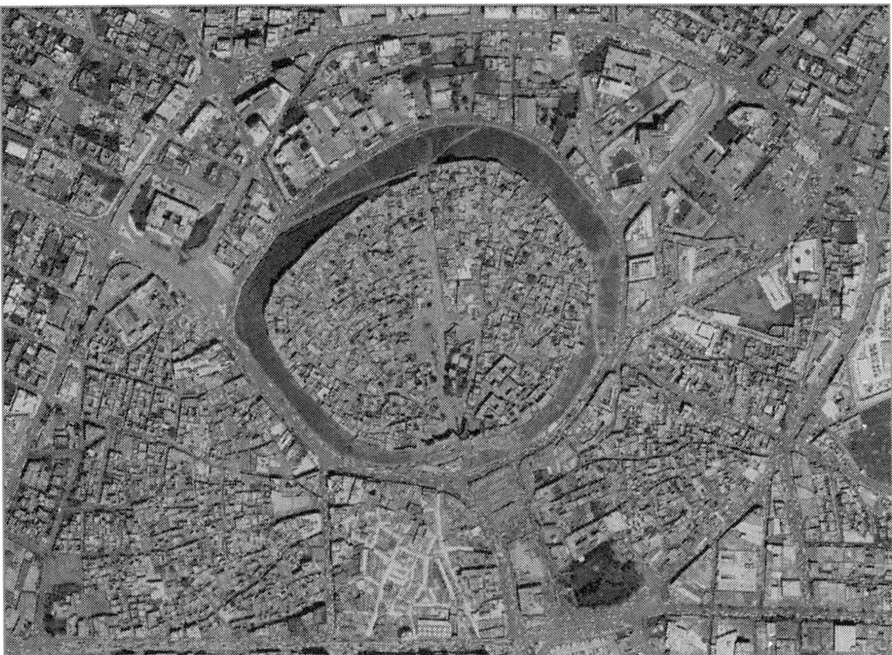

Fig. 1. Arbela (or Erbil) a walled city inhabited continuously since its creation, B. C., up to date. Source Google earth 2011.

That practice of strong separation transcended for centuries, and even though the thinking that something appreciated should be enclosed flourished in managed landscape –or better garden- through the middle ages *orthos conclusus*, and later the green labyrinths, or portions of nature locked up to be enjoyed only by a few people, like an individual property.

Once the social and political circumstances that caused the growth of walled towns had changed and overcrowding became a problem, the surroundings had to be occupied. However, the feeling of being unprotected promoted in some cases the construction of a new peripheral wall that would symbolize security. This second wall, though, was not as fortified as the first.

By the time when walled cities reached their peak in Asia and Europe, in the land that later would be named America, a very different thinking guided its inhabitants relation to the earth. They used to define themselves, and still remnant tribes do, as part of nature. This thought is quite nicely expressed by the Uruguayan writer Eduardo Galeano, who states, even opposed to the Catholic believe of God's Ten Commandments, that God forgot the eleventh commandment: You will love and will respect nature to which you belong. (Galeano, 1994).

The conquering army's power defeated the natives´ thought power and a result of that was the establishment of a walled city with the most extensive fortifications in South America: Cartagena de Indias, which is still the best example today. (Figure 2). Described as the masterpiece of Spanish military engineering in America and located on the Atlantic coast to the north of Colombia, the walled part of the city was declared by UNESCO as a Historic and Cultural World Humanity Heritage in 1984.

Fig. 2. A portion of the present urban area of Cartagena de Indias and, within the circle, the ancient walled city. Source: Google Earth 2012

Surrounded by water, although not rectified or geometrically transformed as had previously happened in other walled cities, in Cartagena de Indias the sea and the swamp offered the right environment to settle an urban core defended by bodies of water, which were reinforced by the infallible wall.

That inherited defensive attitude, that had repercussions on people's perception and interpretation of their relation to landscape, still remains in urbanism practice, particularly within residential units or condominiums and may be seen everywhere in Colombia. This phenomenon results in a kind of landscape that wastes the wide richness of local landscape and hurts the local landscape identity.

2.2 Over passed borders

Over the centuries, the population growth led inhabitants to pass the second wall, when it existed, and to invade the nearest surroundings in a, at the beginning, moderate process, which was later accelerated by the effects of the industrial revolution.

Pablo Arias (2003), while revising urban history concludes that from the Roman city up to the eighteenth century, the formal and physical relationship of the city with its surroundings would remain relatively stable, with a closed city stated in the territory as the central fact and character, without altering the environment in which it was settled.

Traditional agricultural practices accelerated by new technologies, utensils and machinery depersonalized the previous relationship between towns and their territories. New tools formed part of the everyday landscape, and the result of their use, in many cases, homogenized the peripheral and rural landscape next to urban conglomerates.

Following the mentioned author, the difference between the ancient towns and the modern city, in terms of expansion, is the different behaviour in relation to its surroundings. The old historic towns reinforced their identity through the manner in which they were linked with the territory. Modern cities, on the other hand, exert the right to prey on the territory in searching of resources, some indispensable to live such as water and food, and some others necessary for social and economic development, such as roads and factories.

The confusion of this overwhelming texture of networks and frames depersonalized the old heritage sense of the city image in its territory; it is one of the most significant losses in the current city (Cano 1985)[1]

The city of Adelaide (Figure 3) represents an interesting example that illustrates a historical border that persists despite the later strong urban sprawl. Founded in 1836, the origins of this planned city have very little to do with walled towns, but the observed plan shape tells the story of a historical centre, a surrounding fringe and the later irregular sprawl. Adelaide was planned under *Light*[2] *Vision*, and the fringe –the Adelaide Parklands- that initially acted as the growth limit, contention and definition of the inside and the outside, now represents a great advantage. The needed green areas, usually desired when the population increases, were already there, bordering the old town. Although their general shape does not follow the Torrens river flow or other natural features that surely were there before the city construction, that green area represents an outstanding environmental and landscape resource that has a clear balancing effect.

Besides the environmental damage widely analysed under the concept of ecological foot print (Rees & Wackernagel 1994), the growth without borders or, better, without control, triggered by the conjunction of diverse forces that result in an invasive stain that spreads on the natural support to blot out all traces of what it was before.

---

[1] Cited by Arias 2003
[2] Colonel William Light, planner of the city

Fig. 3. Adelaide, South Australia. Source: Google Earth 2012

2.3 Demographic explosion effect

Facing the need of expansion due to economic growth, cities used to be thought of in a centrifuge manner. The borders moved faster than the planning authorities attempted to solve or even understand problems. The growth had been predictable or at least reachable by remedial strategies for centuries, but the demographic explosion of the 60´s made an abrupt change on the previous inertia, mainly in the named "developing world". In this part of the earth, the situation has not just been severe because of the rapid urban rendering, expansion and consequent deterioration of places, but it has been aggravated by social unbalance, socio-political complex situations, and extreme environmental damage as well as consequent landscape disfigurement.

Facing the growth from this time onwards, planning authorities were at the beginning focused to solve issues from a single functional point of view, ordering and distributing land uses, as if the habitat were independent of inhabitants. That was the time of "zoning", a technical exercise that minimized the importance of the human behaviour of the diverse groups of population and communities and the significance of natural determinants of the territory.´

The evident dysfunction of that planning system and the increasing social set of problems drove planners attention to society; that is to say, to the collective human factor. Aspects such as education, health, the right to work, social security, among others became even more important than the assignation of uses to the land. In Colombia the beginnings of this kind of planning attitude could be placed in the 80´s decade, to be followed in the 90´s by the environmental worry.

Environmental issues have been much treated since the Stockholm Conference in 1972. But in the developing world it started to be important enough to be incorporated in local law two decades afterward. In general, the complex environmental problem seems to be increasing in a geometrical tendency, while the solutions increase in an arithmetical way.

Meanwhile, economical factors and land speculation go over the common sense of preserving resources and to treating them in a real sustainable manner. Words such as ecology, green, and sustainable, have lost their actual meaning and are used without measure. The "green wash" has invaded contemporary discourses hiding the real environmental question posed by the urban expansion.

To complete the spectrum, of zoning + social + environment, and rooted in its agglutinative and unifying role, many signals seem to point to this time being that when the integrator par excellence: THE LANDSCAPE occupies the deserved place as an important determinant in planning decision and purposes. As Sir Geoffrey Jellicoe thought *The world is moving into a phase when landscape design may will be recognized as the most comprehensive of the arts* (Jellicoe 1982).

To deal with the complex issue of indiscriminate urban expansion and moving peripheries a strategic coordination of many actors and factors is necessary. Of course the landscape design discipline is not enough but its contribution is indeed necessary first in helping to understand and balance the multiplicity of facets of the urban-rural border phenomena, and second to promote integrated answers to the complex trouble.

3. A research attempt to planning landscape in borders

As mentioned before, research is being carried out at the Universidad Pontificia Bolivariana related to the landscape in borders in Medellín. The aim of this work is to contribute to the local authorities' acquaintance for better political decisions through the production of landscape guidelines applicable in any urban, civil, architectonic or infrastructural intervention that takes place on borders. This is in order to respect and understand the abundant, and by now abused, streams that run down the valley where the city is settled as a landscape structure on its borders. A very important circumstance comes with the city administrator team recently elected, one of whose major interests is focused on the urban rural borders.

3.1 Background

Historians have traced Medellin´s act of foundation in four different dates in between 1575 and 1675, and also the site of foundation has been placed in different points. During the 18th century life in Medellín elapsed in a more rural than urban environment and even today this city shows a closer relationship with the rural environment and traditions than other Colombian towns.

The city was founded later than many other Colombian cities, for example those on the Caribbean coast or the capital city of Bogotá. Probably due to the location of the Aburrá valley in the middle of an intricate set of rough mountains of difficult access, formed by the Colombian Andes, in Antioquia province a region where the central and west branches of the mountain system get closer, before of descending to the Atlantic Coast swampy savannas. Nevertheless, during the 19th century, Medellin underwent a population growth much higher than the rest of the country (Álvarez, 1996)[3]

While the British industrial revolution expanded to the rest of the First World between 1750 and 1850, South American territories were just being colonized. Industrialization arrived in 1930 in the form of Medellin's first factory, which as was the case in Manchester, was a textile factory.

The twentieth century brought an extreme increment of population to the city. Even before the demographic explosion the industry progress attracted workers from villages around that exerted a considerable population increase during the 40's. Since then the occupation process has been markedly informal and only from the second half of the 20th century, when the city had around 360.000 inhabitants, there were important planning efforts (EAFIT, 2010, pg. 50).

About natural resources and their landscape shape, EAFIT Department of Geology, cites Parsons (1997), who mentions a cut on the rock of the southern strait (of the valley), which deviates the river Medellín[4] to be able to explode the gold alluvium. And also mentions that communication between the diverse urban cores of Medellín was difficult because that they were separated by a wide muddy swath (EAFIT 2010 pg. 53). Nowadays, the river is completely channelized and the old muddy swath has been completely occupied. Many industries, administrative buildings and residential units are placed there, separated from the river by main roads that interrupt a sound relationship between people and the main natural landscape feature present in the city.

In the same source the following affirmation is founded: "The covering of the Santa Elena stream (the main affluent of the Aburrá river in urban area) started in 1926 and was completed in 1940 by the construction of the Nutibara square and Hotel. The rectification and the channelling of the river was started in 1912 and after several stage it finished with the construction of the Metro in 1985".

La Violencia[5] in Colombia began in 1948, having an effect on the main cities, but particularly on Medellin, due to the attractiveness represented by work opportunities in a booming industry. That violence would derive in a hard urban violence that has represented an immense obstacle to a sound development and to a healthy relationship to the landscape.

During the 70's the city suffered the very negative effects of mafia and drugs and it was sadly named as *the most violent city in the world*. Nevertheless, Medellín bears other titles that better account of the actual reality and landscape identity: *The mountain capital, The city of everlasting spring or The silver cup*. It is, as well, the only city in Colombia using the Metro

---

[3] Cited by EAFIT 2010, pg. 51

[4] Two names are used for the river either Aburrá or Medellín. The first corresponds to a tribe of pre-Columbian inhabitants' name, and the second to the name that the Spaniard conquerors gave to the city in resemblance to a village in Spain.

[5] La Violence (1948-1958) Tensions between the two traditional parties, the Conservatives and the Liberals, led to a civil war. The violence, which left between 100 000 and 200 000 dead, ended with the establishment of the National Front (1958-1975), an agreement in which the conservative and the Liberals decided to share power even alternation in the presidency every four years, thus excluding all leftist movements (Rozema, 2007).

transport system, the first one to establish passengers' transport by cable and also a well-known touristic destination, positioned to host congresses and conventions.

3.2 Medellín borders on steep slope

Medellín is located in a basin with, proportionally quite little flat land, which of course was rapidly occupied. This relative topographical difficulty is compensated by a very friendly climate and humidity in between the range of human comfort through the whole year.

3.2.1 Current situation

The topographical complexity is overlapped by a quite difficult social situation derived from rural forced displacement that makes the population and consequently the settlement grow not only from inside to outside but also by groups coming from distant places attracted by the urban imaginary, but stopped at the fringe. Medellín and nine more minor municipalities occupy the Aburrá river valley in a closed conurbation. This physical relationship is reinforced by the geographical identity of belonging to the same watershed (Figure 4) The steepness of the slope lets the permanent sight from one side to the other, from bottom to up, vice versa, and in general overpowering panoramic views (Figure 5).

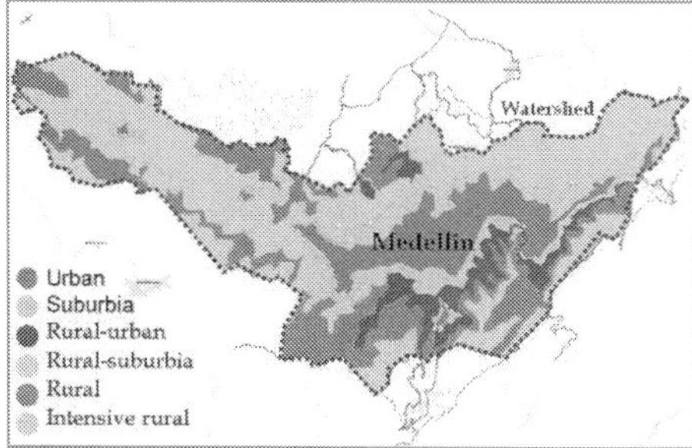

Fig. 4. Aburrá valley watershed

Fig. 5. Centre-east side of Medellín

Simultaneously, with a nice view to observe, those panoramas show the uncontrolled and worrying climbing of urban occupation on the hills (Figure 6). The picture changes day-by-day and in contrast to the Adelaide case presented previously, there are not enough open areas reserved. This is worst because the numerous streams that run down are being buried and their adjacent watersides completely occupied and deteriorated. To aggravate the situation, this invasion happens not only by informal occupation (Figure 7) but also by planned housing developments by high-income neighbourhoods (Figure 8).

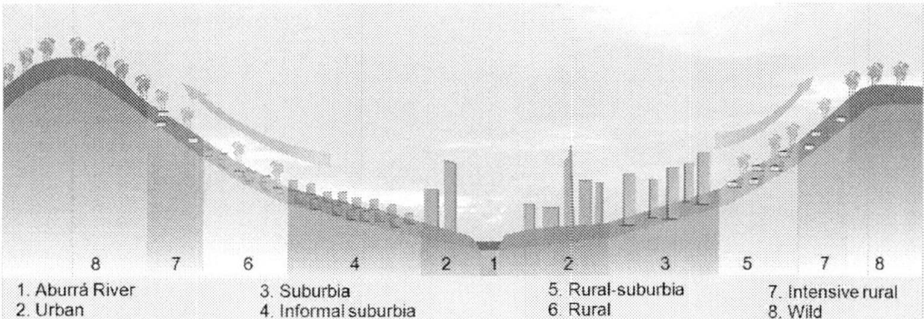

Fig. 6. A sketch section of the urbanized Aburrá river valley, the border moves upwards constantly and menaces with colonizing the edge

Fig. 7. Urban growth out of control

Fig. 8. High-income neighbourhoods

3.2.2 Solution attempts from the urbanism perspective

Local authorities have carried out laudable efforts that have propitiated a wide recognition of the recent development of Medellín throughout the world. There are two outstanding programs that exemplify those efforts: social urbanism and PUI.

Social urbanism is an innovative urban development strategy oriented to solve the conflicting circumstances that had driven the city to a critical situation. It consisted in to investing in grassroots communities in order to pay the "historical debt" that society owed to these urban links of the ignored city. It is not just intended to solve the underlying problems related to housing, employment and poverty. By building metro cables, parks, libraries, schools in high-quality architecture, public spaces, and other projects with a high aesthetic and social impact (Figure 9) it seeks not only to "make the best architecture, which raises pride and self-esteem of the community, an architecture that generates a sense of belonging", but also implement projects to" lead a profound social transformation. "(Mayor of Medellin, 2008).

Fig. 9. España Library, Metrocable transport system, the bamboo bridge. Three of the many works recently developed as part of the program "social –urbanism", developed in the east side slope of Medellín.

Integral Urban Projects (known as PUI, in Spanish). "PUI is an urban intervention instrument that gathers physical, social and institutional matters, with the aim of solving specific problems on a defined territory. In this way, the City Hall, using all development tools in a planned and simultaneous manner, gets that actions oriented to development reach vulnerable zones"(Mayor of Medellín 2009).

According to local planning and development authorities the PUI components, in order, are:

- Community participation
- Coordination of diverse institutions plans and efforts

- Housing promotion
- Improvement of public space and mobility
- Alignment and construction of community facilities
- Environmental recovery

A prominent official plan on the topic is the Borders Master Plan carried out by the Urban Development Agency (known as EDU in Spanish). The plan represents a good institutional effort to develop an interdisciplinary exercise, focused on two particular sectors of borders, and that resulted in specific projects, already executed and of good reception from the community. It must be credited to this process the involvement, of a moderate contribution from the landscape discipline. This circumstance opens the door to professional landscape participation in projects leader by public institutions.

The advances in Medellín to improve the urban habitat in all senses deserve all admiration, but it has to be said that landscape has not been attended as much as it ought to have. Many times it is considered a superfluous activity that may be present or not, depending on the budget remaining and that could be solved by planting some trees. That is the landscape professionals' challenge.

3.3 Borders from the landscape perspective

Similar to many places in the world, Medellín has followed the planning process characterized by: functional emphasis, social emphasis, environment regards. The last, up to a level that could not be properly considered an emphasis jet.

Landscape is the core of the research, variable in itself, and in the case of the matter, that variation depends on other research variables like border, hillside and streams.

The "border" from institutional consideration, is usually seen as a line that the planner draws on a map, attending the use conditions more than the natural realities. As a line that, even in the recent past, some governors have pretended to identify by a particular colour (difficult or even impossible to materialize, but with a good reception from a naive point of view), to be seen from as many places as possible. A line almost without thickness that it is sometimes referred to as a "membrane"; a border as fragile as an administrative division that ignores or contradicts natural limits such as watersheds.

From the landscape point of view, this is to say, from a perspective that gathers natural dynamics, values and forms, signs of permanent occupation or in consolidation process, affective and appropriation relationships to the site, borders are not a line, and not even a fringe. It is an elongated space, composed by fragments or subspaces that aren't anything else but micro-basin portions occupied and deformed or mistreated. These portions are curiously placed in a perpendicular position to the basin axis: the streams. This is, in terms of the reinforcement and prevalence of this site as a settlement, beyond its natural calling and shape identity.

As an academic work and counting on the experience, even brief as it is, of the Master in Landscape Design programme, an approach from the landscape discipline corresponds. The landscape, as people perceive it, is the result of the interaction of natural and human factors with an eye on ecological, social, functional and economic values. That means, as it was mentioned before, an integrated interdisciplinary focus.

The nature of a site's landscape is the result of many interactions; that is why the approximation to its complexity makes it necessary to split it up in parts that aren't just elements or physical subdivisions, they're systems or layers, in a cartographic language. With this premise, we proceed to identify the different parts that in this case are defined as components and subcomponents (Table 1). Such disintegration allows a precise analysis, a settled research and a well-balanced result of the weight assigned to each part, in the general definition of work. In this matter, once there's a certain grade of clarity about the circumstances and the meaning of each analysis component, we proceed to confront ones with others and to identify its interferences. This constitutes the first step to a new and necessary aggregation. Gradually, after retrospective revisions, that aggregation was consolidated, as complex as certain limits allow it to be, to conclude with a work of an integral proposal.

The work, according to the expressed methodology, was organized selecting the following landscape components and subcomponents that in the context of this research were considered the most relevant:

LANDSCAPE COMPONENTS AND SUBCOMPONENTS	
Components	**Subcomponents**
• Natural	Relief (topography) Flora Fauna
• Hydrologic	
• Social	Actor Legal situation Use of space and resources Effects on space and resources
• Morphologic	Natural Urban
• Normative	Environment, Water, Biodiversity, Landscape — International, National, Regional- Local
• Spatial/perceptual	Panoramic Middle distance Experiential

Table 1. Disaggregation in components and subcomponents for landscape diagnosis.

It is necessary to point out that even when the hydrologic matter is part of the nature component, in attention to the streams importance in the work it was decided to develop the hydrologic component apart.

3.4 Landscape principles

To start, it is pertinent to select some concepts that express our interpretation of landscape and simultaneously support the landscape axis of the research. For such purpose, some substantial reference parts on the matter have been extracted, and are presented as follows.

The first reference is taken from La Gro´s Ten perceptions of landscape meaning (La Gro 2008, pg.157) and adapted to drive the basis for the intended guidelines (Table 2). The aim of replying to the author ideas was to establish a starting point that would compromise the research team attitude from the beginning.

SOME PRINCIPLES OF LANDSCAPE DESIGN		
Landscape as:*	**Associated concepts:***	**How we should reply****
Nature	Fundamental, Enduring	*Profound responsibility and axis of design*
Habitat	Adaptation , Resources	*Creativity*
Artifact	Platform, Utilitarian	*Respect for nature respect, and social and common sense*
System	Dynamic, Equilibrium	*Interdisciplinary strategies*
Problem	Flaw, Challenge	*Intelligence, sensibility*
Wealth	Property, Opportunity	*Social justice*
Ideology	Values, Ideas	*Re-discovering own roots*
History	Chronology, Legacy	*Tracing back past and heritage*
Place	Locality, Experience	*Facilitating perception*
Aesthetic	Scenery, Beauty	*Discovering, enhancing, enjoying*

Table 2. Some principles of landscape design. *Taken from La Gro James A. Jr. 2008. **Proposed by the author of this chapter.

The second reference comes from Lucia Costa who says that landscape and city are destined to a permanent complicity relationship and supports herself on Lawrance Halprin (1981) when the landscape designer argue that the most interesting cities are those that allow to reveal that complicity (Costa 2006). These ideas encompass what we are daring to pursue through our work.

The third reference is taken from Anne W. Spirn who referring to her book The Granite Garden, says: after its publication in 1984, I was surprised how many people, including scientists and naturalists who refused to accept or ignored the evidence that human settlements, including cities, are part of the natural world. I have found that those ideas about nature and what is natural come from very deep feelings and beliefs. These views are personal and varied, and to change them is not simply a matter of some verbal arguments convincing, but to reach both the mind and the heart of people. Photography and landscape architecture are powerful ways to help people to feel, as well as reflect on the place of humans in nature. (Spirn, 2006). This reference drives our thought into the sensible and human side of landscape that could not be absent in a serious and complete landscape project.

From a more practical point of view it was agreed by the research team that the landscape of borders has to be seen at least in three scales:

- The panoramic scale that shows the fact and let appreciate it in terms of composition: line, texture, colour and form It is particularly evident in the situation of the valley always present wherever the observer is positioned (Figure 10).

Fig. 10. Panoramic view on the west side of the Aburrá valley basin. Photo C D Montoya

- The middle distance scale where other aspects of landscape, in addition to the composition itself, emerge as the identification of local particularities. Here the rugged relief besides the unfeeling inclusion of buildings and civil works make landscape reading even more intricate (Figure 11) and 12).

Fig. 11. Middle distance view of a place where a stream is hidden.

Fig. 12. Civil works on a natural environment

-   The experiential scale, in which the other senses, besides the sight, gain prominence in the perception of the place (Figures 13 and 14).

Fig. 13. Experiential landscape

Fig. 14. Extremely hard stream treatment

3.5 Research process

As a basis, support was founded in the PIOM (Integral Plans of Ordering and Management) applied by now to seven streams in the area of Medellín. Those were analysed in searching their landscape approaches, if any, but also because of the wide and up to date basic information that they could provide, avoiding invert efforts in a job already accomplished.

The AMVA (Environmental Authority of the Aburrá valley) has produced a PIOM Methodology to be followed in the case of every one of Medellin's streams. It is worth to annotate that landscape factor is absent of all considerations, as it is frequent in local public documents. In the scarce cases that landscape is mentioned it is addressed as the vegetation piece of urban design, or in any case as a secondary matter.

In their development, some PIOM make emphasis on one component while others do on others. In order to get a balanced approach, the established components were analysed on the same basis.

The hydrologic component was analysed under general concepts of water cycle and functioning, while the major and minor watersheds were revised in the territorial ordering context[6]. Within Medellin's jurisdiction[7], more than 400 streams and 60 minor watersheds are reported; those conform 6 watersheds that tribute to the Aburrá river major watershed, a considerable water presence. In a closer analysis conflicts between streams, resource use, urban uses or mistreatment were studied.

After visiting several cases, observe, perceive experimenting and talking to local people, the analysis from each component was captured on a template designed to gather in an organized way the contributions from each sub-component to the landscape diagnosis. One of the records obtained from the spatial/perceptual landscape analysis is shown ahead as an example, selected because in some way it gathers other topics and because this specific topic could be more relevant in the context of this book. (Figure 15)

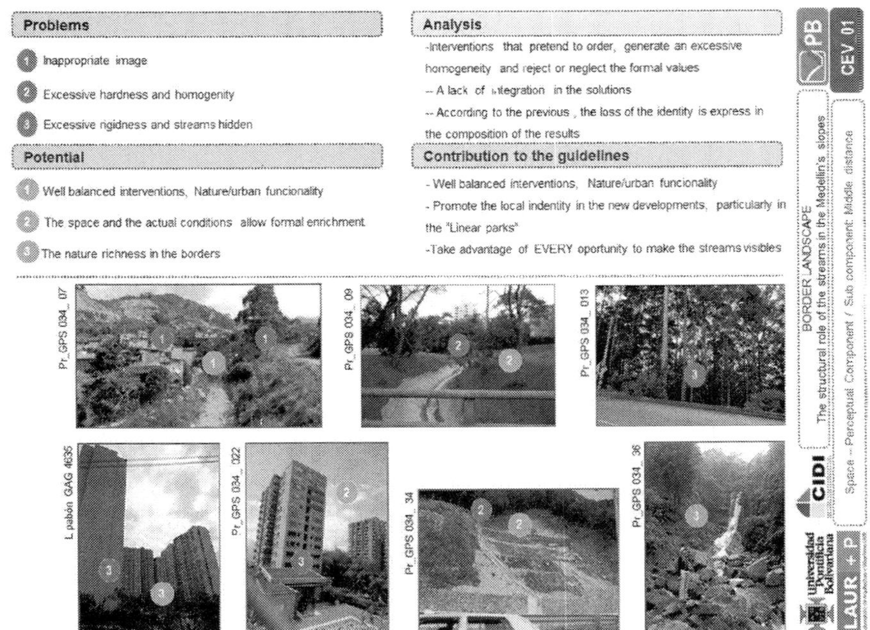

Fig. 15. Example of landscape analysis template, to be applied to each landscape component

To be able to identify the streams in the panoramic scale would contribute to recognize those landscape features, and would help to local and visitors to orientate themselves by the reading of the territory. It is particularly evident in the situation of the valley always present wherever the observer is positioned and because the skyline, a significant visual resource, sometimes is menaced by the urban expansion. That identification also would help local inhabitants to get familiar with the streams, as well as other landscape features, and even though to be proud of them.

---

[6] Colombian Law 2002, Decree 1729
[7] An extension of 380 Kms.2

In Figure 11, where middle distance scale is illustrated it is difficult to guess that a stream exists, because the scene is all starring by the buildings, road and bridge. An appropriated forestation of the waterside, completely missing, would help to identification and orientation of users and also to a harmonious equilibrium of the scene. The case illustrated in Figure 12 shows how sometimes public works go ahead of urban occupation against a beneficial use of natural features in the landscape.

In the experiential scale (examples shown in Figures 13 and 14) landscape is perceived as more dynamic than the other two scales mentioned, because of the detail that could be experienced and because that the atmospheric phenomena could be felt more strongly. Hearing, smell and touch become more noticeable and discomfort could become exacerbated.

3.6 Results

Research is still on the process but partial results have been obtained as preliminary guidelines, taking as reference the Manual of Environmental Guidelines for the design of Infrastructural Projects in Bogotá D.C[8].

The first result has been the construction of general criteria to facilitate the applicability of landscape concepts and assessment on the identified reality. These criteria intend to cover, in balance, the broad conceptual basis of landscape: natural, human perception, social function, and environment. As it may be identified in the following list, criteria from 1 to 3 correspond to natural basis, from 4 to 6 correspond to human perception, from 7 to 9 correspond to social function and the last one to environment.

1. Hydrological functionality
2. Promotion of biota
3. Contribution to environmental conditions
4. Valuation of riparian landscape on hillside
5. Conscience of the historic present
6. Responsible appropriation of streams in hillside
7. Minimization of the risk
8. Recreational benefit
9. Educational benefit
10. Environmental sustainability

The second result is a set of guidelines for intervention in the landscape, feasible to be adopted in local politics and regulations that foster the streams as landscape structuring entities. Each one of the three tables produced addresses a feature of those present in the streams: one on the course, another on the waterside, and the last one on intersections with roads or other infrastructure items. Only the second one is presented ahead (Table 3).

The guidelines have been organized according to progressive stages of a project: planning, design and intervention. The first stage is addressed principally to public decision makers, the second to designers and the third to constructors. Crossing the mentioned criteria with these stages, a matrix was obtained to register, as guidelines, the ideas discussed and agreed

---

[8] Developed in 2003 and revised in 2006 by the author of this chapter, for Bogotá local planning and environmental authorities

LANDSCAPE GUIDELINES FOR STREAMS IN URBAN-RURAL EDGES

WATER SIDE

CRITERIA	PLANNING	DESIGN	INTERVENTION
1. Hydrological functionality	• To respect and increase catchment surfaces • To identify and protect associated water flows	• To guarantee appropriate permeability and drainage • To make the water visible to generate appropriation and responsibility to it.	• To guarantee the natural flow of the run off • To facilitate the relation vegetation / water
2. Promotion of biota	• To enhance the streams by means of water side management. To constitute the water sides in green corridors as bird habitats • To provide identity by sectors, through design diversity	• To take advantage of the biodiversity and enrich it in a sustainable way • To recover degraded areas. • To give continuity to existing environments.	• To establish diversity of vegetation associations attractive for birds. • To attend multilayer and natural succession • To restore the aquatic ecosystems
3. Contribution to environmental conditions	• To respect the character of every border area. • To design and carry out every border with similar quality • To recover ecosystems and balance opportunities in the city	• To improve watersides to generate environmental richness • To promote them as transition or ecotones between different uses. • To respect the existent natural functioning	• To apply phytoremediation techniques • To improve degraded areas through appropriated vegetation
4. Valuation of riparian landscape on hillside	• To enhance water sides as attractive and relaxing spaces, climatic stabilizers and landscape articulators.	• To reinforce site or corridor identity • To condition the design to natural landscape and follow its image, without imposing capricious geometry	• To take advantage of relief, rocks (if there are some) and vegetation formal qualities • To avoid interruption on user and water relation relationship
5. Conscience of the historic present	• To contextualize planning of these areas into landscape premises, instead of the application of foreign urbanism patterns	• To recover local cultural and natural values and features.	• To value and reintroduce artisanal principles and techniques
6. Responsible appropriation of streams in hillside	• To favor understanding of hydrological functionality by the community • To coordinate positive rapprochement from inhabitants to streams • To involve them in assignments that lead to the stream enjoyment	• To keep 80% of the area in soft surface to make drainage easier • To take creative advantage of area natural morphology • To provide generous spaces for permanence and slow walks.	• To involve local residents in the construction works. • To use signposting to inform and welcome visitors
7. Minimization of the risk	• To constitute the water sides in shock-absorbing of environmental risk • To avoid assignment of uses in risky places	• To design respecting morphology and natural functionality of the area • To harmonize risk mitigation strategies with previous landscape • To define spaces and use vegetation allowing visibility of critical points	• To mitigate risks with natural structures • To allocate signposts and marks for a correct orientation
8. Recreational benefit	• To constitute water sides in routes and stay spaces of visual recreation • To improve recreational offer for free time enjoyment	• To compose subspaces and transitions harmonically. • To avoid drastic divisions • To facilitate the development of group activities	• To establish vegetation attending their functional and esthetical qualities • To make good use of local vegetation qualities to provide comfort
9. Educational benefit	• To study the behavior and wealthy appropriation of water sides in diverse social groups • To empower educational institutions to use and care for water sides.	• To facilitate the rapprochement and knowledge of natural richness • To provide spaces for informal education and observation	• To preserve, compensate and label the existing vegetation • To allow participation of the community in construction works
10. Environmental sustainability	• To respect the environmental offer at the intervention and improve its role as part of the ecological net and air cleaners • To avoid the uses that may stimulate pollution	• To take advantage of underestimated spaces • To guarantee the clean air necessary to breath while exercising • To propitiate shock-absorbing of noise in favor of fauna • To guarantee a healthy environment to sane users development.	• To value and take advantage of marginal spaces • To take advantage of functional and formal qualities of vegetation • To propitiate self-sufficient maintenance and management of bio-mass wastes

Table 3. Example of the guidelines matrix set, applied to the streams waterside.

by the research team, each one of the members' standard-bearer of her or his component of responsibility.

In synthesis, summarizing the three described tables there are around 200 ideas that could be taken into account when facing a project in borders, near a stream, for a better landscape in projects responsible and respectful of natural resources, people's feelings, society needs, and environmental consciousness.

4. Conclusions

Borders have been always a special issue for settled communities. The landscape of borders on steep slopes is a very dynamic and complex fact that deserves, on one side a deep analysis and, on other, creative solutions to cope with preservation of natural resources, satisfaction of social needs and development of cultural identity.

Certain Rogers´ statements have been confirmed: Cities have become pests in the landscape, vast bodies that absorb energy from the planet for their maintenance: relentless consumers, relentless pollutants. (Rogers, 2000, pg. 27). Although many efforts have been carried out, the evidence shows that those are not enough to counteract the environmental damage. A healthy environment is the basis of a sound landscape so the two issues have to be attended together for both body and spirit heal.

There will be no sustainable cities up to the moment that urban ecology, economics, sociology be integrated into urban planning (Rogers, 2000, pg. 32). To complete this comprehensive statement it is also necessary to apply and benefit from the integrative function of the landscape approach and its perceptual issues, to accomplish, not just well constructed or equipped, but enjoyable cities that reach both mind and hearth of people.[9]

An intense responsible work has to be undertaken, at least in the developing world, to situate landscape matters in the authorities and managers' minds and hearts. This is part of the academy responsibility to develop a strategic way to evidence the significant importance of landscape matters and put them closer to public decision makers to position them on par with infrastructure, housing, mobility or industry in search of better cities for happier people.

5. References

Alcaldía Mayor de Bogota D.C. (2003). *Guia de Lineamientos Ambientales: Para el diseño de proyectos de infraestructura en Bogota* D.C. Bogota D.C.: Alcaldía Mayor de Bogota D.C.

Alcaldía Mayor de Bogota D.C. (2007). Revisión *Guia de Lineamientos Ambientales: Para el diseño de proyectos de infraestructura en Bogota* D.C. Bogota D.C.: Alcaldía Mayor de Bogota D.C.

Arias, P. (2003). *Periferias y nueva ciudad: el problema del paisaje en los procesos de dispersión urbana.* ISBN 84-472-0805-2. España Sevilla: Universidad de Sevilla.

Benevolo, L. (1977). *Diseño de la ciudad.* ISBN 84-252-1026-7. Barcelona: Gustavo Gili.

---

[9] According to Spirn quotation in page 14 of this chapter

Camargo, G. (2008). *Ciudad Ecosistema: Introducción a la ecología urbana*. ISBN 985-97488-7-2 Bogota D.C.: Universidad Piloto de Colombis

City of Tucson, Department of Transportation, Stormwater Management Section. (2005). *City of Tucson Water Harvesting Guidance Manual*. Tucson, Arizona. Retrieved from http://dot.tucsonaz.gov/stormwater/downloads/2006WaterHarvesting.pdf

Costa, L. M. (2006). *Rios e paisagens urbanas: Em cidades brasileiras*. ISBN 978-8588721388. Rio de Janeiro: Prourb-FAU-UFRJ.

EAFIT, Centro de Estudios Urbanos y Ambientales (2010). *Medellín medio - ambiente, urbanismo y sociedad*. ISBN 978-9587200744. Medellin: Fondo editorial Universidad EAFIT.

Galeano, E. (1994). *Úselo y tírelo: El mundo del fin del milenio, visto desde una ecología Latinoamericana*. ISBN 978-9507425424. Buenos Aires: Planeta Biblioteca de ecología.

IFLA. (2008). *Transforming with water*. ISBN 978-90-8594-021-0. The Netherlands: Blauwdruk and Techne Press.

Jellicoe, S. G., & Jellicoe, S. (1982). *The Landscape of Man: Shaping the Environment from Prehistory to the Present Day*. ISBN 978-0500278192. GDR: Van Nostrand Reinhold. Thames & Hudson.

LaGro, J. (2008). *Site Analysis: A Contextual Approach to Sustainable Land Planning and Site Design*. ISBN 978-0471797982. Hoboken, New Jersey: John Wiley & Sons, Inc.

Larsen, E., Girvetz, E., & Fremier, A. (2007). Landscape level planning in alluvial riparian floodplain ecosystems: Using geomorphic modeling to avoid conflicts between human infrastructure and habitat conservation. *Landscape and Urban Planning, 79*, 338-346. Retrieved from http://escholarship.org/uc/item/8rs5v2kp

Palmer, M.A., Bernhardt, E.S., et alt. (2005). Standards for ecologically successful river restoration. *Jurnal of Applied Ecology, 42(2)*, 14 de Marzo, 208-217. Retrieved from http://onlinelibrary.wiley.com/doi/10.1111/j.1365-2664.2005.01004.x/full

Pflüger, Y., Rackham, A., & Larned., S. (2010). The aesthetic value of river flows: An assessment of flow preferences for large and small rivers. *Landscape and Urban Planning, 95*, 68-78. Retrieved from http://www.elsevier.com/locate/landurbplan

Qviström, M., & Saltzman, K. (2007). *Ephemeral Landscapes at the Rural-Urban Fringe*. Lisbon: Edições Universitárias Lusófonas. Retrieved from http://tercud.ulusofona.pt/publicacoes/Book/12.pdf

Rogers, R. (2000). *Ciudades para un pequeño planeta*. ISBN 978-8425217647. Barcelona: Gustavo Gili.

Rozema, R. (2007). *Paramilitaresy violencia urbana en Medellín, Colombia*. ISSN 0185-013X. Foro Internacional, XLVII, 535-550. Mexico, Mexico.

Spirn, A. W. (1 de marzo de 2006). *Ser uno con la naturaleza: paisaje, lenguaje, empatía e imaginación*. Retrieved junio de 2010 from Ciudades para un futuro más sostenible: http://habitat.aq.upm.es/boletin/n38/aaspi.html

Spirn, A. W. (1984). The Granite Garden: *Urban Nature and Human Design*. ISBN 0465026990. USA: BasicBooks.

The Landscape Partnership Ltd. (2007). *Peterborough Landscape Character Assessment -Urban Fringe Landscape Sensitivity Study*. Peterborough: Peterborough City Council.

Wang, H., Gu, Y., & Li, H. (2007). *Landscape Ecological Planning of the Urban Fringe.* China: Hebei Agricultural University. Retrieved from http://webcache.googleusercontent.com/search?q=cache:http://www.seiofbluem ountain.com/upload/product/201004/2010lyhy04a7.pdf.

5

Urban Green Space System Planning

Bayram Cemil Bilgili and Ercan Gökyer
¹Çankırı Karatekin University, Çankırı
²Bartın University, Bartın
Turkey

1. Introduction

Today's changing world, values and standards of human were changed with urbanization. In this change people was differentiated existing uses and created new areas. These changes are different from country to country to the extent of economic, cultural and geographical reasons. In addition, these areas were determined to same principles basis for human uses.

Life style was changed with urbanization. In this process rural areas were transformed to urban areas. These areas are dominated by mass of concrete. In these areas there are small green areas at a micro level. In the process of rapid urbanization was created an unnatural environment. In the developed countries, urban areas were effected physical and mental development of people. This effect was adversely. With this change in urban areas, people entered into a yearning for natural areas. At beginning, green areas have been established to resolve natural longing of people. Urban green spaces have become the indispensable elements of ecological, aesthetic and recreational value. Establish of urban green space systems has become a necessity in today.

Urban green areas were not established for recreational needs. At the same time urban green spaces are ecological based requirement (Bilgili, 2009). Urban green space and green space systems were reviewed in this section.

2. What is urban green space?

Urban green spaces are urban areas which were occurred that, natural or semi natural ecosystems were converted urban spaces by human influence. Urban green spaces provide the connection between urban and nature. In this context, green areas are reflection in the urban spaces of natural or near natural areas surrounding the cities. The green fields are continuation of mostly landscapes around the city. Besides, urban green areas provide lots of ecological benefits which were established especially needs of urban people.

The increase of spare time of urban resident's and pressure of work and study enhances their demand of green space. Generally, the determining of necessary work time and to have more time make the leisure activities, which help people engage in self-creation activity and relaxation of body and soul, possible. This special time pay attention to two kinds of activities: people deal with nature; the other that between people. People's desire for fresh air, natural views and natural attractions, which reflects people's natural perception; the latter reflects their social behaviour. During the long historical time period, human has used a kind of ability to appreciate nature, to get the flavour of life and formed some psychological processes dependency on nature. This kind of feeling and perception through the realization of a better build and strengthen self-identification. In addition, these feelings reinforce mutual understanding and trust, strengthen the relationship with each other and may be responsible. All this really help to achieve self-worth. This is the most powerful reason why the communication in green area can never replace that in open public area. People begin to realize the crucial mechanism of urban green space system is to transform the active mechanic space into ideal state, i. e. form the value of environment mechanism, in relation to people's life (Wuqiang et al., 2012).

This urges urban spatial pattern to develop a kind of diversity system to relate other spatial shapes and itself can provide city with ecological safety value (Wuqiang et al., 2012). The requirement of green space, one of the main drives of world city system: most of the multinational corporations will choose the areas of headquarter and branches by comparing the urban environment and landscape of many cities. And of course, the favourable urban system, i. e. embodiment of the urban spatial pattern based on the integrate green space system, can attract more attention of the investors (Baycan-Levent and Nijkamp, 2009; Wuqiang et al., 2012)

Regional green space is based on the protection and optimization of natural ecological system and actually refers to continuous suburban green space of large size. It not only improves the whole ecological environment of the city region and its neighbours, and provides important support of urban environmental improvement. Furthermore, introduction of suburban green space into city also acts as the base of ecological balance. In practice, problems of urban woods and citied agriculture should be paid sufficient attention (Wuqiang et al., 2012). Green space systems require improvement of the spatial pattern of urban green space. To identify potential improvements, we compared the predicted development of planned cultivated and natural green spaces (Kong et al., 2010). Urban green space systems includes protection of existing green spaces, creation of new spatial forms, and restoration and maintenance of connectivity among diverse green spaces. To maintain or restore connectivity, planners must identify the best habitat and potential corridors by considering distances and the barriers between habitats (impedance) posed by the landscape and land use (Kong et al., 2010)

Urban green spaces provide many functions in urban context that benefits people's quality of life. There is therefore a wide consensus about the importance and value of urban green spaces in cities towards planning and constructing sustainable or eco-cities of 21st century. Steadily growing traffic and urban heat, especially in the developing countries is not only damaging the environment but also incur social and economic costs. The ecological benefits bestowed in green spaces which range from protecting and maintaining the biodiversity to helping in the mitigation of change cannot be overlooked in today's sustainable planning. Inner-city green spaces are especially important for improving air quality though uptake of pollutant gases and particulates which are responsible for respiratory infections. Green spaces also help in reduction of the energy costs of cooling buildings effectively. Furthermore, due to their amenity and aesthetic, green spaces increase property value. However, the most sought benefits of green spaces in a city are the social and psychological benefits. Urban green spaces, especially public parks and gardens provide resources for relaxation and recreation. Ideally this helps in emotional healing (therapeutic) and physical relaxation. In order to meet social and psychological needs of citizens satisfactorily, green spaces in the city should be easily accessible and in adequately optimal in quality and quantity. Green spaces need to be uniformly distributed throughout the city area, and the total area occupied by green spaces in the city should be large enough to accommodate the city population needs (Haq, 2011).

The provision, design, management and protection of urban green spaces are at the top of the agenda of sustainability and liveability. Urban green spaces play a key role in improving the liveability of our towns and cities. The quality and viability of cities largely depend on the design, management and maintenance of green as well as open and public spaces in order to provide their role as an important social and visual way. Urban green spaces are not only an important component in housing areas, but also in business, leisure, retail and other commercial developments (Baycan-Levent, 2002).

3. Urban green space systems

Urban Green spaces refer to those land uses and land cover that are covered with natural or man-made vegetation in the city and planning areas. It has been long argued about the definition of green space system. Different disciplines have used various definitions from their own professional concept, such as Horticultural Greenland System, Urban Greenland System, Ecological Greenland System, and Urban Green Space and Green Open Space (Manlun, 2003).

Common is that they are primarily linear or networks of linear lands designated or recognized for their special qualities Table 1. 1 (Hellmund and Smith, 2006)

TERM	OBJECTIVE OR CONDITION	EXAMPLES
Biological corridor (biocorridor)	Protect wildlife movement and accomplish other aspects of nature conservation.	Mesoamerican Biological Corridor through Central America; Chichinautzin Biological Corridor, State of Morelos, Mexico
Bioswale	Filter pollutants from storm runoff (usually at the scale of a site).	Numerous examples in various localities. See, for instance, the bioswales that are part of the City of Seattle Public Utilities' Street Edge Alternative (SEA) project in northwest Seattle.
Conservation corridor	Protect biological resources, protect water quality, and/or mitigate the impacts of flooding.	Southeast Wisconsin environmental corridors
Desokota	Blend rural and urban areas in a dense web of transactions, tying large urban cores to their surrounding regions in the same landscape. (From the Indonesian words "desa, " for village, and "kota, " for town. Also known as the McGee–Ginsburg model.)	Indonesia and China

Dispersal corridor	Facilitate migration and other movement of wildlife.	Owl dispersal corridor in the Juncrook area of the Mt. Hood National Forest in Oregon; Marine dispersal corridors for blue crab in the Chesapeake Bay
Ecological corridors (eco-corridors)	Facilitate movement of animals, plants, or other ecological processes.	North Andean Patagonian Regional Eco-Corridor Project
Ecological networks	Facilitate movement or other ecological processes.	Pan-European Ecological Network for Central and Eastern Europe
Environmental corridor	Protect environmental quality.	Southeastern Wisconsin environmental corridors
Greenbelts	Protect natural or agricultural lands to restrict or direct metropolitan growth.	City of Boulder, Colorado, greenbelt; London, England, greenbelt
Green extensions	Put residents in contact with nature in their day-to-day lives through a system of residential public greenspace, shaded sidewalks, and riparian strips.	Nanjing, China
Green Frame	Provide a network of greenspace for a metropolis or larger area.	San Mateo County, California, Shared Vision 2010 for the county's future development green frame; Addis Ababa, Ethiopia, green frame
Green heart	Protect a large area of greenspace that is surrounded by development. Originally referred to a specific area in the Netherlands, but now more widely used.	The agricultural open space surrounded by the Randstad, Holland's urban ring, consisting of the cities of Amsterdam, The Hague, Rotterdam, and Utrecht
Green infrastructure	Protect greenspace for multiple objectives on equal grounds with gray infrastructure (i. e. , roads, utility lines, etc.).	Maryland Greenprint Program; Chatfield Basin Conservation Network— Denver, Colorado, metropolitan area
Green fingers	Purify stormwater through bioswales.	Buffalo Bayou and Beyond for the 21st Century Plan, Houston, Texas, area
Green links	Connect separated greenspace.	Green Links initiative to connect isolated patches of habitat throughout the lower mainland of British Columbia
Greenspace or green space	Protect lands from development.	Countless systems (usually called "open space") across North America
Green structure or greenstructure	Connect separated areas of greenspace and provide a structure around which development may occur. Term is commonly used in Europe.	Greater Copenhagen Green Structure Plan

Table 1. Urban green space systems in the different countries (Hellmund and Smith, 2006)

Green Belt land is contributing to the healthy ecosystems which underpin many natural processes supporting a range of services including pollination, soil fertility, flood defense, air filtration and carbon capture and storage. Without the Green Belt designation it is likely that a proportion of this land would have been lost to urban development and associated infrastructure. Green Belt landscapes have been fragmented by development in a number of locations over time, however, and there may be a correlation between this and the relative lack of large and/or nationally important nature conservation areas. Green Belt land needs to be recognized as an integral part of ecological networks, forming healthy, functioning ecosystems to benefit wildlife and the people who live in adjacent towns and cities (Anonymous, 2010).

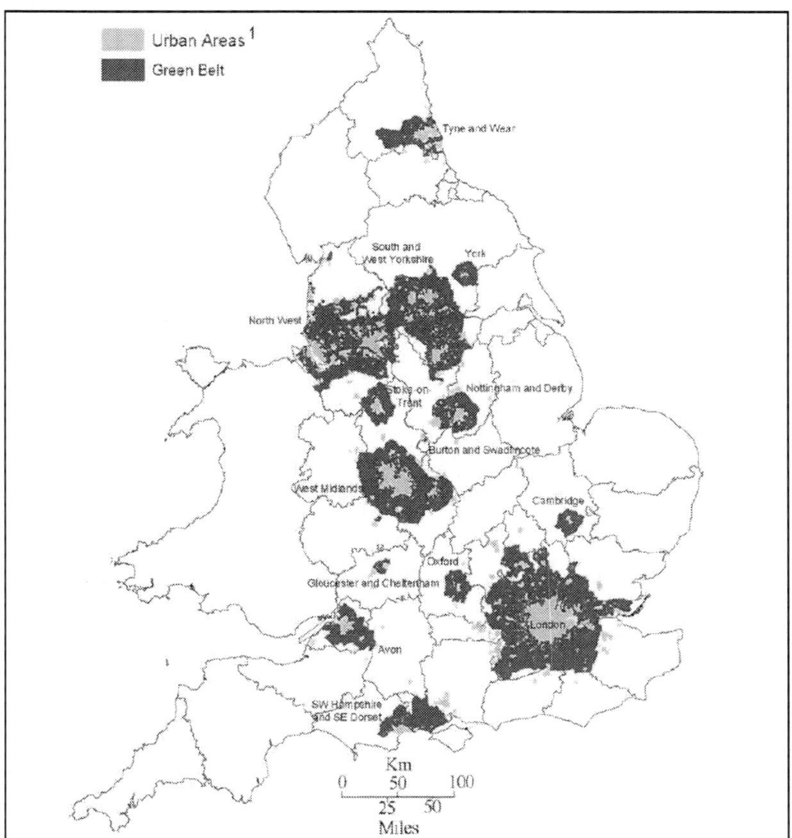

Fig. 1. Green belts in England (http://www. buildinglanduk. co. uk/greenbelt-land-uk. htm)

Greenways are being designated as green network in cities and countryside throughout North America and elsewhere. Sometimes these conservation areas are a response to environmental problems, such as flooding or degrading water quality. Other times their creation is an act of pure vision— people imagining a better community— one where people and natural processes coexist more closely. Often, despite this recent popularity, people fail to recognize the full range of contributions greenways can make to society and the environment. It is as if open spaces, especially in metropolitan areas, have been thought of as just so much generic greenery, mere backdrops for people's activities. In this chapter we suggest why greenways are deserving of their newfound popularity and how their functions can be enhanced, but also consider their limitations. We discuss how the greenway concept came to be, how it has been defined, and how its spatial form and content have varied. We also highlight the significant social and ecological functions of greenways, in advance of a fuller discussion of greenway ecology and design in subsequent chapters (Hellmund and Smith, 2006).

A network of green spaces which supply life support functions including food, fiber, air to breathe, places for nature and places for recreation. The Green Infrastructure approach seeks to use regulatory or planning policy mechanisms to safeguard natural areas. Multifunctional green infrastructure refers to different functions or activities taking place on the same piece of land and at the same time. For example, a flood plain providing a repository for flood waters, grazing land, a nature reserve and a place for recreation (Anonymous, 2010).

4. Classification of urban green/open spaces types

There are different ways to classify urban open space and greenspace, such as its size, how people use it, its intended function, its location etc. (Byrne and Sipe, 2010). Types of green spaces that serve different uses over the city, green space systems can be created as a result of efficient organization. In this context, urban green areas were classified different categories, according to the spatial characteristics, service purposes and state of property. Classification of green spaces is seen in the figure 1 according to the property.

4.1 Parks/public open space

Nowadays, in the cities, there are limited green areas. Parks or public open spaces are very important in the life of urban people. People who lives in the cities want to go outside (especially green areas) whenever they have spare time. They go parks or public open spaces. Parks are designed different type, size, and functions. In the parks, people can do lots of activities.

Typically classification types are based upon the size of the park, its deemed function, it geographic location and the types of facilities present within the park and sometimes the degree of naturalness of the park. Parks can be variously described as urban parks, nature parks, pocket parks, district parks, community parks, neighbourhood parks, sporting fields, urban forests and the like. But there are other ways of classifying parks too. These include factors such as the activities that occur within the park (e. g. cricket oval, skateboard park, bowling green), the agency responsible for managing the park (e. g. national park, state park, city park), the history of the park (e. g. heritage rose garden), the condition of the park, the land use history of the area (e. g. street-corner neighbourhood park), the types of people who use the park, landscaping and embellishments (e. g. dog park, bike park or Chinese garden) and the philosophy behind the park's development (e. g. recreation reserve or civic square). Combining these various factors can result in all sorts of combinations and permutations, rendering a standardized method of classifying parks virtually impossible and rather pointless. Parks are not the only type of urban greenspace though. In most cities while parks comprise a large portion of green and open space, other types of urban greenspace and open spaces are present too including plazas, urban trails and even well-vegetated streets (Byrne and Sipe, 2010).

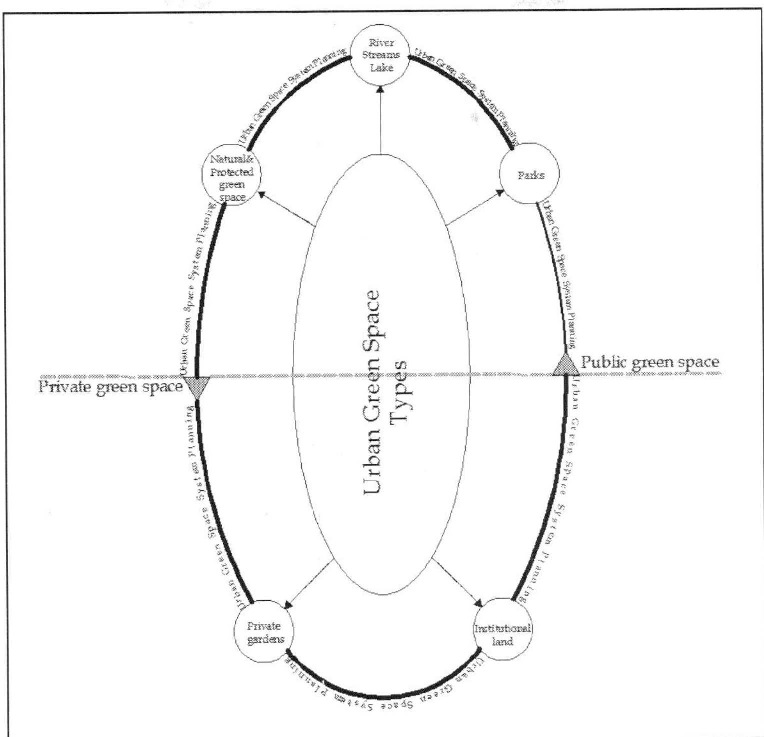

Fig. 2. Classification of green spaces according to the property.

Fig. 3. Urban Park in Ankara, Turkey (http://www.anfaaltinpark.com.tr/)

Fig. 4. Pocket Park in Çorum, Turkey.

Times have changed, somewhat. Sitting in a municipal park, looking at the flowers and listening to the occasional brass band do not feature prominently in surveys reviewing the popularity of leisure-time activities. Rather, people desire access to rich and varied landscapes with scope for many outdoor activities. Park planners responded to the new age by tearing down park railings and planning webs of interconnected green space, originally known as park systems. The diagnosis was correct. The treatment was pathetically oversimplified. Public open space should be planned in conjunction with other land-uses for multiple objectives. New parks and new links should be designed by planning recreational and conservation uses in conjunction with other land-uses: urban reservoirs can make splendid waterparks; ornithological habitats and hides should be designed in conjunction with sewage farms; wildlife corridors should be planned beside roads, railways and streams; flood prevention works can yield canoe courses; public gardens can sit on top of office buildings. New uses and new layers of interest should be brought into public open spaces. Some open spaces could supply firewood and wild food (nuts, berries, herbs); others could infiltrate rainwater back into the ground, instead of allowing the water to accentuate flood peaks; Sunday markets can fit well into parks. Every public open space can have a specialist use, in addition to its general functions. One could be a centre for kite flying one for tennis; one for lovers of herbaceous plants; one for reenacting military battles; one for every special recreational type which has a magazine on your local newsstand (Turner, 1998).

Fig. 5. National Park, Küre Mountains, Bartın, Turkey.

5. Benefits of urban green spaces

Urban green spaces have many functions and benefits. These functions and benefits are important for to improve life quality in the urban areas. Green spaces provide linkage between people (who lives in the urban) and nature. So, these areas are very important for the urban people.

Urban green spaces are important as functions and meanings for (Alm, 2007):

- Urban climate, noise moderation, air cleaning and handle of surface water
- As an indicator of environmental changes
- As a part of the circulation of nutritive substances
- Cultivation of energy plants
- Biodiversity; to save valuable urban species, as refuges for species from rural biotopes and as spreading corridors.
- Social and cultural values; for health, recovering and rehabilitation, to give beauty and comfort, to give room for passivity and activity, as a cultural heritage, as an arena for citizenship, for education.
- Gardening and allotments; as history of urban landscapes, as a social function, for life quality and beauty, providing a reserve.
- Urban design; to give the city an understandable structure, to connect different scales and parts of the urban landscape.

The benefits of urban green areas were described as detailed below under the main headings.

5.1 Environmental benefits

Ecological Benefits

Urban green spaces provide to cities with ecosystem benefits ranging from maintenance of biodiversity to the regulation of urban climate. Comparing with rural areas, differences in solar input, rainfall pattern and temperature are usual in urban areas. Solar radiation, air temperature, wind speed and relative humidity vary significantly due to the built environment in cities. Urban heat island effect is caused by the large areas of heat absorbing surfaces, in combination of high energy use in cities. Urban heat island effect can increase urban temperatures by 5°C. Aside from these human benefits, well designed urban greenspaces can also protect habitats and preserve biodiversity. Greenspaces that feature good connectivity and act as 'wildlifecorridors' or function as 'urban forests', can maintain viable populations of species that would otherwise disappear from built environments (Haq, 2011; Byrne and Sipe, 2010).

Pollution Control

Pollution in cities as a form of pollutants includes chemicals, particulate matter and biological materials, which occur in the form of solid particles, liquid droplets or gases. Air and noise pollution is common phenomenon in urban areas. The presence of many motor vehicles in urban areas produces noise and air pollutants such as carbon dioxide and carbon monoxide. Emissions from industrial areas such as sulphur dioxide and nitrogen oxides are very toxic to both human beings and environment. The most affected by such detrimental contaminants are children, the elderly and people with respiratory problems. Urban greening can reduce air pollutants directly when dust and smoke particles are trapped by vegetation (Haq, 2011).

Noise pollution from traffic and other sources can be stressful and creates health problems for people in urban areas. The overall costs of noise have been estimated to be in the range of 0. 2% - 2% of European Union gross domestic product. Urban green spaces in over crowded cities can largely reduce the levels of noise depending on their quantity, quality and the distance from the source of noise pollution. In the contemporary studies on urban green spaces consider the overall urban ecosystem, conservation of the urban green spaces to maintain natural ecological network for environmental sustainability in cities. For the cities in fast urbanizing and growing economy, country like China should consider the dynamic form of urban expanding to manage effective urban green spaces which will contribute to reduce the overall CO2 by maintaining or even increasing the ability of CO2 absorption via natural eco-system (Haq, 2011).

5.2 Biodiversity and nature conservation

Green spaces do functions as protection centre for reproduction of species and conservation of plants, soil and water quality. Urban green spaces supply the linkage of the urban and rural areas. They provide visual relief, seasonal change and link with natural world. A functional network of green spaces is important for the maintenance of ecological aspects of sustainable urban landscape, with greenways and use of plant species adapted to the local condition with low maintenance cost, self-sufficient and sustainable (Haq, 2011)

Fig. 6. Green areas are important for biodiversity, Samsun, Turkey.

5.3 Economic and aesthetic benefits

Energy Savings

Using vegetation to reduce the energy costs of cooling buildings has been increasingly recognized as a cost effective reason for increasing green space and tree planting in temperate climate cities. Plants improve air circulation, provide shade and they transpire. This provides a cooling effect and contributes to lower air temperatures. A park of 1. 2 km by 1. 0 km can produce an air temperature between the park and the surrounding city that is detectable up to 4 km away. A study in Chicago has shown that increasing tree cover in the city by 10% may reduce the total energy for heating and cooling by 5 to 10% (Haq, 2011)

Property Value

Areas of the city with enough greenery are aesthetically pleasing and attractive to both residents and investors. The beautification of Singapore and Kuala Lumpur, Malaysia, was one of the factors that attracted important foreign investments that assisted rapid economic growth. Still, indicators are very strong that green spaces and landscaping increase property values and financial re-turns for land developers, of between 5% and 15% depending on the type of Project (Haq, 2011).

Fig. 7. Green areas provides aesthetically well placesAnkara, Turkey (http://www. anfaaltinpark. com. tr/).

Fig. 8. Green areas near the housing area, Ankara, Turkey (http://www. anfaaltinpark. com. tr/).

Fig. 9. Green areas near the housing area, Ankara, Turkey
(http://www. anfaaltinpark. com. tr/).

Fig. 10. Green areas near the housing area, Ankara, Turkey
(http://www. anfaaltinpark. com. tr/).

5.4 Social and psychological benefits

Recreation and Wellbeing

People satisfy most of their recreational needs within the locality where they live. Urban green spaces serve as a near resource for relaxation; provide emotional warmth. In Mexico City, the centrally located Chapultepec Park draws up to three million visitors a week who enjoy a wide variety of activities (Haq, 2011)

Fig. 11. Recreational activities on water surface, Altınpark, Ankara, Turkey (http://www.anfaaltınpark. com. tr/).

Fig. 12. People are sitting in a park for recreational activity, Samsun, Turkey.

Human Health

People who were exposed to natural environment, the level of stress decreased rapidly as compared to people who were exposed to urban environment, their stress level remained high. Certainly, improvements in air quality due to vegetation have a positive impact on physical health with such obvious benefits as decrease in respiratory illnesses. The connection between people and nature is significance for everyday enjoyment, work productivity and general mental health (Haq, 2011)

6. References

Alm, L. E., 2007. Urban Green Structure A hidden resource, Baltic University Urban Forum Urban Management Guidebook V, Green Structures in the Sustainable City, Chalmers University of Technology, Editor: Dorota Wlodarczyk, Project part-financed by the European Union (European Regional Development Fund) within the BSR INTERREG III B Neighbourhood Programme.

Anonymous, 2010. A report by natural England and the campaign to protect rural England, PRE/Natural England 2010, ISBN 978-1-84754-160-4, CatalogueCode: NE196.

Baycan-Levent, T. and Nijkamp, P., 2009. Planning and management of urban green spaces in Europe: Comparative analysis. Journal of Urban Planning and Development, 135: 1.

Baycan-Levent, T., 2002. Development and management of green spaces in European cities: a comparative analysis. Research Memorandum, 2002: 25.

Bolund, P. and Hunhammar, S., 1999, 'Ecosystemservices in urban areas', Ecological Economics 29, pp. 293-301.

Bowdler, S., 1999, 'A study of Indigenousceremonial ("Bora") sites in eastern Australia', paperpresented to Heritage Landscapes: Understanding Place and Communities, Southern Cross University, Lismore, November, 1999.

Byrne, J. Sipe, N., 2010. Green and open space planning for urban consolidation – A review of the literature and best practice, Urban Research Program, ISBN 978-1-921291-96-8.

C. Francis, "People Places; Design Guidelinesfor Urban Open Space, " Second Edition, John WileyandSons, Hoboken, 1997.

D. Huang, C. C. Lu and G. Wang, "Integrated Management of Urban Green Space: The Case in Guangzhou China, " 45th ISOCARP Congress 2009.

EDAW and Skyes Humphrey's Consulting, 2008, 'Creating Better Parks: Brimbank Open Space and Playground Policy and Plan', EDAW and Skyes Humphrey's Consulting, Brimbank, Melbourne.

Fernández-Juricic, E. and Jokimäki, J., 2001, 'A habitat islandapproachtoconservingbirdsin urban landscapes: case studies from southern and northern Europe', Biodiversity and Conservation10, pp. 2023-43.

Fernández-Juricic, E., 2000, 'Bird community composition patterns in urban parks ofMadrid: the role of age, size andisolation', Ecological Research15, pp. 373-83.

Haq, S. M. A., 2011. Urban green spaces and an integrative approach to sustainable environment. Journal of Environmental Protection, 2(5): 601-608.

Hellmund, P. C. and Smith, D., 2006. Designing Greenways : Sustainable Landscapes for Nature and People. Island Press, Washington, DC, USA. http://www. iadb. org/sds/doc/ENV109KKeipiE. pdf.

Kong, F., Yin, H., Nakagoshi, N. and Zong, Y., 2010. Urban green space network development for biodiversity conservation: Identification based on graph theory and gravity modeling. Landscape and urban planning, 95(1-2): 16-27.

L. Loures, R. Santos and P. Thomas, 2007. "Urban Parks and Sustainable Development: The case study of Partimaocity, Portugal, " Conference on Energy, Environment, Ecosystem and Sustainable Development, Agios Nikolaos, Greece

Low, S., Taplin, D. and Scheld, S., 2005. Rethinking Urban Parks : Public Space and Cultural Diversity. University of Texas Press, Austin, TX, USA.

M. Neuvonen, T. Sievanen, T. Susan and K. Terhi, 2007. "Accessto Green Areas and the Frequency of Visits: A Case Study in Helsinki, " Elsevier: Urban Forestry and Urban Greening, Vol. 6, No. 4, pp. 235-247.

M. Sorensen, J. Smit, V. Barzettiand J. Williams, , 1997. "Good Practices for Urban Greening, " Inter American Development Bank.

Manlun, Y. 2003. Suitability Analysis of Urban Green Space System Based on GIS. International Institute for Geo-information Science and Earth Observation. Masterthesis. Enschede, The Netherlands.

Manlun, Y., 2003. Suitability Analysis of Urban Green Space System Based on GIS, University of Twente, Netherlands, 90 pp.

Melles, S., Glenn, S. and Martin, K., 2003, 'Urban bird diversity and landscape complexity: species environment associations along a multiscale habitat gradient', Conservation Ecology 7, pp. [online].

P. Bolund and H. Sven, 1999. "Ecological Services in Urban Areas, " Elsevier Sciences: Ecological Economics, Vol. 29, pp. 293-301. doi:10. 1016/S0921-8009(99)00013-0

P. Grahn and U. A, 2003. Stigsdotter, "Landscape Planning and Stress, " Urban Forest: Urban for Urban Green, Vol. 2 pp. 001-018.

Swanwick, C., Dunnett, N. and Woolley, H., 2003, 'Nature, role and value of green space in towns and cities: An overview', Built Environment 29, pp. 94-106.

Turner, T., 1998. Landscape Planning and Environmental Impact Design. Routledge, Florence, KY, USA.

V. Heidt and M. Neef , 2008. "Benefits of Urban Space for Improving Urban Climate, " Ecology, Planning and Management of Urban Forests: International Perspective

Ward, S., 2004. Planning and Urban Change (Second Edition). SAGE Publications Inc. (US), London, GBR.

Wilby, R. L. and Perry, G. L. W., 2006, 'Climate change, biodiversity and the urban environment: a critical review based on London, UK', Progress in Physical Geography30, pp. 73.

Wuqiang, L., Song, S. and Wei, L., 2012. Urban spatial patterns based on the urban green space system: A strategic plan for Wuhan City, P. R. China Shi Song.

6

Tourism Planning in Rural Areas and Organization Possibilities

Tuğba Kiper and Gülen Özdemir
*Namık Kemal University, Faculty of Agriculture,
Department of Landscape Architecture, Department of Agricultural Economics
Turkey*

1. Introduction

Along with changing life conditions, various orientations have emerged in tourism. Instead of conventional tourism centers, quiet, natural and authentic places are being preferred. However, the economic output provided by this initiative has not been able to reach to the expected level and sustainability due to lack of a planned and rational basis from the beginning. Uncontrolled pressure generated from this might have irreversible impacts on natural and cultural landscapes. Therefore, a rational, practical and preservation-based planning should be implemented in order to ensure a sustainable development in regions where no intense tourism activities exist yet but that have significant potential. At this point, first potential a region has regarding the natural and cultural landscape values should be determined; and then applicable tourism types and their implementation areas should be established. Additionally socio-economic development and change to be generated by probable tourism activities should be analyzed carefully. Moreover opinions of local people regarding this subject must be considered throughout the decision-making process. Here cooperatives and organizations in rural areas step in. Planning decisions, which are made without considering the balance between usage and conservation and provide a temporary solution, would damage authentic sources of a rural area in an irreversible way. Rural areas are settlements which represent reserved areas remained by natural areas and contain a unique rural life culture. Therefore, among tourism types they are not only regions suitable for development of rural tourism, but they are also considered to be spaces of many tourism types which share this environment.

In this study, it is aimed to reveal rural tourism concept and types, analyzing rural tourism possibilities implemented in the world and in Turkey; and to create rural tourism planning and organization models. Within this scope, we focused on rural tourism's definition, its objectives, the reasons of its emergence and development, its principles, its types, its environmental, social and economic impacts; and on the examination of approaches to tourism at rural areas in Turkey and Europe.

In this section, the subjects below will be discussed.

- Definition of Rural Area
- Definition of Rural Tourism and its types

- Relationship between Rural Tourism and Natural - Cultural Environment
- Rural Tourism-Rural Development relationship
- Rural Tourism Policy in Turkey
- Rural Tourism – Examples of Implementation
- Rural Tourism Organizations
- Conclusion

2. Definition of rural area

The definition of rural has been in dispute for decades (Gilbert, 1982), many different definitions of the rural have been given, each focusing on a different specialized aspect: in turn, statistical, administrative, built-up area, functional regions, agricultural, and population density.

Many commentators define rural areas as those with less than 10-20 per cent of their land areas covered by the built environment. There are three important implications here. These areas will be dominated by agrarian and forest-based economic activities. They will be, to a large extent, repositories of the natural world and wild-life. For the visitor, they will give an impression of space, and a traditional non-urban, non-industrial economy. Their economies will be strongly influenced by the market for farm and forest products. Although the labour force required for farming and forestry has declined rapidly in recent years, rural areas still show a strong bias towards jobs in the farm/forest sector. Additionally, they usually exhibit low female activity rates outside the home because of the shortage of job opportunities for women in many rural areas (Organisation For Economic Co-Operation and Development [OECD], 1994).

The OECD Rural Development Programme uses a pragmatically based series of indicators: while at local level a population density of 150 persons per square kilometre is the preferred criterion, "at the regional level geographic units are grouped by the share of their population which is rural, into the following three types:

- predominantly rural (> 50 per cent),
- significantly rural (15-50 per cent),
- predominantly urbanized regions (< 15 per cent)" (Organisation For Economic Co-Operation and Development [OECD], 1994).

The frequently quoted definition of rurality based on population density criteria as used by the OECD is a typical example of this type of definition. Besides, an EU classification of rural areas (integrated rural areas, intermediate rural areas and remote rural areas) based upon socio-economic trends, such as population growth, land use change and employment conditions (European Commission, 1988), belongs to the descriptive definitions (Elands & Freerk Wiersumi 2001)

This point is also illustrated when examining the size of settlements classified as rural by a selection of various states, given below: (Table 1)

According to the European Commission 1997., approximately 80% of the territory of the European Union can be called 'rural'. These rural areas or countrysides. include a great variety of cultures, landscapes, nature and economic activities that shape a palette of rural identities (Huigen et al., 1992; Slee, 2000).

Country	The criterias to define the rural area
Austria	Towns of fewer than 5 000 people.
Australia	Population clusters of fewer than 1 000 people, excluding certain areas
Canada	Places of fewer than 1 000 people, with a population density of fewer than 400 per square kilometre.
England and Wales	No definition -- but the Rural Development Commission excludes towns with more than 10 000 inhabitants.
Denmark	Communities with fewer than 200 households
Ireland	The separation between total urban areas and rural area are determined 100 settlements.
Italy	Settlements with fewer than 10.000 people.
France	Towns containing an agglomeration of fewer than 2 000 people living in contiguous houses, or with not more than 200 metres between the houses.
Norway	Communities with fewer than 200 households
Portugal	Settlements with fewer than 10.000 people.
Scotland	The local authority areas less than 100 people pers q km.
Spain	Settlements with fewer than 10.000 people.
Switzerland	Settlements with fewer than 10.000 people.
Turkey	Provincial and district centers outside settlements with fewer than 20.000 people.

Table 1. The criterias to define rural area in different countries (Roberts & Hall 2003; Gülçubuk, 2005, UN Demographic Year-books and Robinson, 1990)

According Agricultural and Rural Development Operational Program [Ardop], (2006) It covers areas of population density of or under 120 persons/km² or settlements with a population under 10,000 inhabitants. From this array of varying definitions, two clear points stand out. Rural settlements may vary in size, but they are small, and always with a population of fewer than 10 000 inhabitants. They are almost always in areas of relatively low population density (Organısatıon For Economıc Co-Operatıon and Development [OECD], 1994).

Typically rural areas have low population densities: this is a result of small settlements, widely spaced apart. The natural and/or the farmed/forested environment dominates the built environment.

This point is various definiations common features of rural space are (Ashley and Maxwell 2001):

- Spaces where human settlement and infrastructure occupy only small patches of the landscape, most of which is dominated by fields and pastures, woods and forest, water, mountain and desert
- Places where most people spend most of their working time on farms
- Abundance and relative cheapness of land
- High transaction costs, associated with long distance and poor infrastructure
- Geographical conditions that increase political transaction costs and magnify the possibility of elite capture or urban bias

Rural areas generally suffer high levels of poverty, and are also characterised by lower levels of non-farm economic activity, infrastructural development, and access to essential services. They may also suffer from depopulation of the able-bodied, and lack of political clout.

3. Definition of rural tourism and its types

The concept of rural tourism is multidimensional and there are several different definitions about rural tourism. The followings are some examples: according to the Organization of Economic Cooperation and Development [OECD], rural tourism is defined as tourism taking place in the countryside.

It has been argued above that rurality as a concept is connected with low population densities and open space, and with small scale settlements, generally of fewer than 10 000 inhabitants. Land use is dominated by farming, forestry and natural areas. Societies tend towards traditionalism: the influence of the past is often strong.

Government policies lean towards conservation rather than radical or rapid change.

It follows, therefore, that rural tourism should be:

- Located in rural areas;
- Functionally rural, built upon the rural world's special features: small scale enterprise, open space, contact with nature and the natural world, heritage "traditional" societies and "traditional"practices;
- Rural in scale -- both in terms of buildings and settlements -- and, therefore, usually small scale
- Traditional in character, growing slowly and organically, and connected with local families. It will often be very largely controlled locally and developed for the long term good of the area;
- Sustainable -- in the sense that its development should help sustain the special rural character of an area, and in the sense that its development should be sustainable in its use of resources. Rural tourism should be seen as a potential tool for conservation and sustainability, rather than as an urbanizing and development tool;
- Of many different kinds, representing the complex pattern of rural environment, economy, and history.

According to Lane (1994a) rural tourism should: be located in rural areas, functionally rural, rural in scale i.e. usually small-scale; be traditional in character; grow slowly and organically; be connected with local families; and represent the complex pattern of rural environment, economy, history and location.

Descried rural tourism as multi-faceted activity rather than farm-based tourism only. It not only includes farm-based holidays but also comprises special interest nature holidays and ecotourism, walking, climbing and riding holidays , adventure, sport and health tourism, hunting and angling, educational travel, arts and heritage tourism and in some areas, ethnic tourism (Baramwell& Lane, 1994).

Rural tourism refers to those traveling activities that aim at pursuing natural and humanistic attraction with rurality in rural area (Jingming& Lihua, 2002).

Macdonald and Jolliff (2003) introduced the concept into this patticular study and defined it as cultural rural tourism. By this definition rural tourism refers to a distinct rural community with its own traditions, heritage, arts, lifestyles, places, and values as preserved between generations. When tourists visit these areas, they are well informed about the culture and experienced folklore, costoms, natural landscapes, and historical landmarks. They may also enjoy other activities in a rural setting such as nature, adventure, sports, festivals, crafts, and general sightseeing.

The connection between rural tourism, agricultural tourism and farm tourism is synthesized in figure 1. In relation to the scope of rural tourism, McGehee & Kim (2004) provide examples of tourism types as illustrated in Figure 1 below:

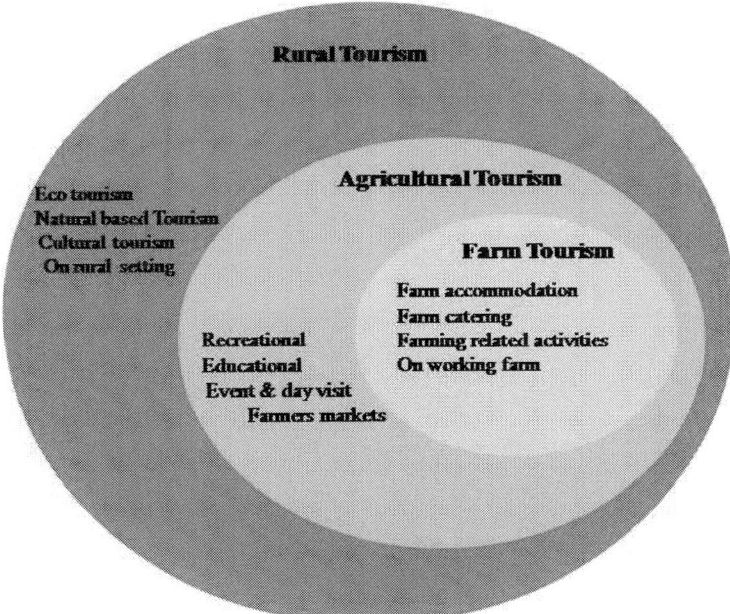

Fig. 1. The classification of different tourism activities in rural areas (McGehee & Kim, 2004)

Rural tourism, or rurally-located tourism, can include the above but also campsites, lodges, safari drives, craft markets, cultural displays, adventure sports, walking trails, heritage sites, musical events indeed any tourist activity taking place in a rural area (Table 2). Rural tourism is a kind of rural activities and its characteristic is natural and humanistic (It includes customs, scenery, landscape (about local country and agricultural) and other attractions, Its types of activities basically are leisure, sightseeing, experience and learning, and so on (Jingming & Lihua 2002; Deqian, 2006; Holland, Burian & Dixey, 2003). According to Nilsson, rural tourism is based on the rural environment in general whereas farm tourism is based on the farm and farmer. This means that within the framework of rural tourism, farm tourism enterprises are more closely related to agriculture than other rural tourism operations.

Zone	Activity	Facilities
Areas for picking wild vegetables, sightseeing	Picking wild vegetables	Explanations and maps on signboards, paths, restrooms
Areas to experience agricultural life and culture of mountain people	Triditional agricultural activities, tours	Explanations and maps on signboards, paths, restrooms
Trout fishing	Ecology of trout fishing	Parking area, landscape beautification
Exhibition of agricultural products	Selling agricultural products, traditional culure of mountain people	Service center for seling agricultural products and tradional crafts
Meals made from local specialtles	Tasting and knowing how to cook with local special ingredients	Noticeboards, parking area, service center, landscape beautification
Farmstays	Staying with local people	Noticeboards, parking area, landscape beautification, improvement of accommodation facilities
Waterfall areas	Sight-seeing at water- falls, often with butterflies	Noticeboards, landscape beautification, parking areas
Cultural area of mountain people (Buson tribe)	Exhibilition of culture, dancing, traditional festivals	Explanations and maps on signboards, exhibition center, museum, festivals, parking areas
Natural landscape area	Climbing, hiking	Paths, noticeboards to signal route and warm of any dangers
Heritage area	Heritage interpretation, telling of folk storles	Explanations and maps on signboards, setting out of tourist route

Table 2. Activities and facilities in different zones (Hong, 1998)

Agri-tourism is when the purpose of the visit has a specific agricultural focus such as being with animals, enjoying a vineyard. Tourism on the farms enables farmers to diversify their activities while enhancing the value of their products and property. Farm tourism also helps to reconcile farming interests and environmental protection through integrated land management in which farmers continue to play a key role. Tourists who choose farm accommodation rather than other kinds of accommodation facilities look for genuine rural atmosphere where they can share intimacy of the household they live in, learn traditional crafts and skills with their hosts, make friends which is a quality, modern times have almost forgotten and above all enjoy home made food and drinks. Some specific food labels can help consumers establish a local produce and can be used as a selling point to tourist who want to taste home grown quality food and drink. Agritourism "is a hybrid concept that merges elements of two complex industries—agriculture and travel/tourism—to open up new, profitable markets for farm products and services and provide travel experience for a large regional market (Wicks & Merrett, 2003). Agritourism helps preserve rural lifestyles and landscape and also offers the opportunity to provide "sustainable" or "green" tourism (Privitera, 2010). Agritourism can be defined as a subset of rural tourism (Reid et al., 2000), and "includes a range of activities, services and amenities provided by farmers and rural people to attract tourists to their area in order to generate extra income for their businesses" (Gannon, 1994).

Farm tourism is when accommodation for rural tourists is provided on farms. The core activity is in the wider rural area (walking, boating) but the vast majority of visitors are accommodated on farms, either working farms or farms converted to accommodation facilities. Farm tourism activities can include farm markets, wineries, U-Picks, farming interpretive centers, farm-based accommodation and events, and agriculture-based festivals.

Heritage and cultural tourism in rural areas comes in a wide range of forms most of which are unique to an individual local and a valuable component of the rural tourism product. Heritage and cultural tourism includes temples, rural buildings but may be extended to local features of interest including war remnants, monuments to famous literary, artistic or scientific people, historic remains, archeological sites, traditional parkland etc.

Eco tourism; many tourists visit rural areas for the purpose of bird and animal watching and learning about local flora and fauna. Rural tourist destination as a product is definitely very fragile in ecological, social and cultural sense. Its development requires very specific approach that could help it remain sustainable in the long term. In many rural regions, tourism is accepted as a natural part of the socio-economic fabric juxtaposed with agriculture.

Rural tourism is among the most polymorphous of all forms of Special Interest Tourism (SIT). The diversity of attractions included within rural tourism embrace: Indigenous and European heritage sites, Aspects of culture (agriculture), Industrial tourism (farm practices), Education tourism, Special events, Ecological attractions, Adventure tourism and Wine tourism.

Such diversity represents major opportunities for rural areas that have turned to tourism as a means of supplementing diminished incomes (Douglas & Derret 2001).

Lane, (1994) identifies four necessary features for the sustainable rural tourism strategies as:

- It is important that the person or team formulating the strategy is skilled not only in tourism development but also in economic, ecological and social analysis
- Wide consultations amongst all interest groups are essential. These consultations will include trade and business, transport, farmers, administrators, and the custodians of the natural and historic assets of the area
- Tourism relies more than any other industry on local goodwill. The local population must be happy with their visitors and the secure in the knowledge that the visitor influx will not overwhelm their live, increase their income hosts and impose new and unwelcome value systems on them
- The strategy-making process should not be a once-only affair. It has to be an evolving long-term enterprise, able to cope with change, and able to admit to its own mistakes and shortcomings. It is the beginning of a partnership between business, government and cultural and conservation interests

In recent years, rural tourism has been developing rapidly. In order to promote the development of rural tourism, the local government paid more and more attention to the planning of rural tourism. Rural Tourism (RT) has long been recognized in certain parts of Europe as an effective catalyst of rural socio-economic regeneration for over a hundred years.

Rural tourism can therefore encompass a wide range of rural- based attractions, events and services that can provide the context for economic diversification and a mediating factor for sustainability. Rural tourism can promote heritage appreciation and resource conservation, contribute to social-economic change, and provide the context for interactionbetween local rural peopleand the tourists. On the other hand, rural tourism development can promote undesirable changes in the landscape, negatively influence the social-cultural values of a region, and promote inauthentic representations of local customs and ways of life. In the case of farm tourism, the sustainability of the tourism product can also be influenced by the degree to which operators have developed managerial skills, such as product and market development and customer service skills (Colton & Glyn, 2005)

Rural tourism planning process begins with choosing qualified villages to attract tourists. In fact the first step of rural tourism planning management is to select potential villages for rural tourism. Finally the managerial operations must be implemented in order to maximize the benefits of mral tourism processes (Mahmoudi, Haghsetan, Meleki, 2011).

Rural tourism planning can be organized as shown (Figure 2);

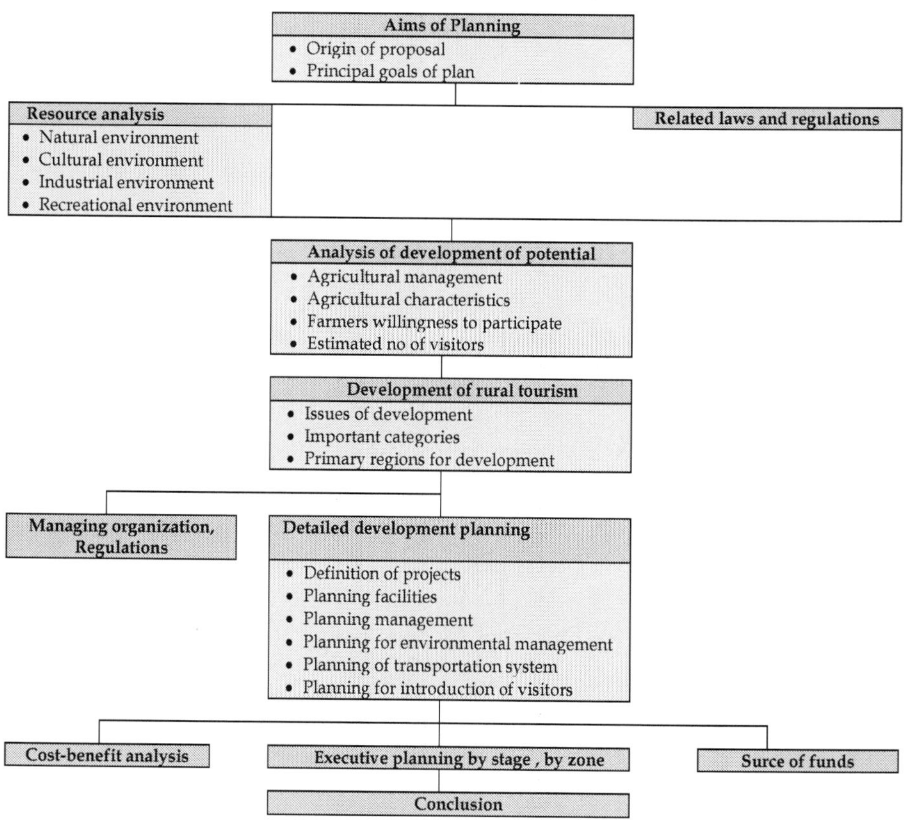

Fig. 2. Planning procedure for rural tourism (Hong, 1998).

4. Relationship between rural tourism and natural-cultural environment

Tourism and environment are representing common relationship concepts. Successful tourism activity needs clean and orderly environment (Kiper 2006). Rural tourism is both entagled with rural settlements and is based on natural resources (Soykan 2003).

Rural tourism, natural resources, cultural heritage, rural lifestyle and an integrated tourism is a type of local economic activities. Therefore, rural tourism in rural areas was carried out with a number of elements in their natural landscape and cultural landscape (water, vista, topography, vegetation, clean air), as well as in the variety of recreational activities suitable for all kinds of environments. Therefore, rural tourism and its natural assets and raw materials to create, as well as directing people to travel is an attractive force (Kiper, Özyavuz & Korkut 2011).

Natural and cultural landscape values form a basis for rural tourism. These values are geographical position, micro-climatic conditions, existence of water, natural beauties, existence of natural vegetation, existence of wildlife, surface features, geomorphologic structure, local food, festivals and pageants, traditional agricultural structure, local handicrafts, regional dress culture, historical events and people, heritage appeals, architectural variety, traditional music and folk dance, artistic activities and so on (Gerry, 2001; Lane, 1993; Lanquar, 1995; Soykan, 1999; Brıassoulis, 2002; Catibog-Sinha & Wen, 2008; Mlynarczyk, 2002; Drzewiecki, 2001, Kiper, T. 2006; Kiper, Korkut & Yılmaz, 2011).

Relationship between rural tourism and natural - cultural environment in Figure 3.

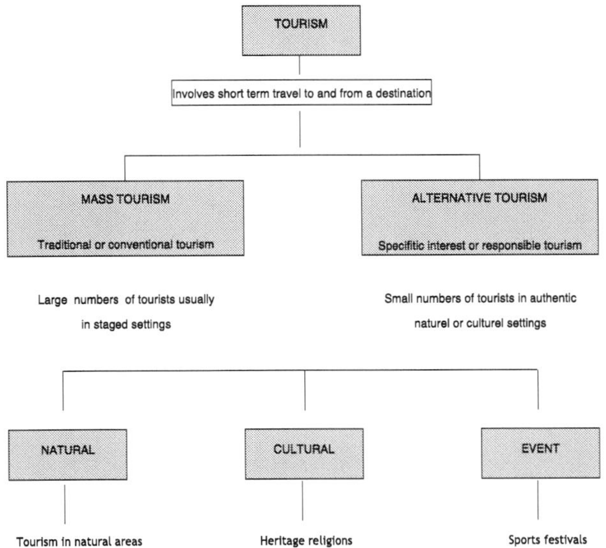

Adventure- emphasis on activity
Nature based- primarily viewing of natural landscape
Wildlife- primarily viewing of wildlife
Ecotourism- includes educative and conservation supprting elements
Rural tourism- primarily viewing of natural and cultural landscape

Fig. 3. An overview of tourism (Dowling, Moore, Newsome 2002)

According to Weaver and Opperman (2000) tourism attractiveness is form of two: natural and cultural.

Natural attractiveness: the topography (mountains, canyons, beaches, caves, volcanoes, fossil sites, etc..), (Fig.4), climate (temperature, rainfall, humidity and so on.), hydrology (lakes, rivers, waterfalls, hot springs, etc..), wildlife (mammals, birds, insects, etc..), the natural vegetation structure (Fig.5, Fig.6)

Cultural attractiveness: Archeological heritage (castle, cave, bridge etc.), religious structures (mosque, monastery etc.), conventional architecture, folkloric values, gastronomy, etc.

Rural tourism resources vary with the geographical, it mostly with natural style, toiling morphology, farm life and tradition primarily. It is affected by the climate and the seasons, so can satisfy tourist various needs. Rural tourism cultural resources include folk festivals, crafts, folk architecture, folk art, marriage customs, fun legends, and so on (Fan &Yang, 2011).

According to Tane and Thierheimer tourism in natural environments is mostly practiced by nature lovers that prefer isolation instead of urban crowds, and silence instead of the noise of tourist resorts. Being directly connected to rural tourism, it takes place in partially or completely isolated areas (mountain, sea, delta), being mostly destined to young or adult tourists.

Fig. 4. Topographic structure

Fig. 5. Natural vegetation cover

Fig. 6. Natural vegetation structure

Tourism in natural environments creates the possibility for tourists to practice climbing, fishing, hunting or other sports like dirt track, canoe racing, sky diving etc.

The relationship between rural tourism and other tourism forms practiced in the rural environment is extremely important, because it establishes the connections between natural, human, economical and social parameters, in order to insure all factors that lead to tourist growth and durable development (Tane Thierheimer, 2009).

According to Butler and Hall, to speak about rural tourism in a place, an economically viable rural population sustaining rural culture and identity through being engaged in rural activities is a prerequisite.

For, the target group will be the local community who is both the most influenced by tourism in the region and is the most influential on it. Certain states such as participation and embracing are prerequisites for sustainability (Kiper, 2011).

5. Rural tourism-rural development relationship

Tourism has long been considered as a potential means for socio-economic development and regeneration of rural areas, in particular those affected by the decline of traditional agrarian activities. Peripheral rural areas are also considered to be repositories of older ways of life and cultures that respond to the postmodern tourists' quest for authenticity (Urry, 2002).

Rural Development is an instrument which requires an integrated approach regarding economic, ecological, cultural and social way. Rural tourism is one of the most important instruments of Rural Development

According to Keyim, Yang & Zhang, 2005; Rural tourism has long been considered the means of accelerating economic and social development, and has become a development tool for many rural areas.

More specifically, the development rural tourism offers potential solutions to many of the problems facing rural areas, These may be summarized as:

- Economic growth, diversification and stabilization through employment creation in trades and crafts; the creation of new markets for agricultural products; and a broadening of a regions economic base, because local residents with a few can readdily work as food servers, retail clerks and hospitality workers. Tourism also skills can serve as a vehicle for attracting potential investors, as todays tourist may spend their life after retirementor start a business there.
- Socio-cultural development, including the maintenance and improvement of public services, the revitalization of local crafts, costoms and cultural identities, increases opprtunities for social contact andexchange.
- Production and improvement of both the natural environment built infastructure. Tourism, whichs generally considerered to be a relatively clean industry, may support local environment conservation (Keyim, Yang& Zhang, 2005).

According to Mahmoudi, Haghsetan, Meleki (2011); decrease in rural area's population and increase in the urbanity rate in recent years is a result of poverty and the absence of proper access to resources of welfare and livelihood services. Optimum usages of environmental, economical and production potentials of rural areas for improving income and welfare can be efficient in reaching the goal of reducing the rural immigration. Rural tourism is a part of tourism market and is a source of employment and income. Also it can be presented as a significant tool for the socio-economical development of rural areas.

Declining economic activity, restructuring of the agricultural sector, dwindling rural industrialisation and out-migration of higher educated youth, has led to the adoption, in many western nations, of tourism as an alternative development strategy for the economic and social regeneration of rural areas (Pompl & Lavery, 1993; Williams & Shaw, 1991; Hannigan, 1994; Dernoi, 1991; Wickens, 1999)

Rural tourism in one of the forms of sustainable development that through promoting productivity in rural zones, brings about employment, income distribution, preservation of village environment and local culture, raising host community's participation and presenting appreciate methods to conform beliefs and traditional values with new circumstances (Kanaani, 2005).

Sustainable development is a process having economic, social, cultural and environmental–ecological dimensions. This process is perceived as a development in all respects for both urban and rural societies. Yet, in most of the developing countries rural population is gradually diminishing, notwithstanding the agricultural lands that are losing productivity are increasing. While this situation primarily results in increasing impoverishment of rural society, it also causes problems such as deforestation, erosion and productivity loss with the misuse of resources. On the other hand, damaging the natural resources emerge problems such as migration, poverty and hunger. These problems primarily affect rural people. Most affected ones by these problems are women and children. Overcoming these problems would be possible by sustainable planning and management of rural areas in accordance with their resource potential (Golley and Bellot, 1999).

Rural tourism has a positive impact on agricultural development, farmers' incomes and the standard of living in rural areas. However, one important issue is how to reach a consensus among local government, farmers' associations, and individual farmers on what kind of projects to promote, and how they should be funded and operated (Hong, 1998).

Agro-tourism represents a real opportunity for the local economy, the main motivation in training and development initiatives, the traditional activities that have long been neglected, of crafts, the strengthening and development of local artistic creations, linking to friends, material needs and spiritual needs of tourists, local economic activities stimulated life.

Rural household translates offer accommodation and services and creates motivation agro household to prepare and arrange inside and outside the household to obtain revenue, stimulate peasant to invest in their own household, to develop complementary activities of its concerns. The villager will compete, will carry from now on business in a competitive environment and will be forced to become competitive, increase the quality of their services and products to be applied (Munteanu, 2007).

Today, Villages are one of the resources, attracting the attention of tourism planners more than ever. Those with some specific cultural, natural or social appeal have a very strong potential for attracting tourists from close or remote areas and this can have significant role in rural development (Mostowfi, 2000).

Rural development is a strategy for improving economical and social life of poor villagers and a multilateral endeavor to reduce the poverty. That will especially be possible through increase in production and promote productivity in rural environment (Yadghar, 2004).

Certain rural areas depending considerably on stock-breeding and forestry have retrogressed rapidly with the technological developments. This change has rebounded on the economic life styles and agricultural production. Therefore, unifying agricultural activities with recreation and tourism and carrying out plans all together matter within the scope of both enlivening agricultural activities, of the prevention of using agricultural for non-agricultural purposes and of the reunion of people, who left nature and production, with production processes (Kiper, Özdemir & Başaran, 2011).

Since the 70s of 20th century, tourism activity in rural areas has remarkably increased in all the developed countries worldwide, which has played a key role in the development of rural areas that were economically and socially depressed (Perales, 2002).

Rural Tourism (RT) has long been recognized in certain parts of Europe as an effective catalyst of rural socio-economic regeneration for over a hundred years (He, 2003).

In Europe, the rural tourismhas beenwidely encouraged, promoted and relied on as a useful means of tackling the social and economic challenges facing those rural areas associated with the decline of traditional agrarian industries (Wang, 2006; Soykan, 2000). In countries such as France, Austria, and the United Kingdom rural tourism already represents a significant factor and has a growing demand (Pevetz, 1991).

Nowadays, it is seen that there is also a new tendency for rural tourism at local level through local initiatives in Turkey. Although there is still no governmental regulation for rural tourism activities, political and practical developments demonstrate that the sector should be evaluated as a planning element for Turkey.

6. Rural tourism policy in Turkey

The main source which strengthens rural economy is what rural area has regionally (agricultural development, natural resources, historical and cultural assets, traditional values). Here it is seen that they should be integrated within recreational uses which will reveal rural characteristics of the settlement and will develop potential of natural and cultural resources. However, what is important here is the necessity of preservation natural and cultural resources the rural area has through dealing with them within the frame of ecological principles.

In developing countries through the World Bank and United Nations Development Programs the various projects implemented in rural areas. In Turkey, implemented rural tourism projects for rural development in the planning period.

Rural tourism policy in Developments Plans in Turkey (Table 3).

Developments Plans	Policies
I. Development Plans for Five Years	The priority regions to development studies are determined (West and Southwest Anatolia Region).
II. Development Plans for Five Years	Giving priority to undeveloped and development regions with less power are in the planning process for preventing the formation of large single-center.
III. Development Plans for Five Years	Planning studies have only been treated as priority regions. In addition, developmental sequence based on the detection of natural and cultural resources and socio-economic indicators are considered.
IV. Development Plans for Five Years	Regulatory studies were conducted on the use of public land and strengthening of agriculture for priority regions for development,
V. Development Plans for Five Years	Resource utilization and priority regions for development are dicussed.
VI. Development Plans for Five Years	Targetet to meet the needs of current population and the cities are ranked according to rates of development. • Targetet to protect culture, tourism, history and natural assets for planning region • ÇED is suggested. • Environmental organization plans are purposed for the detection of pre-basin and the surrounding land use.
VII. Development Plans for Five Years	The most extensive form of policies to the regional development are reflected in the seventh plan.
VIII. Development Plans for Five Years	The concepts of socio-economic framework, organizational structure and the current situation in rural ares are given. • Sustainable terms of human development. based on rural development strategies
IX. Development Plans for Five Years	• Promotion and marketing of food products, forestry, agriculture in rural areas. • Development of tourism and recreation, handicrafts, agro-based industries and other alternative production activities, • Rural residential planning principles and criteria are determined according to the needs of rural area and rural society, • Giving priority to tourism regions, developing residential units • EU institutional framework are intended to establish for rural development policies and compliance.

Table 3. The arrangements for development plans in rural areas (Sarıca, 2001; Turhan 2005; Ninth Development Plan (2007-2013)

In the Tourism Strategy of Turkey-2023 and the Ninth Development Plan (2007-2013), it is aimed to utilize natural, cultural, historical and geographical values of Turkey based on conservation-use balance, to increase the share of Turkey from tourism and to promote the attractiveness of regions via alternative tourism types like rural tourism (Tourism 2023 Strategy Plan, 2007; Ninth Development Plan, 2007-2013).

In spite of the fact that Rural tourism that has been taken part in as a income supplement generating activity under Rural Development projects carried out in the different provinces in Turkey, The required attention has not sufficiently been given until now. Rural Turism submeasure is taken place under "Diversification of Economic Activities" of IPARD (Instrument for Pre-accession for Agriculture and Rural Development) Programme which is carried out by Ministry of Agriculture and Rural Affairs in the period covering 2007-2013. Ministry of Agriculture and Rural Affairs with the aim of the related public and private sector representatives is expected to play a vital role in the near future for making significant contribution for economic development of Turkey by being explored of the areas that have never been exploited before by both domestic and foreign tourists (Şerefoğlu, 2009).

7. Rural tourism – Examples of implementation

The rural recreation areas of Europe provide a wide variety of experiences and attractions for the visitor. Domestic and international visitor flows mostly from conurbations are turning to countryside destinations for holidays in increasing numbers (Arzac, 2002).

The rural tourism is outstandingly developed in Austria and France in the European Union. In Austria rural tourism businesses receive government subsidies and they may apply for loans on preferential terms while in France government beyond the financial promotion education plays a significant role. In Hungary the history of rural tourism, similarly to the beginning of the Hungarian tourism, started in the thirties. The new age of rural tourism is 10-15 years old, it started in 1989-90 when local governments recognised the rural tourism as a possibility for economic achievements. Generally businesses of rural tourism were initiated by civil associations. Rural tourism connects tourism products. Rural tourism connects areas of rural leisure activities. Therefore the rural tourism, based on the rural circumstances, is a type of tourism which can be combined with the elements of cultural and active tourism (e. g.: horse riding and hiking). Synthesising the elements of rural tourism the system of definitions of rural tourism is shown in figure 7 (Pakurar & Olah 2008).

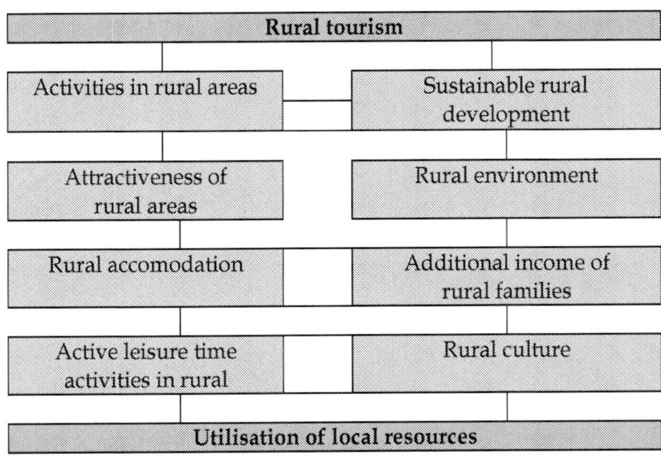

Fig. 7. System of definitions of rural tourism (Könyves, 2001)

- **Examples of Rural Tourism in Italy**

Currently, in Italy agritourism is regulated by a Law (L. 5/12/1985 n. 730 now superseded by art. 14 L. 20/2/ 2006 n. 96; sentence Court of Cassazione 2/10/2008 n. 24430) that defines agritourism as: "accommodation and hospitality activities carried out by farmers..., through the utilization of their own farms in connection with the activities of the cultivation of the land, of forestry, and of the raising of animals". Italy is the only country in the European Union that has specific laws regulating agritourism, whereas elsewhere this particular type of accommodation is included in the more general sector of rural tourism.

The phenomenon of agritourism has reached a consistent level of diffusion in Italy. The most recent official statistics refer to about 18,500 authorised enterprises nationwide, with an annual increase of 4.3%. In particular, in the fiveyear period 2004–2008, the entire sector shows an overall growth regarding accommodation, in terms of farms (32.5%), catering regarding farms (30.7%), tasting of local products and other activities that concern a growing number of enterprises that offer services that are on the increase (nature watching, courses); some of them have increased considerably. The Italian enterprises are almost all family-run; very few tend to take on employees from outside the family or join cooperatives and associations (Privitera, 2010).

- **Examples of Rural Tourism in Taiwan**

Rural tourism is becoming popular in Taiwan among both farmers and visitors. It is becoming a new type of agricultural management for farmers. In view of the impact on farm incomes from international free trade, it is important to help farmers increase the profitability of their farms, as well as improving the quality of rural life. Looking at the effect of rural tourism in Nantou County, not only has there been a development of tourist farms around Luku and Shinyi, but also local communities have been revitalized. The potential of rural tourism to attract visitors is strong, especially now that Taiwan is implementing a policy of two days off every other week. In future, the 35 recreational spots of Nantou County will be connected into a complete network which will integrate farmers, rural villages, and agriculture (Hong, 1998).

- **Examples of Rural Tourism in Denmark**

Fieldwork was conducted in three Danish rural municipalities: Rangstup in inland South Jutland, Ryslinge in the heart of Funen, and coastal Stevns in South-East Zealand. Rangstrup offers plenty of accommodation, mainly holiday cottages and campsites, but almost no named sights or attractions. When the authors inquired at the local tourism information, the attractions suggested were all located outside the municipality. In contrast, Ryslinge municipality has almost no commercial accommodation but boasts the mega- attraction of Egeskov castle, complete with moat, extensive gardens, museums and exhibitions, located in a romantic rural landscape, and attracting more than 250.000 visitors in the season. The third municipality, Stevns, faals between the two above extremes in that it displays several attractions, and a reasonable capacity and variety of commercial accommodation, camp sites, hotels and holiday cottages. None of the three municipalities offered many opportunities for farm-stays or the like (Hall, Roberts & Mitchell, 2005).

The municipalities posses the recreational opportunities that private and public forests offer, within the restrictions that public use of such areas is subjected to in Denmark. Furthermore, Rangstrup has golf courses and put-and-take fishing, while Stevens, the anly littoral municipality of the three, has fine beaches. All three municipalities are criss-crossed by regional and national bicycle routes, while a national scenic route designed for motoring passes through Ryslinge and Stevns, but not Rangstrup. In total, the three municipalities reflect the diversified distribution of facilities for rural tourism and recreation in Denmark (Hall, Roberts & Mitchell, 2005).

- **Examples of Rural Tourism in Korea**

Planning for rural development should be based on development capabilities and advantages of each village so as to form the development plan of each village based on its potentials. Hence, regarding the fact that planning capability for rural development of these two villages is tourism, its development plan should be based on tourism. Regarding the close relation of tourism development and bio-environmental, economical, socio-cultural, and legal-political factors, it is necessary to consider the capabilities and limitations of these factors. So development strategies and operational ways about these factors were determined and are shown in tables 4 and 5. For the aim of determining the appropriate strategies, and because the aim has been the removal of the limitations to implement the tourism programs, the basis for this decision-making has been the weaknesses and the threats concerning the study area (Mahmoudi, Haghsetan & Meleki, 2011).

Factors	Strategy	Way
Bio-environmental	Improving flora	• protect from running the flora • evolution pasture and forest lands im talented area
Economical	Improving employment capacity	• create new employment opprtunity • optimize the process of current jobs
Socio-cultural	Controlthe immigration rate	• Increase the welfare servicies capacity • protect the culture genuine of villages
Legal-Political	Partnership development in planning	• make new opportunities for native people • giving the responsibilities to native people

Table 4. Strategies and ways for rural developmant in Kore Shahbazi Village (Mahmoudi, Haghsetan & Meleki, 2011)

Factors	Strategy	Way
Bio-environmental	Bio-environment management for natural resources	• control the soil erosion • manage the forest and water resources
Economical	Developing the infastructure (infastructure development)	• impove the accessibility roads to villages • remove the infastructure obstacle
Socio-cultural	Control the immigration rate	• increase the welfare services capacity • creserve the culture gerotect the culture genuine of villages
Legal-Political	Partnership development in planning	• make new opportunities for the presence of native native partnership in decission making • give the responsibilities to natives people

Table 5. Strategies and ways for rural developmant in Tossouj Village (Mahmoudi, Haghsetan & Meleki, 2011)

- **Examples of Rural Tourism in United Nations**

Rural tourism appeals to many travelers: 62% of all U.S. adults took a trip to a small town or village in the U.S. in the past three years, according to a travel poll by the U.S. Travel Association (USTA, www.ustravel.org). This translates to 86.8 million U.S. adults. A majority of these trips (86%) are for leisure purposes, and the most popular reason overall for traveling to a small town or rural area is to visit friends or relatives (44%). Baby Boomer travelers are more likely than younger or older travelers to visit small towns or villages for reasons other than visiting friends and relatives. More than half (55%) of travelers to rural locales travel with their spouse; 33% travel with children. Six percent (6%) of rural travelers go with their parents; 17% travel with other family members. Some travel with friends (11%) or as part of a group tour (3%). Another 11% travel alone (The 2011-2012 Travel & Tourism Market Research Handbook, 2011).

The following are trip activities on most recent trips to a small town or village, ranked by percentage of travelers engaging in each activity: Dining: 70%, Shopping: 58%, Beach/lake/river: 44% Historical sites: 41%, Fishing/hunting/boating: 32%, Festival/fair: 29%, Bike-riding/hiking: 24%, Religious service: 23%, Camping: 21%, Sporting event: 18%, Winery/working farm/orchard: 15%, Gambling/gaming: 12%, Visiting, Native-America community: 11% THE 2011-2012 (The 2011-2012 Travel & Tourism Market Research Handbook, 2011).

- **Examples of Rural Tourism in Turkey**

Turkey has a favourable geographic, cultural and humanity structure for rural tourism. Through all these rich resources, the increasing demand for rural tourism can be met, a considerable amount of share can be gained and many advantages can be taken. However, as in all types of tourism, it is essential to express that local analysis be carried out while dealing with rural tourism. The determination of local facilities for rural tourism is very important in that strategies of tourism all across the country should be identified in a more realistic way. It is necessary that the strengths and weaknesses, threats and opportunities of each region having the potential to develop in terms of rural tourism be exhibited in detail, creating a roadmap.

8. Rural tourism organizations

According to Soykan (2000); rural tourism is a planned an organization.

In the contemporary world, an increasing number of enterprises, including rural tourism businesses, employ marketing methods and knowledge in their activities. Rural tourism has an exclusive link with nature. For this reason its services became very popular. Its development is furthered by the right marketing system and an expedient EU and national support. Rural tourism becomes a new field of activities, which makes good income and returns in rural areas and enables to change from agricultural production to service trade (Ramanauskienë, Gargasas & Ramanauskas, 2006).

Morrison (1998) identified the impotance of co-operation in the tourism sector particularly for those who are located in a peripheral region or area. She defines co-operation as that which is 'between one or more tourist product providers, whereby each partner seeks to add to its marketing competencies by combining some, but not all of their resources with those of its partners for mutual benefit'.

Regarding an organization to be founded, mostly subjects such as creation of opportunities regarding marketing, provision of a good price and utilization of state support and aid are notable. Subjects of education and cooperation follow these.

Answers given to the question of in what areas the organization to be founded for ecotourism would be effective were included in the cluster analysis in terms of the answers.

These are:

- Effective marketing of products related to rural tourism,
- Increase of the value of products,
- Provision of technical support regarding the issues of farming, cultivation, operation, marketing and export and of education facilities,
- Utilization of nature and culture effectively,
- Provision of government subsidies and aids,
- Active cooperation with other firms.

Agro-tourism is an innovative operation that is not bound to providing accommodation and catering services, but also gives the local community the opportunity to develop, maintain its folklore, bring back to life long-forgotten skills and crafts and produce traditional products (woven items, embroidery, preserves, jams, pasta, aromatic herbs etc). It also helps in the revival of local customs and the organization of traditional events. In achieving the ambitious aims of agrotourism, the contribution of women is of primary importance. Women develop various agrotourism activities through private companies (individual or corporate enterprises) or cooperatives (Aggelopoulos, Kamenidou & Pavloudi, 2008).

Women's organizations of rural tourism and the related projects and educational studies in Turkey are evaluated and some selected significant information is presented.

- **Karaburun Women Agro-Tourism Cooperative**

Being the first agricultural tourism project of Turkey, Karaburun emerged as a product of the efforts of Winpeace, Turkey-Greece Women's Initiative towards bridging the women in respective countries. Established by a group of women inhabiting Küçükbahçe, Sarpıncık and Parlak villages of İzmir's Karaburun district, "Karaburun Women Agro-Tourism Cooperative" presents a very diverse understanding of tourism. The women in these three villages seek to make a contribution to home economics by turning their houses into guesthouses. Guests who can participate in activities such as cooking bread, preparing traditional dishes, carpet weaving, horse riding, trekking and picking up olives pass a nature-friendly holiday. In addition, agricultural tourism tries to preserve the traditional products subject to be forgotten such as jam, handicrafts and aromatic plants.

For the education of women in Karaburun, similar cooperatives in Greece were visited and educational trips were organized to the villages in Turkey that are excelled in running guesthouses. This project aims at achieving the economic independence of women villagers who previously worked as unpaid family workers and their social enhancement. The project in Karaburun employs 18 people in total. The Karaburun example clearly shows that once the project extends in the whole of Turkey, it would be greatly beneficial in every respect (Anonymous, 2005).

- **Mersin-Erdemli District Kösbucagı village agricultural tourism development project**

The project location, Kösbucağı village, is in 12 km distance from Erdemli and it has some Roman and Byzantine ruins. The main source of income of the village is agriculture and animal husbandry. The project aimed at contributing to the development and extension of agricultural tourism, providing employment opportunities other than agriculture, increasing household income by direct acquisition of local products to the consumers, developing collaboration in the locality, and preserving and utilizing the agricultural product diversity, natural resources and cultural assets. The target group of the project is principally women since women have the potential to fulfill the essentials of agro-tourism. It endeavored to make women obtain income by using the education provided. Social status of the women with income tends to increase and consequently contributes to the domestic decision making mechanism. With a thorough marketing, this project is expected to cause major social, cultural and economic changes (Anonymous, 2009a).

- **DATUR Uzungöl Tourism Development Project**

DATUR is the Eastern Anatolia Tourism Development Project initiated by Efes Beverage Group, part of Anadolu Holding, with the support of United Nations Development Project (UNDP) and Ministry of Culture and Tourism.

Carried out between 2003 and 2006, the project aimed at reducing the regional unevenness in the Çoruh Valley in North-Eastern Anatolia by developing community-based tourism as an economic activity that could both be an alternative to agriculture and increase the value added of agricultural sector and at making the local community benefit from the produced income to the utmost. The activities within the scope of the project concentrated on rendering the Çoruh Valley, which has a cultural heritage and natural wealth, an ecotourism destination point in demand by both the domestic and foreign tourists (Anonymous, 2009b). The project team provided consultancy in issues such as organization, legislation forming, and determining an action plan to the 4 non-governmental organizations that were in the establishment stage within the project (Uzundere Sapanca Association, İspir Sıra Konaklar Tourism Association, İspir Women's Association, Uzundere Women's Labor Association).

With the mobility created by the project, 18 new guesthouses went into operation and the local community opened up their houses to foreign tourists as guesthouses in the Çoruh Valley, extending from Uzundere to İspir to Yusufeli. The local youth and the women actively participated in the tourism sector also thanks to the provided educational training. The foot paths between Uzundere-İspir-Yusufeli-Ayder (Çamlıhemşin) were merged to determine trekking tour routes in the form of a spider web and 23 local youngsters were trained as guides specifically.

- **Ecological Agriculture Tourism Education Project**

Ecological Agriculture Tourism Project is a European Union support project with a budget of 62.000 Euro. 9 educators from Samsun 19 Mayıs University Faculty of Agriculture were assigned to the project and the training took place in Uzungöl with the participation of 28 women and 12 men. Since mostly women are involved in agriculture, running guesthouses and hotels in the town, the majority of the participants were women. Not only the participants did have training, but they also went to ecological agriculture farms in all around Turkey to examine them on site. Some facilities that are engaged in both ecological agriculture and tourisy accommodation in Kuşadası and Fethiye were visited. Moreover, course certificates were provided (Anonymous, 2009c).

- **Halfeti Eco Tourism Project**

Initiated by GAP Regional Development Administration, the project aimed at, among others, establishing eco tourism infrastructure in Halfeti in the province of Şanlıurfa in order to give prominence to characteristics and cultural richness of the rural area and its inhabitants, and to provide socio-economic development and employment by creating income earning employment opportunities for the local community. In addition, contributing to the revitalization and the utilization of the regional tourism and helping protect the environment are among the goals.

The following were included within the scope of the project: founding Halfeti Eco Tourism Women's Cooperative, restoration of some selected unique Halfeti houses to be used for accommodation, promotion of traditional product production (food, medicinal herbs, handicrafts etc.) to endow the local community with skills, provision of theoretical and practical trainings on "running guesthouses", "traditional product production and presentation" and "eco tourism" to raise the awareness of the local community (Anonymous, 2009d).

Rural Development within the concept of sustainable economic, social, cultural and environmental development is considered. In addition, rural development is an important factor in increasing levels of prosperity in rural areas. At this point, all this is carried out with rural tourism organizations. An established co-operative organization, especially in rural tourism is used as important requirement in achieving these objectives. Participation and democracy in cooperatives play an important role for sustainability development of the local people. In addition, in rural cooperatives creates employment for rural development in rural areas.

Activating the local potential, getting women to take part in their organizations and generating activities for income increase significance of these organizations (Özdemir, Kiper &Başaran 2009).

9. Conclusion

Rural tourism is a good opportunity for agricultural based communities but the setting of objectives and the final tourism development plan needs caution. For better results the whole range of the stakeholders have to participate in the planning stage. Slow and stable steps needs for this kind of planning in order conflicts and mistakes to be avoided (Douglas & Derrett, 2001; Mathieson & Wall, 1992; Butler, Hall& Jenkins,1998; Richards & Hall, 2000).

In the broadest definition, planning is organizing the future to achieve certain objectives' (Inskeep, 1991). In other words planning may be explained as deciding for the future by the knowledge of past and now. Planning action includes various aims including economic development planning, urban and regional planning, land use planning and infrastructure planning etc.

The benefits of tourism planning for a country (Gürsoy, 2006):

- By the tourism plans, precautions and tools that are deemed necessary for tourism development are identified through the identification of future objectives
- Tourism plans provide a disciplinary order for the achievement of the objectives
- Tourism plans provide assurance for the financial resources and opportunities that the activity requires
- Tourism plans provide the dispersion of the responsibilities and the control on implications

If rural tourism is considered to have economically, socio-culturally and environmentally positive impacts on the region, it is clear that it is definitely necessary to utilize rural tourism potential. Traditional production styles should be arranged according to needs of rural tourism and should be integrated with rual tourism.

The issue is sustainable use and conservation. In the attempts of conservation; the views, opinions and active participations of the local people must also be obtained. This is because a step taken without considering conditions of the local people or the practices influencing their economic and sociological life styles might lead to the worse results. Starting from these, in all these practices, policies and investments taking the priorities of the local people into consideration and making them an active actor would make protective and sustainable precautions more active. Forests have not contributed the district economically. Within this scope, some recommendations have been made to increase ecotourism opportunities to be developed in rural areas.The forest areas should be preserved and evaluated within the scope of rural tourism. Old houses should be preserved via lodging and boutique hotel identity in tourism.

In order to hand down the natural and cultural heritage of the local people, who pursue their traditional life styles without losing cultural infrastructures, to the next generations, and to preserve and evaluate within the scope of rural tourism, awareness raising programs should be carried out. In this context, the fact that ecotourism is an economic opportunity should be emphasized in order to preserve the local people and culture and to hand down a preserved nature to the next generations. Besides, it should be emphasized in the awareness-rising programs that an ill-planned, underdeveloped and unorganized tourism development might cause disturbance of natural landscapes of rural areas, might threaten wildlife and biological diversity, may cause poor quality of water sources, may leads to the immigration of local people and the erosion of cultural traditions. With the objects of diversification of the sources of income of the people and of offering alternatives, side income generating sources in the areas especially like rural tourism, organic agriculture; agricultural product processing and merchantable plants growing should be put in practice.

Rural tourism plans are the parts of a general rural region development plans and planning study is the duty of the official administrative units which have the responsibility of rural development or tourism development. It should be a national or regional administrative unit but it may be a tourism or agriculture administration institution according to the role of the agri-tourism development of the national aim. Rural tourism should be developed with the aim of enhancing agricultural production or tourism development; therefore agriculture and the tourism ministries are the primary responsible institutions for rural tourism planning. Especially in developing countries, 'it is necessary the national intervention for development in any kind of tourism'. Whether agricultural or tourism ministry manages the process, coordination among the all administrative units, also forestry and environment ministries take role in the process, is the must.   Network regarding conservation works among the government agencies, universities, NGOs and private sector should be constituted and this network should be strengthened. Task sharing should be made among these institutions and organizations and the coordination should be ensured.

Rural development carried out by rural tourism organizations and cooperatives increases the levels of welfare people living in rural areas.

In conclusion, rural tourism activities which are not performed according to the purpose, the principles and the characteristics cause the disturbance in environmental, economic and socio-cultural fields due to over-intensification to be occurred especially in sensitive ecosystems like rural areas. Therefore, in order to provide sustainability in the rural tourism, it is necessary to know environmental, social and economical effects of rural tourism activities and to consider these effects during the planning. From this point of view, informing the local people, who are the most affected group by rural tourism, about the effects caused by the rural tourism to be developed in their region is primarily important.

10. References

Anonymous, (2005). http://www.milliyet.com.tr/2005/06/26/business/bus05.html.
Anonymous, (2009a). http://www.yurthaber.com/haber/kirsal-koylerin-kalkinmasi-icin-agro-turizm-modeli-151493.htm.
Anonymous, (2009b). www.datur.com,
Anonymous, (2009c).
           http://www.porttakal.com/haber-uzungol-un-eko-turizmci-yengeleri-56510.html.
Anonymous, (2009d). www.gap.gov.tr

Anonymous, (2011). Travel & Tourism Market Research Handbook12th Edition; May 2011; 513 pages,i www. rkma. com /2011travelTOC.pdf

Ardop, (2006), Agriculture and Rural Development Operational Programme, 2004-2006. Budapest: Ministry of Agriculture and Rural Development. Retrieved September 25, 2007 from .

Aggelopoulos, S., Kamenidou , I. & Pavloudi, A., (2008). Women's business activities in Greece: The case of agro-tourism. Journal Tourism. Vol. 56, No. 4, 371-384 p.

Arzac, S., (2002). Environment and Rural Tourism in Bustamante, Mexico. Monitoring and Management of Visitor Flows in Recreational and Protected Areas Conference Proceedings ed by A. Arnberger, C. Brandenburg, A. Muhar , 384-389p.

Ashley , C. & Maxwell, S. (2001). Rethinking Rural Development, Development Policy Review, Vol. 19, No. 4, 395-425p.

Bramwell, B. & Lane, B. (1994). Rural tourism and Sustainable Rural Development In: Proceedings from the Second International School of Rural Development. London: Channel View Books

Briassoulis, H. (2002). Sustainable Tourism and The Question of The Commons. Annals of Tourism Research, Vol:29, No.4, pp. 1065-1085, Elsevier Science Ltd, Printed In Great Britian.

Butler, R. W. & Hall, C. M. 1998. Conclusion: The sustainability of tourism and recreation in rural areas. In Butler, R.W., Hall, C.M. and Jenkins, J. (eds). Tourism and Recreation in Rural Areas. John Wiley&Sons, Chichester, pp. 249–258.

Butler R., Hall M., Jenkins J., (1998). Tourism and recreation in rural areas, Wiley, Chichester.

Colton, J.W. & Glyn, B., (2005). Developing Agritourism in Nova Scotia: Issues and Challenges. Journal of Sustainable Agriculture, Vol. 27, No.1, 91-112p..

Catibog-Sinha C. & Wen, J. 2008. 'Sustainable Tourism Planning and Management Model for rotected Natural Areas: Xishuangbanna Biosphere Reserve, South China', Asia Pacific Journal of Tourism Research, 13:2, 145 − 162.

Deqian, L., (2006). Several Discriminations in Rural Tourism, Agriculture Tourism and Folklore Tourism [J], Tourism Tribune. Mahmoudi, B., Haghsetan, A., Meleki, R., (2011). Investigation of Obstacles and Strategies of Rural Tourism Development Using SWOT Matrix. Journal of Sustainable Development. Vol. 4, No. 2, 136-141p.

Dernoi, L. (1991). Prospects of Rural Tourism: Needs and Opportunities. Tourism Recreation Research, Vol. 16, No. 1, 89–94p.

Douglas, N. & Derrett, R., (2001). Special Interest Tourism, Wiley, London.

Dowling R., Moore S.A. & Newsome D. (2001). Natural Area Tourism, Ecology İmpacts And Management, Channel View, New York .

Drzewiecki, M. (2001). Bases of Agritourism. – Bydgoszcz: Centre of Organizational Progress

Elands, BH. M. & Freerk Wiersumi, K. (2001). Forestry and rural development in Europe: an exploration of socio-political discourses, Forest Policy and Economics, Vol.3, 5-16.

European Commission, (1988). The Future of the Rural Society. European Commission, Brussels.

European Commission, Directorate General for Agriculture, (1997). Situation and Outlook: Rural Developments; A CAP 2000 Working Document. Office for Official Publications of the European Communities, Luxembourg.

Fan, X. & Yang, P., (2011). The Survey on the Negative Planning mode of Rural Tourism Planning, Remote Sensing, Environment and Transportation Engineering (RSETE), 2011 International Conference on (24-26 June 2011), pp. 906-909, China.

Gannon, A. (1994). Rural Tourism As A Factor İn Rural Community Economic Development For Economies İn Transition. Journal of Sustainable Tourism, Vol. 2(1 + 2), 51–60p.

Gerry, R. (2001). Cultural Attractions an European Tourism. CABI Publishing, Newyork.

Gilbert, J. (1982). Rural Theory: The Grounding of Rural Sociology. Rural Sociology, Vol. 47(4), 609-33p.

Golley, F.B. & Bellot, J. (1999). Rural Planning froman Environmental SystemPerspective, 376 p., NewYork, USA, Springer Verlag, 4–6p.

Gülçubuk, B., (2005). Rural Construction and Rural Development in EU and Turkey, Ankara, p.2.

Gürsoy, S., (2006). Regional Tourism Planning in Terms of Central and Local Governments, Unpublished Master Thesis, Ankara University Institute of Social Sciences, Ankara.

Hall, D., Roberts, L. and Mitchell, M. (2005). New Directions in Rural Tourism, 54-67 p., UK.

Hannigan, J. (1994). A Regional Analysis of Tourism Growth in Ireland. Regional Studies, Vol. 28(2), 208–214p.

He, J. M. (2003). A Study on Rural Tourism Overseas. Tourism Tribune, Vol. 18, No. 1, 76-80p.

Holland, J., Burian, M & Dixey, L. (2003). Tourism in Poor Rural Areas : Diversifying the Product and Expanding the Benefits in Rural Uganda and the Czech Republic. PPT Working paper/No: 12. (online) Available from http://www.propoortourism.org.uk/12_rural_areas.pdf (Acceseds 21 Januar 2011)

Hong, W-C. (1998). Rural Tourism: A Case Study of Regional Planning In Taiwan. (online) fftc.imita.org/htmlarea_file/.../eb456.pdf (Acceseds 21 Januar 2011).

Huigen, P., Paul, L. & Volkers, K., (1992). The Changing Function and Position of Rural Areas in Europe. Netherlands Geographical Studies no. 153, Koninklijk Nederlands Aardrijkskundig Genootschap, Utrecht.

Jingming , H. & Lihua, L. (2002). A Study on the Conceptions of Rural Tourism, Journal of Southwest China Normal University (Philosophy & Social Sciences Edition), (9).

Kanaani, E. (2005). Tourism and Impact on Rural Societies, *Dahati journal, Vol.70*, 38-42.

Keyim, P., Yang, D., Zhang, X., (2005). Study of Rural Tourism In Turpan, China. Chinese Geographical Science Vol. 15(4), 377-382p., Beijing, China.

Kiper, T., Özyavuz, M. & Korkut, A. *(2011).* Impact of the Development of Rural Tourism Natural Landscape Features: Sample of Şarköy County, Tekirdağ, Journal of Tekirdağ Agricultural Faculty, *Vol. 8, No. 3, 22-34p.*

Kiper, T., Özdemir, G. & Başaran, B. (2011). Applicability of agricultural tourism and the role of women. A case study of Şarköy-Mürefte. Journal of Environmental Protection and Ecology, Vol.12, No.3, *1146–1159 p.*

Kiper, T., Korkut, A. & Yılmaz, E. (2011). Determination of Rural Tourism Strategies By Rapid Rural Assessment Technique: The Case of Tekirdag Province Şarköy County. *Journal of Food, Agriculture & Environment,* 9 (1): 491-496, (2011).

Kiper, T. (2006). An Assesment Of The Landscape Potential Of Safranbolu- Yörükköyü In The Frame of Rural Tourism. Universty of Ankara Graduate School of Natural and Applied Sciences Department of Landscape Architecture (Ph. D. Thesis), Ankara /Turkey. p. 367.

Kiper, T.. (2011). The determination of nature walk routes regarding nature tourism in north-western Turkey: Şarköy District. *Journal of Food, Agriculture & Environment,* 9 *(3&4):622-632.*

Könyves E. 2001, A falusi turizmus szerepe Jász-Nagykun-Szolnok megye vidékfejlesztésében, Doktori Disszertáció, Debrecen.

Lane, B. (1993). Sustainable Rural Tourism Strategies: A Tool For Development and Conservation in Bramwell. Proceedings of the Second International School on Rural Development , 28 June-9 July. Ireland.

Lane, B., (1994). Sustainable Rural Tourism Strategies: A Tool *for Development and Conservation*, Journal of Sustainable Tourism, Vol 2, 12-18p, 199.

Lanquar, R. (1995). Tourisme at Environnement on Mediterranean. Economica, Paris.

Mahmoudi, B., Haghsetan, A. & Meleki, R., (2011). Investigation of Obstacles and Strategies of Rural Tourism Development Using SWOT Matrix. Journal of Sustainable Development. Vol: 4, No: 2.

Mlynarczyk, K. (2002). Agritourism. Olsztyn: Publishing of Warmia and Mazury University

MacDonald, R. & Jolliffe, L. (2003). Cultural rural tourism: evidence from Canada. *Annals of Tourism Research,* Vol. 30, No.9, pp.307-322.

McGehee, N., & Kim, K. (2004). Motivation for agri-tourism entrepreneurship. *Journal of Travel Research.* Vol. 43, 161-170p.

Morrison, A. (1998). Small firm co-operative marketing in a peripheral tourism region. *International Journal of Contemporary Hospitality Management.* Vol. 10. No. 5, 191-197p.

Mostowfi, B. (2000). Agrotourism and Sustainable Development, Case Study: Landscape Design for Karyak Village, MSc thesis, Environment Faculty of Tehran.170pp

Munteanu, A. (2007). Agroturism Şi Tehnologia Informaţiei, Editura Eurostampa, Timişoara.

Newsome, D. & Moore, A.S., Dowling, K.R., (2002). Natural Area Tourism: Ecology, Impacts and Management. Cromwell Press, Britain, 341 p.

Ninth Development Plan (2007-2013)., http:// ekutup.dpt.gov.tr/ plan/ plan9.pdf

Nilsson, P. A. (2002). Staying on Farms — an Ideological Background. Annals of Tourism Research, Vol. 29, No. 1, 7–24p.

Organisation For Economic Co-Operation and Development, [OECD], 1994. Tourism Strategies and Rural Development, 94p.

Özdemir, G., Kiper, T. & Başaran, B. (2009). Women's organizations In Turkey's rural tourism. The International Conference of BENA. Pollution Management and Environmental Protection. (16-20 September 2009), (Abstract Book) (Poster), p.171, Tirana, Albania, (2009).

Pakurar, M. & Olah, J., (2008). Definition of Rural Tourism and its Characteristics in The Northern Great Plain Region. Analele Universitatii Din Oradea Fascicula: Ecotoxicologie, Zootehnie Si Tehnologii De Industrie Alimentara, Vol. 7(7), 777-782.

Perales, R. M. Y. (2002). Rural tourism in Spain. Annals of Tourism Research, Vol. 29, No. 4, 1101-1110p.

Pevetz, W. (1991). Agriculture and Tourism in Austria. Tourism Recreation Research, Vol. 16, 57-60p.

Pompl, W., & Lavery, P. (1993). Tourism in Europe: Structures and developments. Wallingford: CAB International.

Privitera, D. (2010),The Importance of Organic Agriculture In Tourism Rural, In Applied Studies In Agribusiness and Commerce (Apstract), Vol. 4, No.1-2, 59-64 p.

Ramanauskienė, J., Gargasas, A., Ramanauskas, J. (2006). Marketing Solutions In Rural Tourism Development In Lithuania. Ekonomika, Vol. 74, 38-51p.

Reid, D.R., Taylor, J., Mair, H., (2000). *Rural* Tourism Development: Research Report. School of Rural Planning and Development, University of Guelph: Guelph, Ontario.

Richards G. & Hall D., (2000), Tourism and sustainable community development, Routledge, New York.

Roberts, L., & Hall, D., (2003), Rural Tourism and Recreation: Principles to Practice. CABI Publishing. UK

Robınson, G. (1990), Conflict and change in the countryside, London

Richards, G., & Hall, D., (1998). Tourism and Sustainable Community Development. London

Sarıca, İ., (2001). Regional Development Policies and Projects in Turkey, Mediterranean Journal of Faculty of Economics and Business, 154-204, Antalya.

Slee, B., (2000). Methods for measuring the contribution of forestry to rural development. International MCPFE Seminar on The Role of Forests and Forestry in Rural Development _ Implications for Forest Policy, Vienna, 5_7 July 2000.

Soykan, F., (2003). *Rural Tourism and its Importance for Turkish Tourism. Aegean Geographical Journal.* Vo. 12, 1-11p.

Soykan, F., (2000). Rural tourism and the experience gained in Europe. Anatolia Journal of Tourism Research, Vol. 11 (March June), 21-33 p.

Soykan, F. (1999). A type of tourism integrating Natural Environment and Rural culture: Rural Tourism, Anatolia Journal of Tourism Research, Vo. 10 ((March June), 68-75.

Soykan, F. (2003). Rural tourism and its importance for Turkish tourism. Aegean Geographical Journal, Vol 12, 1-11p.

Şerefoğlu, C., (2009). Role Of Tourısm In Development- Place, Role And Expected Targets In The Ipard Rural Development Programme Carrıed Out In Between 2007-2013 In Turkey. Ministry Of Agriculture And Rural Affairs Deputy, Expertise Thesis, s. 176.

Tane, N., Thierheimer, W. (2009). Research About the Connectıon Between Different Rural Tourısm Types And Forms. 4th Aspects and Visions of Applied Economics and Informatics March 26 - 27. 2009, Debrecen, Hungar, 902-906 p.

The 2011-2012 Travel & Tourısm Market Research Handbook, (2011). Travel & Tourism Market Research Handbook12th Edition; May 2011; 513 Pages) Www. Rkma. Com/ 2011**travel**toc.Pdf

Tourism 2023 Strategy Plan (2007). Ministry of Culture and Tourism, pp.60..

Turhan, M.S., (2005). Application Process in Rural Development Measures in Turkey towards membership of the European Union, Ministry of Agriculture and Rural Affairs Department of Foreign Affairs and European Union Coordination, Master Thesis, p.145, Ankara.

UN Demographic Year-books, (1998). www.unis.unvienna.org/ unis/.../ pi225.html

Urry, J., (2002). The Tourist Gaze, second ed. Sage, London.

Wang, Y. C. (2006). The new form and model of development of rural tourism in China. Tourism Tribune, 21(4), 6-8.

Weaver, D. &Opperman, M. (2000). Tourism Management. John Wiley and Sons, Sydney, Australia.

Wen-Ching, H. (1998). Rural Tourism: A Case Study of Regional Planning in Taiwan, Agricultural Bureau, Nantou County Government, Chung Hsing Rd., Nantou City, Taiwan, http://www.agnet.org/library/ac/1998b (01.12.2011.)

Wickens, E. (1999). Tourists' Voices: A Sociological Analysis of Tourists' Experiences in Chalkidiki, Northern Greece, Unpublished Doctoral Thesis, Oxford Brookes University.

Williams, A. M., & Shaw, G. (Eds.), (1991). Tourism and economic development: Western European experiences (2nd ed.). London: Belhaven Press.

Wicks, B. E & Merrett, C. D. (2003). Illinois Institute for Rural Affairs, Agritourism: An economic opportunity for Illinois, *Rural Research Report*, Vol.14, No.9, 1-8p.

Yadghar, A. (2004). Changes process and challenges of rural reclamation and development in Iran, *Journal of rural researches*, 48, 71-90

Photos belong to Kadir Düşünen

7

Agriculture and Rurality as Constructor of Sustainable Cultural Landscape

Juan Gastó[1], Diego Subercaseaux[1], Leonardo Vera[2] and Tonci Tomic[3]

*[1]Laboratorio de Ecosistemas, Facultad de Agronomía e Ingeniería Forestal,
Pontificia Universidad Católica de Chile, Santiago,
[2]Facultad de Agronomía, Pontificia Universidad Católica de Valparaíso,
Centro Regional de Innovación Hortofrutícola de Valparaíso (CERES), Valparaíso,
[3]Consultor Ambiental, Corporación Nacional del Cobre (CODELCO), Santiago,
Chile*

1. Introduction

Is the XXI century an age of changes or is it a change of age? We must assume that new and more complex challenges are as necessary as deep culture and paradigmatic modifications. Growing complexity and the present-day territorial degradation has made it necessary that we transform the dominant science paradigm to face the sustainability problems. A new science is essential to improve the understanding of ourselves and our environmental life [1-4].

The evolution of new technology, about ten thousand years ago, gave birth to the artificialization of nature and agriculture. The way said artificialization and management of the natural resources was determined basically by the factors and cultural tendencies. Culture can be defined as a learned system that produces an action and the way we relate with the world [5]. It is a set of subordinated suppositions and beliefs shared by a group or society, influencing their behavior [6].

The cultural landscape concept has emerged as a systemic transdisciplinary study object. To understand the present context it is basic to understand the cultural landscape concept, representing an expression of cultural activities in a territory, and as such, it is a key factor for the sustainability study [7]. Cultural landscape is a XXI century integrative concept.

Depredating conditions and trends of ecological-territorial systems and their effects on planetary life require an urgent change of the present dominant artificialization style and cultural landscape construction. We are part of the unique and interdepending web of life [8]. Complementary couplings of our construction cultural landscape style and nature organization result in healthy and sustainable cultural landscapes [9].

Starting from a historical ecological-territorial footprint and facing the relationship between agriculture, rurality and cultural landscape, the main objective of this work is to state the fundamentals to understand, develop, and construct a sustainable model adaptive of our age.

2. Agriculture from nature

2.1 Nature

Nature is the set of all entities and forces that constitute the territory. It is the natural world without mankind or civilization [10]. The natural world is the background matrix where humans have evolved during a long period of time, leading to rurality and urbanity as a complement to wildland [11, 12]. Since the presocratic time of Anaxagoras, it is stated that nothing is born or dies, but that everything emerges from preexisting entities and elements; just as it happens in Nature, which when artificialized, is transformed into a cultural landscape. Natural resources are the supply source of our civilization and act as the life support for our domain of existence [13]. This is the reason why the resources should be sustainably managed and maintained, turning the agricultural activities into a main component. Complementary, ecosystem is a concept that allows placing and integrating the various disciplines that transform the agronomic sciences into a transdisciplinary dialog. Recently, cultural landscape emerges as a strong concept. It develops from the territorial stakeholders in a certain cultural context integrating the various sustainability and development dimensions. All of this arises from a social-cultural coevolutionary process with nature, and from the stakeholders with their surroundings.

Territory may be conceived as a "land or aquatic volume or area belonging to a farm, county, province, region or nation" [14]. The territory is used by society, originating from the interaction of three main components: nature, society and technology. Nature comes before man, what grants it a different evolutionary meaning. Man develops culture as a way to establish a relationship with the world, gradually organizing growing and complex structures integrating ethnics, politics and labor, among others, generating as a result the social structure. From the resulting integration of nature with social structure emerges technology as an articulating component for both. This process gives birth to a territorial system which in time becomes an integral unit [14, 15].

Cataldi, an italian mathematician and designer during the XVI-XVII centuries, states that man modifies nature until finally transforming it into a cultural landscape. As a result, he generates a sustainable or unsustainable system depending on the behavior of the people, and ultimately, on the type of activities carried out by the stakeholders.

2.2 Agriculture: Definitions and formulation

Agriculture *sensu lato* can be defined in various ways. Lawes [16] and Prado [17], defined it as a process of artificialization and decision taking about nature, with some specific human purposes, such as producing food, fiber, leather, wood or landscape beauty. It is, therefore, a process of transformation with a given objective, involving nature, stakeholders and technology. In this context, agriculture *sensu lato* includes numerous activities related to multiple land use for production purposes (vegetable garden, forestry, aquaculture, livestock, etc.), protection (of soils, fauna, banks, landscapes, etc.) and recreation (agrotourism, camping, sports, entertainment, and so on).

During the seventies, when hard productivity technologies were being enforced, agriculture was defined as putting a harness to solar energy through plants for human purposes [18]. An earlier definition, 1814, describes it as science of managing farmland [19]. The latter definition is consistent and complements, as well as, integrates the above ones; it combines Nature and its artificialization with land management, organized around rural properties. In all these definitions the ecosystem is essential and a priority.

On the other hand, agriculture can be defined as an economic activity related to the sustainable production of crops and its transformation into elements which can be consumed by man. Many people perform this activity as a way to live [20]. This definition expresses the policy approach of the farmers' associations and some of the Ministries of Agriculture, who tend to consider agriculture as a mere business, when it should be seen as a central component for integral rural development.

In recent decades, agriculture has been looked upon only as an agribusiness, which takes away much of its significance and meaning, leaving it only as a minor branch of the economy [21]. In Chile, in general this began mainly during the second half of last century and continues today. Agriculture has been restricted to crops, economy and enterprise, overlooking its farm dimension and in many cases causing the degradation of the natural resources of the country [11, 22, 23] Unlike this, the traditional large farm (hacienda) for the first 300 years after the conquest and colonization of America was the major territorial, social, economic, and management unit, later complemented with other styles of farm in all its forms [24, 25].

In the early Christian age, at the time of Columella [26, the original paper written during the 1st century aD], there was talk of *re-rustic*, referring mainly to the rurality, which is complemented by the urbanity that takes place in small towns and villages in the territories of Babylon, Greece and Rome [27, 12]. It was necessary to supply the cities with abundant food; thus, it was necessary to develop specialized farms with efficient production processes. The English word farming derives from here, differentiating it from cropping and husbandry (analogous to agriculture in Castilian). Farming can be defined as the arrangement, management and administration, of rural lands, which achievement center on the territory articulated by technological activities related to agriculture *sensu lato* [11, 28].

Ecology is incorporated formally and rigorously since the twenties, adding the ecosystem in the year 1935 [29], and becoming generally known between the sixties and seventies. It is difficult to argue that modern agriculture can develop sustainably without incorporating the ecology as a fundamental paradigm. This due to the agricultural matrix land generated from the artificialization of the natural ecosystem forming rural properties, and due to expansion of the agricultural frontier as generalized as a worldwide phenomenon [1, 2, 23, 30].

Symbolically, the artificialization (A) of nature, or agriculture, can be represented as [15]:

$$A = (\pi_a / \pi_a : \Sigma_0 \rightarrow \Sigma_1) \tag{1}$$

where:

π_a: Set of operators of artificialization for a state of artificialization "a"
Σ_n: State of the ecosystem to time n=0, previous to time n=1
$\Sigma_0 \rightarrow \Sigma_1$: Change of the state of the ecosystem from Σ_0 to Σ_1

From an operational point of view, the farm can be defined as "an organized unit of decision making, an area of renewable natural resources, connected internally and limited externally, which aim is to make agriculture" [31, 32]. Formally, the farm (P) consists of [15]:

$$P = f\ (S,\ \Sigma,\ \phi,\ \sigma_a) \tag{2}$$

where:

S: Space-time, $L^3 \times T$ (length^3 x time)

Σ: Spatio-temporal units of renewable natural resources

Φ: Inter or intra flow of matter, energy or information

σ_a: Answer or output as a function of artificialization

2.3 Ontology and epistemology

Ontology refers to the nature of the reality or phenomenon under study. In this case, it is the agriculture, rurality and cultural landscape in the context of integral and sustainable development based on local and global landscape design, situated in a systemic theory [33], ecological theory [29, 34], as well as the information theory [35, 36], the complex systems theory [8, 37, 38], and cognitive theory [3, 13, 39, 40]. The adaptive flexibility of the cultural landscape is related to the information content of the system [41]. Information and diversity, from the operational point of view, can be considered equal.

Despite the enormous technical advances modern agriculture has undergone, the late twentieth century lacked a unifying theory integrating all the above issues, as well as its thematic and conceptual context. A theoretical framework was needed to locate and frame the agricultural engineering in a holistic, systemic, integrated, and transdisciplinary context, in view of the advance of science and engineering paradigms by the end of the century [13, 42-44]. This theoretical framework arises for agriculture and for several other disciplines from general systems theory, holism, ecology, and new paradigms emerging in recent decades.

The final rationality of stakeholders, as cognitive agents, is to maintain the structural coupling with its domain of existence [13, 39, 40]. In this context, mutual determinations that keep this co evolutionary coupling between the stakeholders and their scenario are of an emotional nature [4]. The stakeholders experience an emotion when confronted with the phenomenon they perceive, determining the action which will generate the landscape, which in turn feeds their perception [8].

According to Röling [13], the cognitive support of collective decision making is sized into four components: value, theory, context and action (Figure 1). According to Lawes´ [16] definition of agriculture, value must be based on ecological rationality given by principles, laws and ecosystem structure, with any style of agriculture. In the theoretical model the value must be constructivist, so it must be generated within an epistemological framework for dialogue and collective subjectivity. Action should be deliberate and collective according to the culture of the stakeholders, associated with their perception and cognition. Finally, the context of agriculture should focus on man as the greatest driving force of the cultural landscape and determining their own future. However, the territorial problems as well as degradation of ecosystems and natural resources, show the lack of an instrument which lets us handle this force [2].

The four dimensions of cognitive support of collective decision making are considerably modified if instead of using the definition of Lawes, we use a definition that increases an agriculture focused on production. The prevailing definition of agriculture determines the paradigm that governs the actions on the cultural landscape and its sustainability.

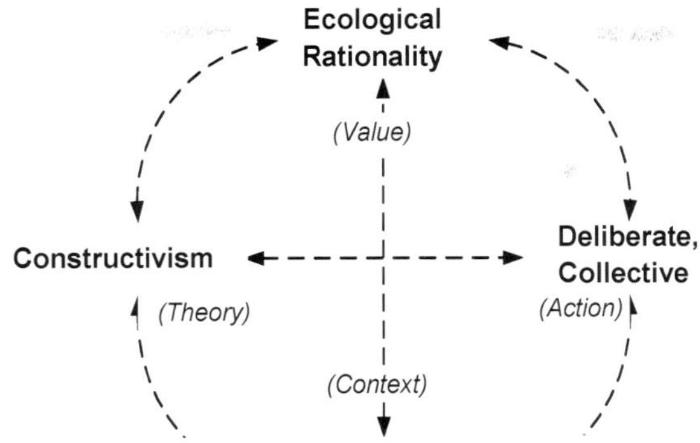

Fig. 1. Decision taking as a function of Lawes definition of agriculture as artificialization of nature [adapted from 13].

3. Rurality, territory and cultural landscape

3.1 Hominid frontier expansion and cultural landscape

Land cropping and the following appearance of the rural cultural open landscape[1], occurs only starting around 10.000 years ago. This is the starting point of the process of landscape hominization [45, 46] and the homind frontier expansion. Each society relates differently to nature and its surroundings, arranging the territory according to its culture, setting the bases of the different cultural landscapes.

The nature artificialization process, and the expansion of the hominid frontier is intended to conquer niches and improve anthropogenic canalization of goods and services, requiring the extraction and insertion of elements into the ecosystem.

As example of the hominid frontier expansion and creation of a cultural landscape, Gastó [47] reports what happened in the range lands of the North American west. After the arrival of settlers there was degradation of the soil and vegetable covering, and as a consequence of this, large stretches of land were abandoned due to low productivity. These settlers didn´t have the necessary knowledge to open up, order, manage and administer the territory. Faced with this, the Government, got involved and establishes the National Park Service (1873), National Forest (1890), Native American Reservations, Wild Life Shelters and the Land Grant College. At the same time, and in order to improve the public land management, the Government set up the Forest Service (1905), Bureau of Land management-BOM (1935), and the Soil Conservation Service (1905). Meanwhile settlers were converting private land into great ranches. The American Society of Range Land Management was created in the 1940´s, with the intention of developing a science based on principles differing from those of agronomy. Currently one of the most important aspects is

---

[1] Rural, etymologically means wasteland, opened by and for mankind. This is within the expansion of the hominid frontier, the place where man can live and generate rurality.

the publishing journal of range management for continuous renewal of concepts, technology, and guidelines, in order to be consistent with the demands of society and maintain the sustainability of the territory [28]. Because this, "rangeland" is an expression of the contemporary American cultural landscape.

Another interesting case of cultural landscape generation is *dehesa*[2], in Mediterranean Spain. Dehesa corresponds to a cultural landscape created and developed by the popular culture. By definition it is a typical natural dense sclerophyllous Mediterranean forest, with a simplified structure and diversity of species achieved by reducing the tree density by pruning and thinning, developing isolated fruit producing trees loaded with acorns and stimulating the formation of a natural prairie in the undergrowth [47, 48]. Two main livestock niches are generated: one of the acorn consuming pig and the other of the ruminant ovine and bovine grass consumers (Gastó 2008). The evolution of the *dehesa*, of its elements and landscapes is deeply related to the development of the transhumant livestock, which has been very important for the Iberian development. According to Gastó [47] the *dehesa* is a sustainable system by generating products of great value while maintaining landscapes of immense aesthetic value with a mixed wintry herbaceous cover and evergreen trees.

In both cases, the rangeland in the United States, as well as, the *dehesa* in the Iberian peninsula, the expansion of the hominid frontier and of the construction of the cultural landscape, created a stable cultural landscape, harmonizing the economic, social and ecological services with a remarkable identity [49].

Easter Island on the other hand has become an emblematic case [12, 50] of a very fast hominid frontier expansion, which extremely modified and depleted a fragile ecosystem (isolated area in the middle of the Pacific, 388 Km[2]). There are various hypotheses, which explain this particular degradation process. Some of them suggest that the deforestation and severe depletion of the ecosystem was the result of the increased demand for logs used for the transportation of the Moais, and that the population of the Island got to be 7.000 [12]; other hypothesis sustain that the disease and slavery brought by the Europeans were the main reason that triggered the population crisis, aside from the introduction of the Polynesian rat that prevented the forest from regenerating and generalized harvest [50]. Nevertheless, the Eastern Island society colonized said territory but failed in its attempt to make it sustainable, producing the depletion of its own ecological support and thus, its own extermination.

In each one of these situations (rangeland, *dehesa* and Easter Island) man colonized a territory, expanded the hominid frontier, artificialized nature and transformed the ecosystem, creating a new cultural landscape to fit their needs, culture and technology, and attaining an improved or poorer system in terms of sustainability and life quality [28, 49].

3.2 Rural, urban and wildland. Territories typologies and components

Before mankind all that existed were natural scenarios based on systemogenic processes and ecological succession guiding the ecosystem to more complex and self-organized stages [49, 51]. People colonize habitats and develop niches; starting the process of nature artificialization and hominid frontier expansion, as well as, the clearing of the wildland and

---

[2] The *dehesa* is a *savannah* Spanish landscape type.

its transformation into rural (wasteland) and, afterwards into urban (built territory); *bann* (abandoned territory) and *agri deserti* (agonizing territory) [49, 52]. When man clears the *wildland*, there is a simultaneous hominid frontier expansion, simplification of the natural ecosystems and input-output of ecosystem elements, shaping the territory on the basis of society´s culture and technology. In this context, the cultural landscape appears gradually as an expression of the sociostructure over the biogeostructure, in a coevolutive context articulated by the tecnostructure [23, 52, 53].

The hominid frontier expansion is followed by a territorial specialization and the emergence of different territory typologies, such as: protected wildlands, rural and urban. These territorials typologies are differentiated by the various proportions of the three territorial components within: *saltus*, *ager* and *polis* (Figure 2). *Saltus* represents the territorial component which is not directly affected by the anthropic action; *ager* is a territorial component cleared with direct artificialization due to the anthropic action in a intermediate level, being land cropping the predominant artificialization style; *polis*, refers to a territorial component with a high level of artificialization, being its main style construction and infrastructure. The protected wildland territories are made up of in large proportion by *saltus* and in lower proportions by *ager and polis*. Urban territories are mainly made up of the *polis* component; and rural territories present a more balanced situation of these three elements: *saltus, ager and polis*.

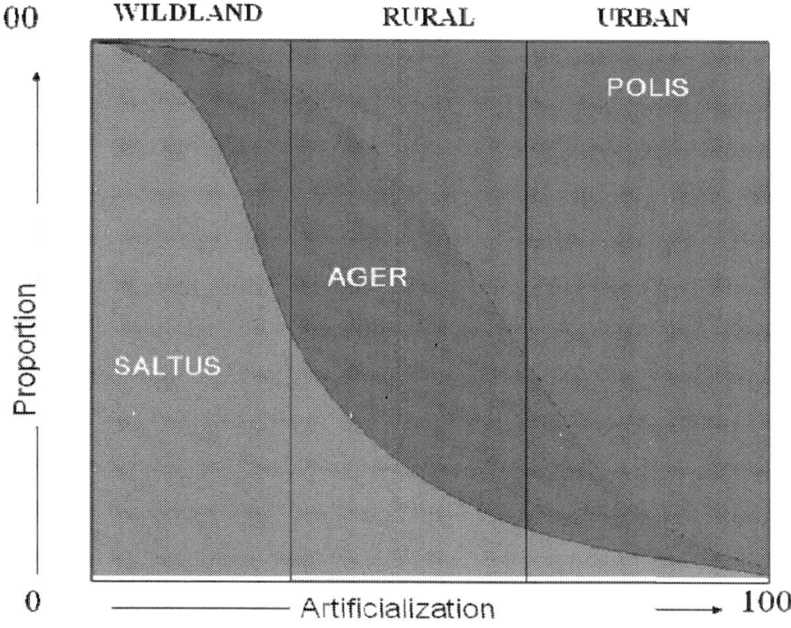

Fig. 2. Relative proportions of territorial components: *saltus, ager* and *polis*, belonging to the territorial typologies: protected wildland, rural and urban, depending on the level and style of artificialization [adapted from 54].

Various farms management typologies appear in these territories: in the urban territories, there are megacities, cities, towns, villages, among others; in rural territories, different kind of farms, vegetable gardens, urban parks, ranches, coexist; an finally, in the natural territories there are National Parks, Biosphere Reserves, Forest Reserves, Nature Sanctuaries, ethnic reserves, and the like.

4. Cultural landscape construction and governance

4.1 Spirit of age and place

Culture has become the main factor to determine the evolutionary dynamic of the ecological-territorial systems of our age, and consequently the construction and resulting cultural landscapes of the stakeholders. The spirit of age was first developed in Germany in the year 1769 by the poet and philosopher John Gottfried von Herder, giving it the name of *Zeitgeist* which means: spirit (*Geist*) and age (*Zeit*). The *Zietgeist* concept is mainly known in relation with the German historical philosophy of the philosopher George Wilhelm Friedrich Hegel. Zeitgeist refers to the predominant cultural tendency at a certain time in the history of mankind. There is a certain vision and behavior during each particular period of sociocultural evolution which is expressed in the ecological-territorial systems and resulting cultural landscapes. This vision and style corresponds to the profile of the age and the conception of the world [55], which would be equivalent to the concept of paradigm in the world of science. Zeitgeist defines in the Hegel approach a certain state of the dialectic evolution of a person, a group of people, society or the whole world. Also important, and complementary with the spirit of age concept, is the spirit of place (*Volkgeist*), which mainly refers to the cultural tendencies of groups or societies in different places. This is related with the Nature's conditions in each place.

A common characteristic of the beliefs from eras preceding the industrial revolution was that human acts were limited to our basic needs and that technology only developed according to them [56]. During the industrial age, development was understood as a rebellion in contradiction to the need governing all societies until the XVIII century, and that progress is the success of said rebellion [56]. This has happened in association with technocracy and economic rationality predominance and with the neoliberal economic model [57]. This world notion and, the related growth model have generated great impact on the ecosystems, natural resources and, have been associated with the unsustainability tendencies of the ecological-territorial systems.

During the last decades of the XX century there have been territorial tendencies damaging sustainability and life quality, motivated by the stakeholders. The predominant sociocultural, economic and territorial tendencies in our time make it necessary to integrate new regulatory and management parameters, as well as new methodological tools to explicit and integrate the ecologic and social approach, methodological frameworks and design tools in addition to ecological-territorial planning [51].

Such unsustainable territorial tendencies have been and are presently generated by issues, such as: the economic strategies and targets that seek principally short term maximization of financial profit; predominance of private short term interest above long term public interest; a sectorial organization and design incapable of integrating the various dimensions of human society development; the non-inclusion of the social and environmental services in the regional o national accounting [58].

The predominant green agricultural industrial revolution known as conventional agriculture is associated with institution, policies and technologies administered form urban centers and markets, which interact with the present-day development model. The green agriculture revolution is based on a great use of capital and exhaustive transformation technology, as well as, laborer reduction, high energy, water and mechanization input, applied in high potential productive ecosystems. Industrial agriculture can be defined as a way of artificializing nature and natural resource management, in pose of agriculture productivity, giving great importance to the economic profit through marketing, and occasional technological processing of highly homogenous products, by means of exogenous inputs into the agro ecosystem, by its artificialization, simplification and destruction of the natural recycling energy and material process [59].

This kind of context and agriculture has generated an important territorial-ecological impact and footprint in rural areas. The main footprints have been: carbon, energy, water and information, which put together, can be considered the agricultural footprint of our era. The agricultural frontier expansion and domestication of nature, both associated to the rural and cultural landscape construction, have developed several ecosystem diseases and affected life quality; such as: soil erosion, desertification, biodiversity reduction, cultural landscape homogenization, loss of niches and habitat diversity, in other words eco-diversity, unstable ecosystems, loss of resilience and stability, etc.

Several studies show the consequences of the great economic importance given to the ecological-territorial management, generating ecosystem dysfunction in maintenance, use and regeneration of resources, as well as, degradation of the ecosystem services [60]. These authors indicate the importance of keeping the pressure on the landscape within the required limits for a stable ecosystem function, key for a sustainable management. Unfortunately these limits are frequently trespassed. This is the case of the Australian grazing system management. The innovation and production goals motivated by the wish for great short term profit in the ranching activities have produced many ways of degradation of the cultural landscape: Diminished natural grange and crop productivity; lower tolerance to drought, salinization, acidification, soil structure and erosion, water salinization, eutrophication of streams and lakes; loss of trees considering the cultural landscape scale; loss of important local and regional plant and animal species; invasion of native and exotic grasses; loss of future potential use of the land (tourism, research, etc.); besides the lower rural life quality [57, 60].

In Latin America there are also many cases. One is the Chilean forestry crop industry, broadly studied in academic and scientific literature and fully examined by Erlwein *et al.*, [61]. The tremendous growth of this industry, explained mainly by the forest plantation territorial expansion starting the mid 60's till the end of the 90´s, and due to the increase of plants and production of cellulose, has triggered effects, such as: the unsustainable rurality; increase of the surface intended for intensive production; extreme production which excludes other uses and activities; reduction of native forest patches and of bio and eco-diversity; separation from land multiple use; resource degradation and production potential; capital concentration and socioeconomic inequity; inconsideration to cultural diversity contrary to social ethics; and cultural landscape uniformity, among many others. In conclusion it has been a sectorial growth which hasn´t incorporated any aspect other than the economic growth (such as the historic, social, ecologic, etc.) nor objectives different from the personal and private ones of the social actors, who have administered the process, and consequently have not stimulated and integral and sustainable territorial development [57].

This has all happened jointly with the emergence and development of the "industrial empires" pertaining to our industrial era. By the end of the XVIII century, with the industrial revolution there is a modification of products, transportation, technology and the demand for elements from nature which start becoming scarce or limited, generating the term natural resources in 1875. Modern industrial empires, such as: USA, United Kingdom, Japan, China, Germany, France, and others. Their natural resource requirement is so high that the commodities are extracted from the rest of the planet, generating the ecological footprint [62] of our industrial age. Said ecological footprint is grater in the countries producing the commodity to fulfill the demand of industrialized countries [53].

According to the ecophilosofer Sigmund Kvaloy two basics kinds of society can be distinguishes as a result of the industrial cultural tendencies and cultural landscape construction style: the Industrial Growth Society (IGS) and Life Necessity Society (LNS). The IGS are orientated towards industrial growth, whereas the LNS to fulfilling vital necessities.

IGS are developed through the interaction of four main dynamic factors [41]: oriented towards the linear or accelerated expansion to the production of industrial goods and services using industrial methods, as massive standardized production, the concentration of a few urban centers, and the specialization; the main force is the individual competition in every field of human effort, including leisure and art; the main resource for expansion and to eliminate competitors is not the mineral, energy, etc. resource control but the applied science control. The leading method to manage everything and perform diagnosis and prognosis is quantification. There is only one historic case of this kind of society: The present one which is becoming global. Most human societies have been of the LNS type. Among them there is a subvariety: the "Life Growth Societies" (LGS). These societies are focused on life improvement and promoting ecological complexity, cultural development and human creativity. This kind of society only can surface as a subspecies of the LNS type [41].

At present, progress is focused on the full understanding of territorial development. In which the territory is not a circumstantial factor of the economic analysis, but a descriptive element of the development processes. For a society to approach sustainability there must be cultural and paradigmatic changes to favor and direct, the integral construction of sustainable cultural landscapes, suitable for a good quality of life. For such paradigmatic changes to take place, there must be a considerable reorientation in the approximations that study these issues. Within the following paragraphs we present the theoretic and conceptual basis for the integral construction of the cultural landscape in the context of our era.

4.2 Change of paradigm

During the last thirty or forty years there has been a paradigm shift due to the postmodern scientific revolution, mainly with the emergence of so called complex sciences, which change the object of study from the parts to the whole [30]. This has meant no longer centering the study in linear and determinant processes, but in non-linear processes organized in hierarchical interrelated networks, in order to identify the main interactions among variables and the processes involved in the study´s objective; this way the processes and tendencies that emerge from these interactions, turn the concepts of complexity, network and hierarchy into fundamental issues [57].

This means changing the fragmentation for integration and complementation of the parts. The intention is to trespass the limits of the traditional scientific knowledge which proposes the objectivity and certainty of scientific truths, recognizing the need for an integral and contextual vision, as well as, and the need to deal with uncertainties [13, 30, 63, 64]. The key of the epistemological property of this paradigm shift is the development of an inter- and trans-disciplinary approach that requires variation in the current scientific reasoning. Röling [13] proposes the evolution of the science paradigm, starting with the simple dynamic structures and mechanical models, passing by the self-regulated models and homeostatic feedback models, towards the complex adapting auto-organizing systems, as well as, the autopoietic cognitive models (Figure 3).

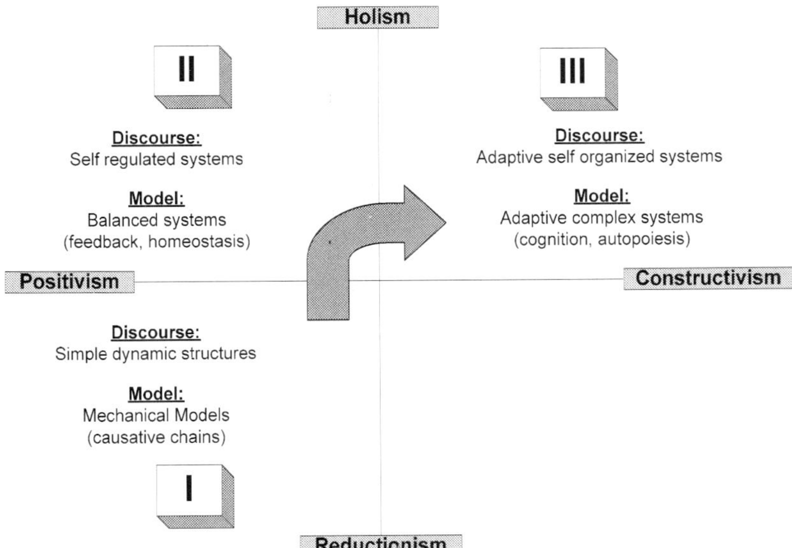

Fig. 3. New scientific paradigm evolution [adapted from 13].

The main difference between positivism and constructivism lies in how you consider epistemologically, the relationship between the observer and the object and phenomena observed. Positivism considers the independent phenomena of the particular observer. Constructivism, on the other hand, incorporates an interaction between the observer and the phenomenon observed, and recognizes that our perception of the world is only an individual and partial one [63]. From the perspective of constructivism, there should be a permanent dialog between the various observers, in order to piece together a group vision of reality, turning this into a collective cognition process. This effective dialogue, resulting from the collective construction is the foundation for the study of the phenomenon from the constructivism perspective.

In the XVII century, the French mathematician Rene Descartes formalized the reductionism perception. According to him, it is necessary to dissect and analyze separately and make precise measurements of the complex phenomena to fully understand it. This approach is synthetized in *Discours de la méthode* (1637). As a consequence, it has created a utilitarian criteria of the truth and a reduction of the phenomenon studied to an instrumental notion [30]. During the same century, the English physicist, Isaac Newton, complemented this approach with mechanical vision of the universe. In this approach the wish to set rules and laws, and even some regularities could be sensed [65].

The holistic approach is based on the system theory, and thus, on the approach which established that the universe is an interrelated system, originating in the aristotelic consideration that the whole is greater than the sum of its parts. All the data is more than the sum of the fragments of information, having to know it all to understand the collective behavior of the parts [30], namely, its combinations, and functional interactions in the construction of the systemic totality. The holistic approach considers that the problems must be tackled from the totalities and considering the contexts, as well as from the qualitative approach which gives meaning and sense to the quantitative.

The first quadrant of Figure 3, shows the reductionist - positivist approach, where each phenomena is perceived and treated independently from every discipline; the second quadrant is still based on positivism, but has evolved from a reductionist to a holistic perspective. There is a partial integration of the positivist – reductionist disciplines, but not enough to develop an integral and operational approach toward transdisciplinary and multidimensional problems.

The third quadrant presents an holistic and constructivist approach. Here the Adaptative Complex System (ACS) is located [*sensu* 64], the cognitive theory [39, 40, 8], the social knowledge based on and intentional and adaptive collective cognition in the design and management of our own destiny [1, 2, 13].

One of the outstanding values of the systemic approach, which is based on second order cybernetics, is that it may overcame epistemological barriers between science and humanity, as well as, between the techno-economical-political areas, where the decision process regarding the management of territories and natural resources take place [13, 30].

The homeostatic systems are related to the model equilibrium paradigm [35, 63], that is to say, they are connected to nature and to the perception of ecosystem as a balanced system. A central issue of this paradigm is the system´s tendency to reach a unique state of stability. In the evolving complex system study emerges a non-equilibrium paradigm [66]. Key aspects of the non-equilibrium paradigm are: the system can reach numerous constant states and keep the organizational pattern; the system has an open relation with its surroundings; it is capable of focusing on the continuous process co-evolutionary coupling [66].

The Adaptive Complex System (ACS) is a concept and model that correspondence a turning point for the study systems of traditional sciences. The main feature of ACS, according to Gell-Mann [64], could be its use for landscape study. Each landscape is an iterative information processing system interacting with its environment; it continuously processes new information from its surrounding environment, generating new adaptive tendencies, coupling and stability. Since, the historical evolving process doesn't couple under the new circumstances and information, it can`t adapt to the system not connect with its surrounding environment, and thus, collapses.

In systems far from equilibrium, such as the ACS's, order and disorder (chaos) are continuously interacting. In the chaotic stage, these systems tend to dissipate energy and generate entropy, creating conditions with new, continuous and iterative, order patterns, and occasionally developing a new organizational pattern and type of system [8, 13, 30]. This perspective is necessary to understand the adaptive evolution of cultural landscapes [13, 30, 67, 68].

The goal of the ACS's is to adapt to variable and changing environments, through different schemes stored in the historic system memory. The self-oriented capacity to adjust is explained by the ACS model. Highlighting human behavior as the main determining factor in the cultural landscapes dynamic and evolution.

4.3 Development approach and models

At this moment in time, the processes of human society development are dominated by the sectorial approach. Each sector pursues optimization according to their own requirements, such as: economic, urbanistic; agricultural, rural, real state, forestry, mining sectors. This approach triggers territorial degradation tendencies as is doesn't considerer territorial integrity. It is a merological approach, reducing the problem to specific problems and interests.

On the other hand, the territorial approach centers the main objective on the landscape planning units and its surroundings, focusing on the integral system development. It is based on the holistic system paradigm, emphasizing a transdisciplinary approach as a key epistemological attribute for human development processes.

The XX century traditional paradigm focuses on three main interacting components: sectors, people and economic efficiency (Figure 4). With this approach, activities and development processes are evaluated as successes or failures, considering mainly the economic parameters, such as IRR (internal return rate) and NAV (real net value) [53].

In the XXI century a new paradigm emerges. The model integrates three dimensions: territory in instead of sectors; stakeholders instead of people; global quality instead of economic efficiency (Figure 4). The global indicator parameter for the sustainable cultural landscape construction and evaluation is related to each specific condition, and is a function of the following variables: ecological, social and economic. It is a determinant based on the interaction of three main axes: economic productivity, social equity and ecological sustainability [66, 69]. This approach and paradigm focus on the sustainable development and life's quality.

TRADITIONAL PARADIGM OF THE XX CENTURY:

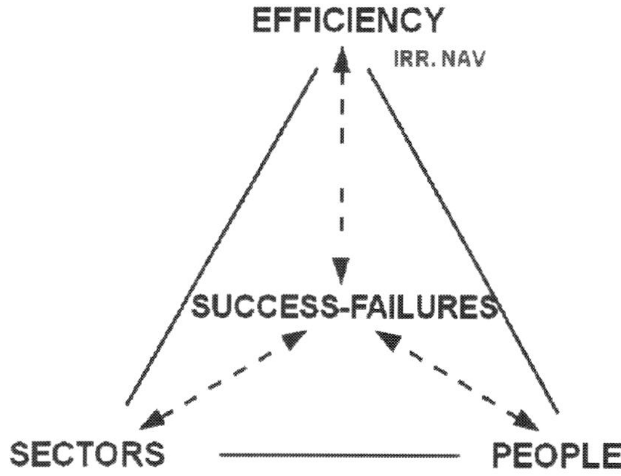

NEW PARADIGM OF THE XXI CENTURY:

Fig. 4. Evolution of the development paradigm. The new paradigm focuses on cultural landscape, integrating sustainable development and life's quality [53].

4.4 Territorial arrangement and cultural landscape design

Territorial arrangement and planning is not only a technological, ecological and political subject, there are also related with the spirits of age and place (*Zeitgeist* and *Volksgeist*). The Territorial arrangements of a country and places are always related with culture [15].

The European Cart of Territory Arragement, emphasized regional territorial balance. They pursued a territorial arrangement with the best distribution of spatial and human activities, to achieve the best combination, as a function of societies´ requirements given its culture, ecological limitations and potential, as well as, life quality optimization and sustainable development. The multiple use principle is a main argument referring to the purpose and management of territorial resources, in order to provide a better use for human requirements without causing ecosystem degradation, as well as, setting up areas for human life and integral development. Thus, multiple use of the territory focuses on different objectives from many sectors and subjects [70].

Watershed is the basic unit for territorial arrangement and where biocenosis (phytocenosis and zoocenosis) interact with the ecotop. Social, economic, institutional and cultural dimensions of the stakeholders administration, resource management and arrangement at the watershed level, are related with α, β, γ diversity.

The design of the cultural landscape is an essential element and operator to reach the goal of balance, the stakeholders need in the landscape context [71]. Presently it is necessary in order to increase territorial services and sustainability, not only to preserve but also to design and construct [13], with an integrative, dynamic, intentional and collective approach. The fundamental dimensions in the cultural landscape design are: ecological, anthropic functionality, life and leisure, and aesthetics.

The ecological dimension refers to system sustainability as a result of cultural landscape nature conservation, ecological connectivity and ecodiversity. It optimizes the positive and negative ecosystem effects, designs the structural cycles (recycling) and ecosystem efficiency, in addition to stability (energy, matter and information). Another key concept is technological receptivity, defined as the amount of technology that could be applied in each particular site to produce a desired sustainable output. Technological receptivity allowed discriminating differences to select the right operator [14].

The functionality dimension is reared towards human actions aspiring to accomplish the activities associated with stakeholders' objectives. The aesthetic dimension deals with symmetry, beauty and landscape perception, which deals with elements such as forms, colors, textures, borders, observations points, etc. Life and leisure dimension are related with resting places for the social actors amusement. Leisure is something highly valuated associated with the creative potential of people and human development [72]. All of this is related with the concept of biophylia, which can be defined as the inherent tendency of humankind to get closer to different kinds of life and natural processes, desirable for a better life quality in step with human evolution during a long period of time.

The landscape design methodology is presented in Figure 5. The first stage is the polithemathic analysis of the landscape's limits, including: zoning, technology, hydrology and the natural matrix in a topological arrangement. The second stage established the threshold of the landscape: functionality, aesthetics, ecology, as well as life and leisure. Then, in the third stage, the territorial components: *saltus*, *ager* and *polis*, and their relative proportions in cultural landscape types: wildland, rural and urban. The last part is directly related with the construction of cultural landscapes by the action of the stakeholders and stockholders.

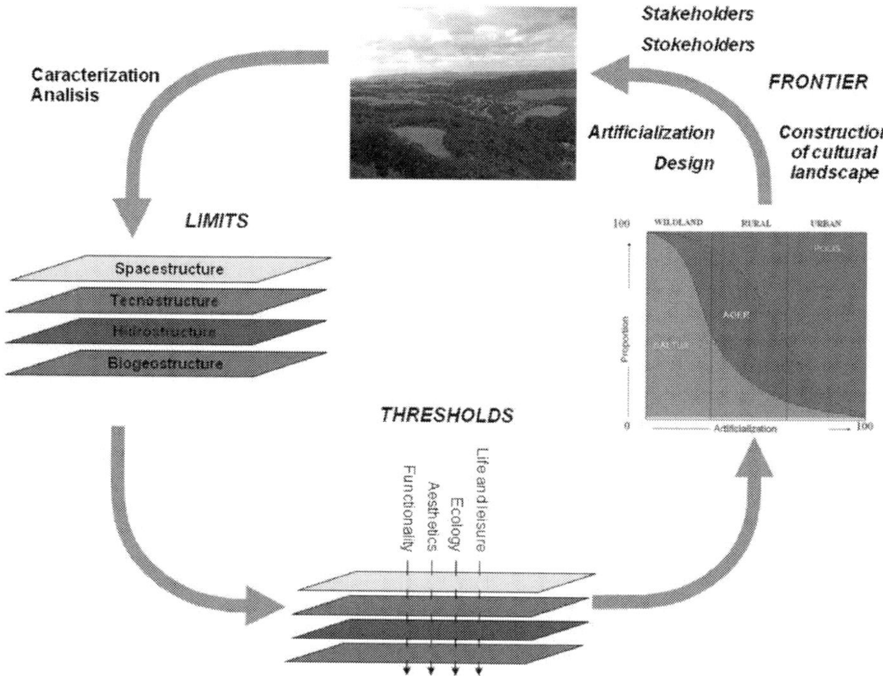

Fig. 5. General methodological model for the cultural landscape design and construction of a sustainable development.

The design and construction of sustainable cultural landscapes include:

- Diversity. Refers in a broad sense to biodiversity (α, β and γ), ecodiversity (niches and habitats) and territorial diversity, the last one related with territorial multiple use.
- Connectivity. Refers to the generation of ecosystemic and territorial networks, including: technological (resource management styles), social, cultural and institutional dimensions. It is a complement of the ecosystem interaction network considering the stakeholder and technology.
- Coupling. The ecological connections aren't enough, they also require energy, matter and information exchange by coupling between system's components. System functionality requires the complementation and integration of their components.
- Location. Technological receptivity and ecosystem resilience is a function of the location of the watershed and biocenosis type [70].
- Recurrence. Design and management of ecological and territorial systems should not be lineal but recurrent. This is equivalent to recycling in natural ecosystems. The recurrent input management is related to achieve adaptation [73]. Agriculture ecosystem design and management is related with connectivity and grater autonomy of agroecosystems.

4.5 Cultural landscape governance

Governance refers to the art and the way of carrying out government, as well as, the executive action. Governance emerges based on the general request for the administration of: natural resources, world ecosystems and territory development. It should be allowed an antrophic control of the phenomena. Governance improves public policies and collective actions to solve problems and take care of the integral development. Nevertheless, it is not possible to predict the future cultural landscape but to simulate and evaluate further scenarios [30]. Some handling capacity can be developed in order to shift to a more specific and desirable situation for a particular culture and stakeholder.

Jentoft [44] states that territorial governance is basically a relationship between two systems: the government system and the governed system. The first is a structure of institutions and control mechanisms. The second is partially social and partially natural: it consists in one ecosystem coupled with its resources, as well as stakeholders, all of them developing institutions and political conditions. Territorial government is related to the connections of both subsystems, by integrating them into only one. In order to make operative governance, both systems should be mutually sensible, combined and coevolving [40].

Jentoft [44] has developed a governance model, where both systems (government and governed) should be efficient (Figure 6).

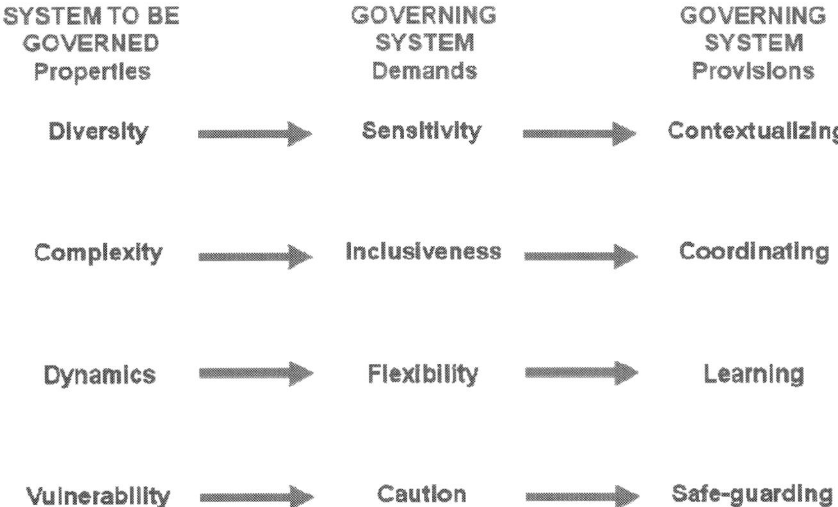

Fig. 6. Territorial governance model. The governed system attributes and the requirements the government system must have [44].

An ecosystem service is a basic element for the territorial sustainable governance that supports life on earth and takes care of the diversity of those services within a varied cultural landscape [74, 75]. These services are necessary for human survivorship and social development [1, 2, 13, 76]. Since ecological services are not tradable in financial markets, there is a shortage of regulatory mechanisms to detect the supply and ecosystem damage [2, 74, 75, 77-79]. Human economy can´t operate without ecosystem services, and thus, the financial value is infinite. Constanza *et al.* [74] present seventeen categories of ecosystem services: gas regulation, climatic regulation, disruption regulation, hydric regulation, water supply, erosion control, soil formation, nutrient cycle, waste treatment, pollination, biological control, shelter, food production, raw materials, genetic resources, recreation, and culture.

It is amazing to notice that conventional productive agriculture only generates two of those seventeen categories: food and raw materials. What is more, the green revolution of industrial agriculture has a negative effect on the other fifteen. However, it´s important to mention that rurality is concerned about all seventeen ecological services.

5. Cultural landscape sustainability

5.1 Universal legality

All human activities linked with artificialization and management of natural resources should be set up on a hierarchical system (Figure 7). The degrees of freedom on each hierarchical level change according to the hierarchical context and wheater the direction is downwards or upwards, in line with the hierarchical theory [30, 43].

Decision making at any level depends on the stages above and below. Political decisions should be subordinate to economic, technologic, social and ecological levels. A right decision should be valid on all the levels of the universal legality.

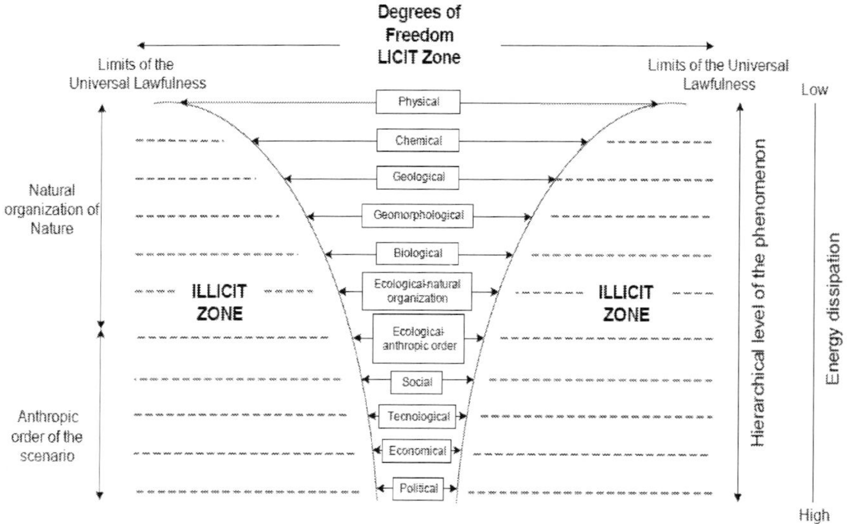

Fig. 7. Hierarchical scheme for landscape decision making and the relative degree of freedom of each level [28].

5.2 Planning and design of sustainable cultural landscape model

One of the main principles for landscape planning is to minimize negative effects, give equal opportunities and maximize the aptitude, all of this in interaction. In order to plan, design and govern the cultural landscape, it is essential to follow the value based model, defined by the particular culture (Figure 8).

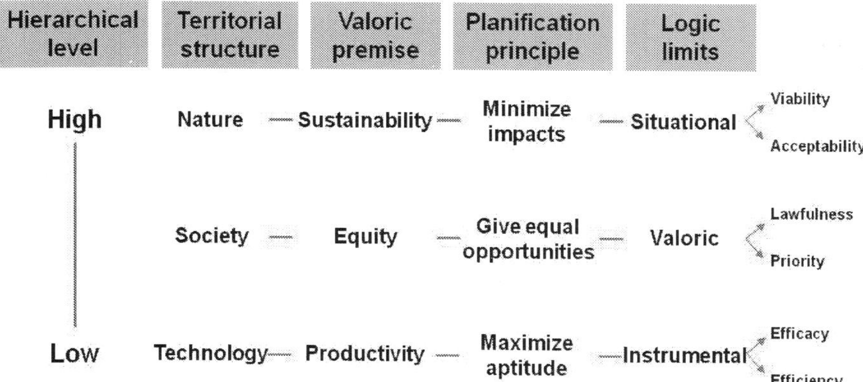

Fig. 8. Model for planning and designing a cultural landscape in order to provide a sustainable governance [51].

5.3 Ecosystem artificialization and sustainability

Artificialization is defined as a way to apply a certain amount and kind of technology to transform the ecosystem. The resulting ecosystem transformation is a function of the technical inputs. Thus, the end result could express the main functions. One of them is the anthropic benefits brought about as a consequence of this transformation. Still, there is a cost associated with the work inputs applied to the ecosystem.

In general, and consistent with the degree of artificialization the cost increases significantly in vulnerable ecosystems. In contrast, the benefits of marginal ecosystems usually increase very little compared with the degree of artificialization. If this case were to happen, both functions would never intersect. As a consequence this extreme vulnerable ecosystem should not be artificialized at all, being necessary to preserve them in a natural state.

Then again, there are highly stables ecosystems where the additional costs to keep them sustainable are insignificant, but the output benefits of artificialization are high. In this case the degree of sustainable artificialization could be immense.

Under usual condition, namely ecosystems which are not extremely vulnerable nor highly stable, there is an intermediate degree of potential sustainable artificialization (Figure 9). At the right side of the figure, the artificialization costs are greater than the benefits, and thus, the degree of transformation should be no higher than this magnitude. In contrast, at the left side, the cost is lower than the benefits, so it is fine to transform the landscape up to this magnitude. This defines the artificialization for the sustainable cultural landscape construction.

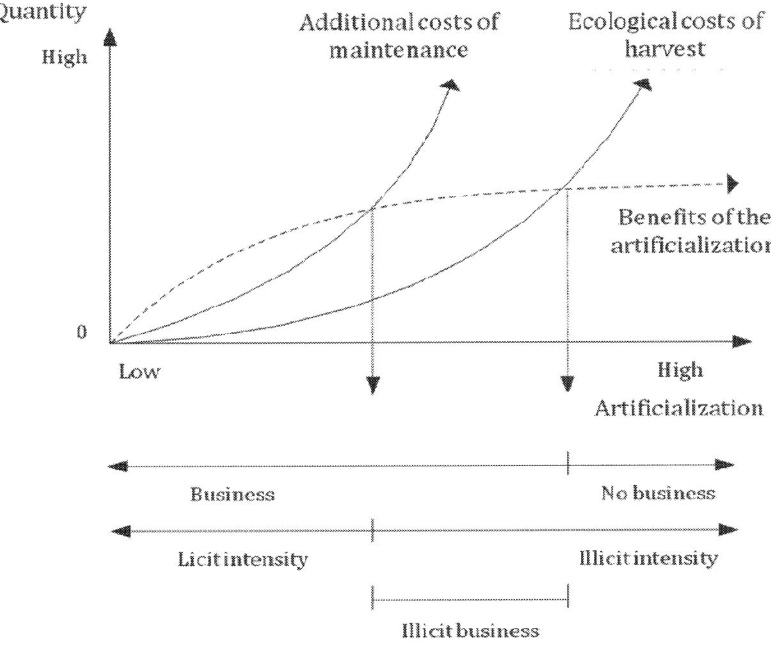

Fig. 9. Cost-benefit relation from artificialization in landscape transformation, for an ecosystem of intermediate vulnerability [23].

5.4 Adaptability and panarchy

The adaptive capacity to environmental changes is key for stability and sustainability of open systems. The adaptability of anthropic ecosystems and cultural landscapes is mainly determined by the stakeholders behavior and management.

Recently the concept of panarchy has been proposed to develop the ACS sustainability theory. Panarchy stems as an antithesis of hierarchy, representing the framework of "nature rules", suggested by the Greek nature god *Pan*. This state is reinforced supported by two main issues. The first one is a four face heuristic model change: exploitation, conservation, creative destruction and renovation, which brings about an adaptive cycle. This is a fundamental model to understand ACS, such as cells, ecosystems, human societies and culture landscapes as a whole.

Three proprieties define the adaptive cycle: potential, which provides and determines the limits of changing capacity; connectivity, which provides the variable internal control and consistency; resilience, which determines the vulnerability of each system's shifting. The adaptive cycle model provides the conceptual bases for understanding hierarchy not as fixed structures but as dynamic entities [80], which is basic for cultural landscape sustainability.

6. Concluding remarks

To come closer to sustainability it is necessary understand the ontology and epistemology of the relation and interaction between Nature and human society, which implies deal with the artificialization of the first one for the stakeholders. This is a central part of agriculture and cultural landscape construction.

The evolution of perception, as well as the interest, stimulus, and priorities of the social actors involved in the construction of agricultural and rural territories is constantly changing. It's related to the spirit of age (*Zeitgeist*), the spirit of place (*Volkgeist*), and certainly, to their culture and the characteristics of their territory.

The main concern associated to agricultural sustainability is discovering the problems affecting the stakeholders and their activities. In this context, technology, nature, society, economy, and ecology are related in different ways and intensities to rurality and agriculture *sensu lato*, in line with the meaning given, as well as where the problem is located and framed.

Landscape is a set of countless ways to characterize, and differentiate a specific area of land. It's a natural and cultural association of society with the components of the land. The cultural landscape is the consequence from the technological activities carried out by the stakeholders in a territory, and its transformation into sustainable or non sustainable agriculture. The cultural landscape concept emphasizes culture as the main dynamic determinant of the territorial evolution, aiming and associating it with the stakeholders' behavior.

Modern agriculture deals with extreme capital use, high technology, reduction of manual labor, high energy and water inputs, as well as, great mechanical labor, all things taking place in high input and high output ecosystems. It's associated with policies, development strategies, institutions, resources and technologies regulated from urban centers and markets. All of this generates, in rural areas, a significant ecological and agricultural footprint. The main ones are: carbon, water, energy, information, and all of this generate a substantial biodiversity loss, as well as, niches, ecodiversity and adaptability reduction. This process, has taken place in combination with a divergence and dissociation of agriculture and the integral rural development.

Rural landscape plays other roles beside those of agriculture, such as: gas regulation, climate stability, water regulation, erosion control, nutrient cycles, biological regulation, recreation, culture conservation, soil formation, as well as, the generation of genetic resources. Agricultural sustainability is a component of the rural landscape and actors. As such, it should also be analyzed in a complementary context, take into account its interaction with the urban areas and the protected wild areas.

Currently, the social and territorial development focuses on the relationship of sustainability with life quality for the collective construction of the territory, associated to the paradigmatic change of science and culture. Several well known schools of thought and intellectual scientific, and philosophical currents approach to this quandary is in a holistic and systemic transdisciplinary way.

The unifying agricultural and rural areas sustainable concept is linked to the territorial governance, limits, regulations, in addition to the development of the rural cultural landscapes.

Only if stakeholders operate with prudence in the artificialización of ecosystems and in the construction of the cultural landscape, according to the universal legality, a sustainable future will be possible. For this we must assume the challenge of design ecological-territorial systems appropriately adaptatives for our age context.

7. References

[1] Vitousek P, Mooney H, Lubchenco J, Melillo J (1997) Human domination on Earth's systems. Science 277: 494-499.

[2] Lubchenco J (1998) Entering the Century of the Environment: A New Social Contract for Science. Science 279: 491 – 496.

[3] Maturana H, Mpodozis J (2000) The origin of species by means of natural drift. Revista Chilena de Historia Natural 73 (2): 261-310.

[4] Plutchik R (2001) The Nature of Emotions. American Scientist 89: 344-350.

[5] Flores L (1994) La Tecnología en el Contexto de la Cultura Latinoamericana." Instituto Latinoamericano de Estudios Transnacionales. Tecnología y Modernidad en Latinoamérica: ética, política, cultura. pp. 19-23.

[6] Hax A, Majluf N (1993) Gestión de empresa con una visión estratégica. Ediciones Dolmen. Segunda edición. Santiago, Chile. 513 p.

[7] Calabuig E. (2002) Prólogo de la edición española. In: Burel F, Baudry J, editors. Ecología del paisaje: conceptos, métodos y aplicaciones. Ediciones Mundi-Prensa, París, Francia. pp. XV-XVI.

[8] Capra F (1996) La trama de la vida. Editorial Anagrama. Barcelona, España. The Web of Life. A New Synthesis of Mind and Matter. Harper Collins Publishers. Londres, Reino Unido. 359 p.

[9] Gastó J, Subercaseaux D (2010) Dimensión ecológica del paisaje cultural en el siglo XXI. Revista de la Escuela de Arquitectura de la Universidad de Talca 4: 60-73.

[10] RAE (Real Academia Española) (1984) Diccionario de la Lengua Española. 20ma. Edición. Espasa-Calpe, Madrid, España. 1417 p.

[11] Gastó J (1980) Ecología, el hombre y la transformación de la naturaleza. Editorial Universitaria. Santiago, Chile. 573 p.

[12] Pointing D (1992) Historia verde del mundo. Paidós. Barcelona, España. 584 p.

[13] Röling N (2000) Gateway to the global Garden: Beta/Gamma Science for Dealing with Ecological Rationality. Eight Annual Hopper Lecture. University of Guelph, Canada. 51 p.

[14] Nava R, Armijo R, Gastó J (1996) Ecosistema. La unidad de la naturaleza y el hombre. Editorial Trillas. México. 332 p.

[15] Gastó J, Rodrigo P, Aránguiz I, Urrutia C (2002) Ordenación territorial rural en escala comunal. Bases conceptuales y metodología. In: Gastó J, Rodrigo P, Aránguiz I, editors. Ordenación Territorial, Desarrollo de Predios y Comunas Rurales. Facultad de Agronomía e Ingeniería Forestal, Pontificia Universidad Católica de Chile. LOM Ediciones. Santiago, Chile. pp. 5-60

[16] Lawes J (1847) On agricultural chemistry. J. Royal Agric. Soc. 8:226-260.

[17] Prado C (1983) Artificialización de ecosistemas. Planteamiento teórico para su transformación y optimización. Tesis de Ing. Agrónomo. Universidad de Chile. Santiago, Chile.

[18] De Wit C (1974) Publication address at the ocasión of the Dies Natalis of the "Landbouwhogeschool". Wageningen University. Holland.

[19] Barnhart R (2004) Chambers. Dictionary of etimology. H.W.Wilson. New York. USA. 1284 p.

[20] Acevedo E (2009) Fisiología de cultivos: intensificación sustentable, captura de carbono y aumento del rendimiento potencial y rendimiento bajo el estrés de los grandes cultivos. Seminario Desafío Científico del Desarrollo de las Ciencias de la Agricultura de Chile. Reunión Académica de Ciencias Agronómicas. Santiago, Chile.

[21] Martínez J (1987) Economía y ecología: cuestiones fundamentales. Pensamiento Iberoamericano 12: 41-60.

[22] Altieri M, Rojas A (1999) Ecological Impacts of Chile's neoliberal policies, with special emphasis on agroecosystems. Environmental, Development and Sustainability 1: 55-72.

[23] Vera L, Gastó J (2011) Expansión de la Frontera Homínida en el Paisaje Cultural. Hominización, restauración y gobernanza de la Cordillera de Los Andes de la Araucanía, Chile. Editorial Académica Española. Saarbrücken, Alemania. 388 p.

[24] Randon R (1994) Haciendas de México. Fomento Cultural Banam. México, D.F. 382 p.

[25] Ohrens O, Alcalde J, Gastó J (2007) Orkestiké. Agronomía y Forestal UC 31: 22-25.

[26] Columela M (1959) (Libro original del siglo I d.C.). Los 12 Libros de la Agricultura. Editorial Iberia SA. Barcelona, España. 458 p.

[27] Hughes J (1975) Ecology in ancient civilizations. University of New Mexico Press. Alburquerque. N. M. 181 p.

[28] Gastó J, Vera L, Vieli L, Montalba R (2009) Sustainable agriculture: Unifying concepts. Cien. Inv. Agr. 36: 5-26.

[29] Tansley A (1935) The use and abuse of vegetational concepts and terms. Ecology 16:284-307.

[30] Naveh Z (2000) What is holistic landscape ecology? A conceptual introduction. Landscape and Urban Planning 50: 7-26.

[31] Ruthenberg H (1980) Farming systems in the tropics. Clarendon Press. Oxford. England. 424 p.

[32] Gastó J, Armijo R, Nava R (1984) Bases heurísticas del diseño predial. Sistemas en Agricultura. IISA 8407. Departamento de Zootecnia. Facultad de Agronomía e Ingeniería Forestal. Pontificia Universidad Católica de Chile. Santiago, Chile. 41 p.

[33] Bertalanffy L von (1968) General System Theory: Foundations, development, applications. George Braziller. New York. USA. 289 p.

[34] Odum E (1953) Fundamentals of ecology. W.B. Sandero. Filadelfia, USA. 598 p.

[35] Margalef R (1974) Ecología. Editorial Omega. Barcelona, España. 951 p.

[36] Margalef R (1993) Teoría de sistemas ecológicos. Publicaciones Universitat de Barcelona. Barcelona, España. 290 p.

[37] Prigogine I (1996) El fin de las certidumbres. Editorial Andrés Bello. Barcelona, España. 226 p.

[38] Solé R, Goodwin B (2000). Sign of Life. How Complexity Pervades Biology. Basic Books. New York, Estados Unidos de América. 322 p.

[39] Maturana H, Varela F (1972) De Máquinas y Seres Vivos. Una teoría de la organización biológica. Editorial Universitaria, Santiago, Chile. 122 p.

[40] Maturana H, Varela F (1987) El árbol del conocimiento: las bases biológicas del entendimiento humano. Editorial Universitaria, Santiago, Chile. 171 p.

[41] Reed P, Rothenberg D, editors (1993) Wisdom in the open air. Univ. Minnesota Press. Minneapolis, USA. 142 p.

[42] Briggs J, Peat F (1994) Espejo y Reflejo: del Caos al Orden. Editorial Gedisa. Barcelona, España. 222 p.

[43] Wu J, David J (2002) A spatially explicit hierarchical approach to modeling complex ecological systems: theory and applications. Ecological Modelling 153: 7-26.

[44] Jentoft S (2007) Limits of governability: Institutional implications for fisheries and coastal governance. Marine Policy 31: 360-370.

[45] Carbonell E, Sala R (2000) Planeta Humano. Ediciones Península. Barcelona, España. 263 P.

[46] Carbonell E, Sala R (2002) Aún no Somos Humanos. Propuesta de Humanización para el Tercer Milenio. Ediciones Península. Barcelona, España. 240 p.

[47] Gastó J (2008) Producción Animal y Paisaje Cultural. In: Junta de Andalucia. Consejeria de Agricultura y Pesca, editor. Pastos, clave en la gestión de los territorios: Integrando disciplinas XLVII. Reunión Científica de la Sociedad Española de Estudio de los Pastos. Córdoba, España. pp.509-523.

[48] Ceresuela J (1998) De la Dehesa al Bosque Mediterráneo. In: Hernández C, editor. Jornadas de Agronomía: La Dehesa. Aprovechamiento Sostenible de los Recursos Naturales. Grupo de Ecologistas de Agrónomos. ETSIAM, Universidad Politécnica de Madrid. Editorial Agrícola Española S.A. Madrid, España. pp.45-52.

[49] Gastó J, Vieli L, Vera L (2006) Paisaje Cultural. De la *Silva* al *Ager*. Agronomía y Forestal UC 28: 29-33.

[50] Hunt T (2006) Rethinking the Fall of Easter Island. American Scientist 94: 412-419.

[51] Gastó J, Pino M, Fuentes V, Donoso S, Gallardo S, Ahumada N, Gálvez C, Gática C, Retamal M, Pérez C, Vera L (2005) Metodologías para la planificación territorial. Ministerio de Cooperación y Planificación, Santiago, Chile. 144 P.

[52] González F (1981) Ecología y Paisaje. Ediciones Blume. Madrid, España. 250 p.

[53] Gastó J, Vera L (2009) Ordenamiento territorial en un mundo centralista. In: Von Baer H, editor. Pensando chiles desde sus regiones. Sinergia Regional, Ediciones Universidad de la Frontera. Temuco, Chile. pp. 455-473.

[54] Gastó J, Gálvez C, Morales P (2010) Construcción y articulación del paisaje rural. AUS (Valdivia) 7: 6-11.

[55] Ferrater J (1965) Diccionario de Filosofía, Quinta Edición. Editorial Sudamericana, Buenos Aires, Argentina. 2017 p.

[56] Illich I (1996) La sombra que arroja nuestro futuro. Entrevista. In: Gardels N, editor. Fin de Siglo. Grandes pensadores hacen reflexiones sobre nuestro tiempo. McGraw-Hill Interamericana Editores. D.F., México. pp. 69-85.

[57] Subercaseaux D (2013) Implicancias Ecológicas de la Priorización Económica en el Paisaje Cultural. Determinante de Orden y Sustentabilidad. Economía, sociedad y territorio, revista de El Colegio Mexiquense A.C. Zinacantepec, México. In press.

[58] Ortega R, Rodríguez I (2000) Manual de gestión del medioambiente. MAPFRE, Madrid, España. 364 p.

[59] Subercaseaux D (2012) Tecnologías Agroecológicas Socialmente Apropiadas y Estilos de Agricultura. Postitulo en Agroecología y Desarrollo Rural Sustentable. Universidad de Santiago de Chile. Santiago, Chile.

[60] MacLeod N, McIvor J (2006). Reconciling economic and ecological conflicts for sustained management of grazing lands. Ecological Economics 56: 386-401.

[61] Erlwein A, Lara A, Pradenas A (2008) Industria de celulosa en Chile: un modelo de desarrollo no sustentable. In: Monjeau A, Parera A, Lacour E, editors. Ecofilosofía. Editorial Universidad Atlántida, Argentina-Brazil, pp. 141-179.

[62] Wackernagel M, Rees W (2001) Nuestra huella ecológica, reduciendo el impacto humano sobre la tierra. LOM Ediciones. Santiago, Chile. 207 p.

[63] Bertalanffy L von (1975) Perspectives of general system theory. Springer Verlag. New York, USA. 253 p.

[64] Gell-Mann M (1995) El quark y el jaguar. Aventuras en lo simple y lo complejo. Tusquets Editores S.A., Barcelona, España. 413 p.

[65] Aguilera F (1992) Posibilidades y limitaciones del análisis económico convencional aplicado al medio ambiente. Ponencia presentada en IV Congreso Nacional de Economía. Desarrollo Económico y Medio Ambiente. Sevilla, España.

[66] D'Angelo C (2002) Marco conceptual para la ordenación de predios rurales. In: Gastó J, Rodrigo P, Aránguiz I, editors. Ordenación Territorial, Desarrollo de Predios y Comunas Rurales. Facultad de Agronomía e Ingeniería Forestal, Pontificia Universidad Católica de Chile. LOM Ediciones, Santiago, Chile. pp. 205-224.

[67] Naveh Z (2001) Ten major premises for a holistic conception of multifunctional landscapes. Landscape and Urban Planning 57: 269–284.

[68] Naveh Z, Lieberman A, Sarmiento F, Ghersa C, León R (2002) Ecología de paisajes. Teoría y Aplicación. Editorial Facultad de Agronomía Universidad de Buenos Aires, Buenos Aires, Argentina. 571 p.

[69] Nijkamp P (1990) Regional sustainable development and natural resource use. World Bank, Annual Conference and Development Economics. Washington D.C., U.S.A. 215 p.

[70] Gastó J, Guerrero J, Vicente F (2002). Bases de los estilos de agricultura y del uso múltiple. In: Gastó J, Rodrigo P, Aránguiz I, editors. Ordenación Territorial. Desarrollo de Predios y Comunas Rurales. Facultad de Agronomía e Ingeniería Forestal, Pontificia Universidad Católica de Chile. LOM Ediciones. Santiago, Chile. pp. 153-169.

[71] Subercaseaux D (2007) Paisaje Cultural: Implicancias Ecológicas de la Priorización del Lucro Económico. Bases Teórico-conceptuales y Planificación del Paisaje Cultural. Tesis de Magister en Recursos Naturales. Pontificia Universidad Católica de Chile. Santiago, Chile.

[72] Max-Neef M (1994) Interview. Por qué un Cristo de plástico acerca más a la gente a la divinidad que un árbol. In: Mendoza M, editor. Todos queríamos ser verdes. Chile en la Crisis Ambiental. Editorial Planeta, Santiago, Chile.

[73] Holling C (1978) Adaptative Environmental Assessment and Management. Wiley, New York, USA. 377 p.

[74] Costanza R, D'Arge R, de Groot R, Farber S, Grasso M, Hannon B, Limburg K, Naeem S, O'Neil R, Paruelo J, Raskin R, Sutton P, van den Belt M (1997) The value of the world's ecosystem services and natural capital. Nature Magazine 387: 253-260.

[75] Costanza R, Farber S (2002) Introduction to the special issue on the dynamics and value of ecosystem services: integrating economic and ecological perspectives. Ecological Economics 41: 367-373.

[76] Chiras D, Reganold J, Owen O (2002) Natural Resource Conservation. Prentice Hall. New Jersey. USA. 640 p.

[77] Boumans R, Costanza R, Farley J, Wilson M, Portela R, Rotmans J, Villa F, Grasso M (2002) Modeling the dynamics of the integrated earth system and the value of global ecosystem services using the GUMBO model. Ecological Economics 41: 529-560.

[78] Farber S, Costanza R, Wilson M (2002) Economic and ecological concepts for valuing ecosystem services. Ecological Economics 41: 375-392.

[79] Villa F, Wilson M, De Groot R, Farber S, Costanza R, Boumans R (2002) Designing an integrated knowledge base to support ecosystem services valuation. Ecological Economics 41: 445-456.

[80] Gunderson L, Holling C, editors (2002) Panarchy. Understanding Transformations in Human and Natural Systems. Island Press. Washington, USA. 507 p.

8

Local Residents' Perceptions of and Attitudes Toward Sustainable Tourism Planning and Management in Amasra (Turkey)

Bülent Cengiz

Bartın University, Faculty of Forestry, Department of Landscape Architecture
Turkey

1. Introduction

Tourism has a prominent role in international economy, and especially for many developing countries, is a major source of income. This economic prominence has led to an increasing competition among destinations (United Nations World Tourism Organization, 2012). Diversification of tourism activities and the quality of local environment play a key role in this competition as tourists seek unique, aesthetically pleasant and culturally attractive places (United Nations Environment Programme, 2009).

For instance, coastal historical settlements have a significant tourism potential as they offer not only cultural and historical values, but also natural landscape through the merger of sea and shore, the prime elements of coastal tourism which is arguably the dominant form of tourism worldwide (Yazgan and Kapuci, 2007; Kelkit et al. 2010). Tourism activities in coastal areas have direct and indirect effects on the local and national economy and the quality of environment (Mason and Cheyne, 2000; Harrill and Potts 2003; Ernoul, 2009). The effects of tourism development vary to a great extent as it has the potential for both positive and negative outcomes, especially at the local level (Lankford and Howard, 1994; Lee et al. 2007). Lack of tourism planning and management causes loss of biodiversity and pressures on natural resources (Puczkó and Rátz, 2000; Israeli et al. 2002; Harrill and Potts, 2003; Welford and Ytterhus, 2004; Ernoul, 2009), air, water and soil pollution, degradation of natural and cultural environment, visual pollution, decline in the quality of life and public health, as well as degeneration of the socio-cultural structure.

The goals of sustainable tourism development are related to the ecological aspects (environmental quality), economic aspects (feasibility), social characteristics of the locality (acceptance by residents), and lastly, to tourist satisfaction (Dymond, 1997; Puczkó and Rátz, 2000). Sustainable tourism enables tourism development that is compatible with the carrying capacity of the ecosystem. It creates recreational opportunities for local residents and visitors alike, and is effective in the protection of historical and archaeological sites (Sertkaya, 2001).

The multifaceted nature of sustainable tourism development requires tourism policy to integrate strategies for sustainable development and poverty reduction and measures related to climate change and biodiversity. Furthermore, to accomplish such multidimensional goals, tourism strategy should engage national, regional and local administrations alike. With respect to the sustainability of tourism benefits at the local level, an integrated approach to management is needed to protect natural resources from diverse tourism pressures (United Nations Environment Programme, 2009).

Participation of local people is important in sustainable tourism planning. In this respect, residents' expectations of tourism, perspectives on tourism and perceptions of the natural and cultural values play an important role in the development of sustainable tourism planning and management strategies.

Ernoul (2009) reports that the local residents' perceptions of tourism development are largely neglected. Several studies focus on special events in order to comprehend the perceptions of the local population regarding tourism (e.g. Soutar and McLeod, 1993; Jackson, 2008). The results of these studies suggest that tourism has a positive impact on the residents' quality of life. According to Lee et al. (2007), research on local populations' perceptions of tourism development is useful for developing measures to overcome problems due to hostility between visitors and local populations (Sethna, 1980) and for making plans to gain resident support for further tourism development.

This study measured the local residents' perception of the necessity of sustainability of the natural, cultural and historical values that constitute the resource of tourism activities in the city of Amasra in order to determine the attitudes toward and participation in tourism development of local residents. With a rational tourism planning approach, it concludes with some proposals for providing opportunities of year-long tourism activities, especially considering its coastal location, gastronomy, local handicrafts and pensions and for contributing to regional economy in terms of sustainable tourism development.

2. Material and methods

2.1 Material

2.1.1 Research site

Amasra is a Western Black Sea coastal town in Bartın Province, Turkey. The town occupies a peninsula with two bays and extends toward north. Located 17 km away from Bartın, Amasra is a historic coastal residential area (Figure 1).

The town of Amasra is influenced by the Black Sea climate. The average annual temperature is 13.8 °C, annual precipitation 1035.22 mm and annual mean relative humidity 69.8 %. Northeaster blows from 15 October to 15 March and the town is also influenced by northerly, southwesterly and northwesterly winds (Anonymous, 2001).

In terms of flora, the research area is located in the sub-category of Eux of the Euro-Siberian region and in the A4 square according to the Davis grid system (Davis, 1988). Yatgın (1996) identified in a study on the flora of the region of Amasra 265 plant taxa belonging to 68 families. The dominant and characteristic species of the region are as follows: *Carpinus betulus* L. (Hornbeam), *Castanea sativa* Mili. (Anatolian chestnut), *Fagus orientalis* L. (Oriental Beech), *Ostrya carpinifolia* Scop. (European hophornbeam), *Tilia argentea* Desf. (Silver Lime). Also common in the Mediterranean maquis shrubland, the following shrubs constitute a major part of the coastal flora: *Arbutus unedo* L. (Strawberry Tree), *Cistus creticus* L. (Pink Rockrose), *Cistus salvifolius* L. (White Rockrose), *Erica arborea* L. (Tree Health), *Juniperus oxycedrus* (Prickly Juniper), *Laurus nobilis* L. (Bay Laurel), *Myrtus communis* L. (Myrtle), *Phillyrea latifolia* L. (Phillyrea), *Rosa canina* L. (Dog rose) and *Spartium junceum* L. (Weaver's Broom). There are also shrub species that occur naturally in forests such as *Cornus mas* L. (European Cornel), *Cornus sanguinea* L. (Common Dogwood), *Cotinus coggygria* Scop. (Eurasian Smoketree), *Crataegus monogyna* Jacg. subsp. *monogyna* (Common Hawthorn), *Corylus avellana* L. (Common Hazel), *Laurocerasus officinalis* Roem. (Cherry Laurel), *Ligustrum vulgare* L. (Common privet) and *Rhododendron ponticum* L. (Common Rhododendron).

Amasra, originally called Sesamos, has a history of 3000 years. Amasra is an important historical town that has been able to preserve the architectural properties of the historical periods and the civilizations it hosted. It is as well an important harbor town that preserves the natural and cultural landscape of many civilizations (Sakaoğlu, 1999; Sertkaya, 2001; Yazgan et al. 2005; Anonymous, 2005; Anonymous, 2007).

According to the population census of 2010, the total population of Amasra is 16.122, with the town population of 6.450 and village population of 9.784 (Anonymous, 2012a). During summer months, the population of the town is almost 40.000 due to tourists. Varying between summer and winter months due to tourism activities, population density of Amasra is 44 people per square kilometer. Population density of the central and coastal villages is more intense than the rest of the villages (Anonymous, 2012b).

The economy of Amasra is mostly based on coal production and tourism, whereas the economy of the inland villages on agriculture and animal husbandry. The coastal ones, in addition to agriculture and animal husbandry, are engaged in tourism (Aşçıoğlu, 2001).

The growth of tourism in Amasra started in the period 1940-1960. In this period, the natural and cultural values of Amasra and its surroundings were discovered by local and foreign visitors. The tourism development period of 1960-1965 was marked with the increase in the number of visitors, and thus, tourism was considered a new source of income. It is also in this period that irregular urban development took place that negatively affected the development of tourism. After 1965, Amasra faced a decline in tourism activities (Sakaoğlu, 1999). Nowadays, Amasra is one of the important regional (the Western Black Sea coastal area) and national point of attraction with its tourism potential.

2.2 Method

This study was conducted with a strengths, weaknesses, opportunities, threats (SWOT) analysis, questionnaires, field work and the review of the relevant literature. The data obtained were analyzed to give some suggestions.

2.2.1 SWOT analysis

The SWOT analysis was conducted to determine the tourism planning and management strategies regarding the natural and cultural landscape values (Kelkit et al. 2005; Ozturk Kurtaslan and Demirel, 2011; Kiper et. al. 2011). The data used in the SWOT analysis were obtained through interviews with the mayor, district governor, employees of public institutions and inhabitants, previous studies conducted in the region or in similar areas, the relevant literature and professional experience, field surveys and in site observations.

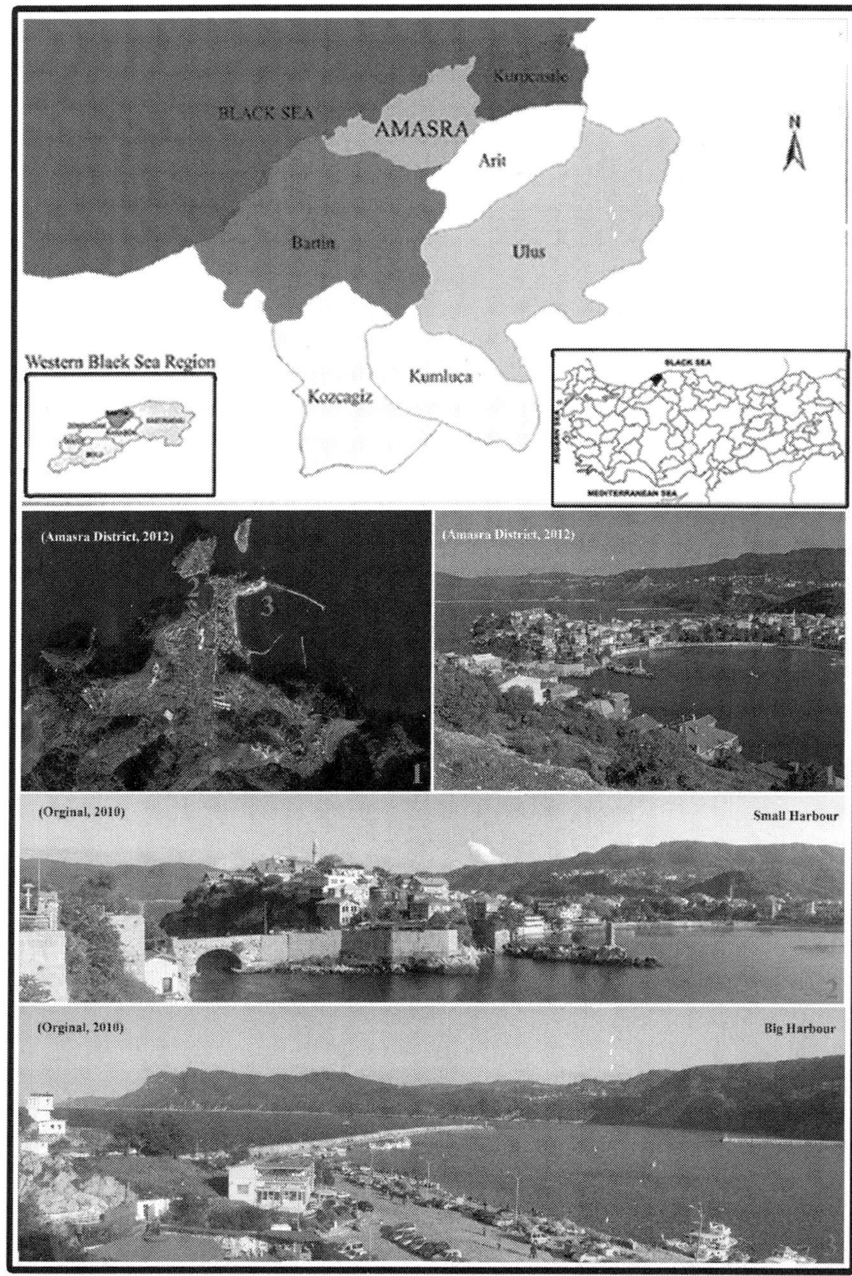

Fig. 1. Geographical location and genereal views of Amasra.

2.2.2 Questionnaire design and implementation

The studies of Kapuci (2004), Pereira et al. (2005), Yaslioglu (2007), and Cengiz et al. (2012) formed the basis for the questionnaire forms.

According to Daniel and Terrell (1995) the following formula, applied for finite populations, was used to determine the sample size (Atıcı, 2012).

$$n \geq \frac{Z^2 \text{xNxpxq}}{\text{Nx } D^2 + Z^2 \text{xpxq}}$$

- **n** = Sample size,
- **Z** = Confidence coefficient (z = 1.96 for 95% confidence level),
- **N** = Population size,
- **p** = Proportion of the sample in the population (as the proportion of the sample in the population was unknown initially and due to the lack of a preliminary investigation, values p and q were taken equal to each other and 50%)
- **q** = 1-p = 0.5,
- **D** = Sampling error (% 10)

$$n \geq \frac{(1.96)^2 \text{x6450x}(0.5)\text{x}(0.5)}{6450\text{x}(0.10)^2 + (1.96)^2 \text{x}(0.5)\text{x}(0.5)}$$

$$n \geq \frac{6194.58}{65.4604}$$

$$n \geq 95$$

According to this formula, the calculated value of "n" was 95. The questionnaire was presented by direct interviews to the local residents on weekdays and weekends within the period January-February 2011. The survey was conducted with 100 randomly selected people.

The questionnaire covered four issues: (i) personal information, (ii) local residents' attitudes toward tourism in Amasra, (iii) local residents' perspectives on the natural and cultural values of Amasra, (iv) local residents' expectations of and suggestions for tourism planning and management strategies. The first set of questions was designed to obtain information on demographic variables like gender, age, residence, educational level and occupation and was composed of nine questions. The second set of questions was composed of 12 questions aimed at determining the local residents' attitudes toward tourism, whereas the third set was composed of 5 questions on local residents' perspectives on the natural and cultural values of Amasra. The final set was composed of 5 questions about the local residents' expectations of and suggestions for the future tourism planning and management strategies. Total number of questions was 31.

The data obtained were analyzed by means of Microsoft Excel and the Statistical Package for Social Sciences (SPSS) Version 16. Contingency tables were constructed and multiple choice questions were analyzed using Chi-square test to measure independence between groups.

3. Results and discussions

3.1 SWOT analysis of the existing natural and cultural values of Amasra

In order to reveal the current status of the area a SWOT analysis was conducted. The results of the SWOT analysis are as follows (Table 1).

Strengths	Weaknesses
• Historical and archaeological sites/ heritage sites • Hills (panoramic views) • Natural protected areas • Natural beaches • Biodiversity • Climate advantages • Visual advantages (opposite shore due to the peninsula) • Gastronomic activities • Fisheries • Coastal features (steep cliffs, caves, interesting coastal formation) • Traditional handicrafts • Folklore • Ports/Harbours (big and small ports) • Tourism activities dating back • 3000-year past • Ancient port town • Established guesthouses	• Topographic obstacles • Traffic and parking problems • Use pressure on carrying capacity in the summer • Lack of promotion • Misuse of protected areas • Factors that cause changes in the shore line (fill practice, etc.) • Degradation in the historic urban pattern • Aesthetic problems in urban and natural environment (visual pollution, etc.) • Lack of infrastructure • Activities of Turkish Hard Coal Authority • Earthquake and erosion risk • Lack of inventory on natural and cultural assets • Lack of tourism education • Insufficient accommodation standards • Financial limitations faced by local administrations • Lack of expertise in local administrations about planning, design and management of natural and cultural areas • Lack of approaches to protection in local administrations • Lack of interest by NGOs and local residents • Irregular urban development
Opportunities	Threats
• Proximity to major cities (Ankara and Istanbul) • Local residents' openness to innovation • Sea tourism • Fisheries open to development • Proximity to Kastamonu-Bartın Küre Mountains National Park, one of the most important natural protection area of Europe for its natural, cultural and historical values • Proximity to the Bartın River, the only waterway available for transport in Turkey • Proximity to the museum city of Safranbolu, which is in the World Heritage List	• Seawater pollution • Anthropogenic pressures on forests • Degradation of archaeological areas • Coastal fill areas • Overcrowding especially in summer months (exceeding carrying capacity) • Increase in building density and height • Urban development in agricultural areas (pressures on agricultural areas) • Cultural degradation • Changes in laws and legislations

• Employment opportunities in tourism for local residents • Active Tourism Vocational School • Accessibility through land, sea and air travel • Being on the current tour routes

Table 1. SWOT analysis of Amasra.

3.2. Survey results

3.2.1 Socio-demographic structure

Table 2 demonstrates the socio-cultural information of the survey participants.

		Percent (%)
Gender	Female	49
	Male	51
Age	15-20	1
	21-30	35
	31-40	25
	41-50	26
	51-60	9
	60+	4
Education	Illiterate	1
	Elementary school	3
	Secondary school	6
	High school	30
	University	60
Occupation	Worker	6
	Civil servant	48
	Self-employed	19
	Retired	7
	Student	5
	Housewife	4
	Unemployed	3
	Other	8
Income (TL)	Less than 500 TL	3
	500 – 750	10
	750 - 1000	7
	1000 - 1500	29
	1500 - 2500	41
	More than 2500	10

Table 2. Socio-cultural information of the survey participants.

48% of the participants are from Amasra. 54% of the parricipants have been living in Amasra for more than 20 years. In terms of source of income, 37% of the participants are engaged in tourism, and 51% in other sectors (Table 3).

		Percent (%)
Being from Amasra	Yes	48
	No	52
For how long they live in Amasra	1 year	7
	1-3 year	12
	3-5 year	6
	5-10 year	10
	10-15 year	6
	15-20 year	5
	20+ year	54
Source of income of the family	No response	2
	Agriculture	1
	Animal husbandry	1
	Commerce	7
	Tourism	37
	Handicrafts	1
	Other	51

Table 3. Participant demographics.

3.2.2 Local residents' attitudes toward tourism

Among the participants whose source of income is tourism, 15% are engaged in running guesthouses, 9% in handicrafts (wood engraving, filigree, etc.), and 8% are restaurateurs (Table 4).

		Percent (%)
Occupation of local residents whose source of income is tourism	No response	62
	Guesthouse management	15
	Handicrafts (wood engraving, filigree, etc.)	9
	Restaurateurs	8
	Hotel management	2
	Other	4

Table 4. Occupation of local residents whose source of income is tourism.

Amongst the primary problems faced by the participants whose source of income is tourism, the first was lack of training in tourism (13%), followed by lack of standards in the tourism sector and lack of awareness among the public about tourism (both 6%) (Figure 2).

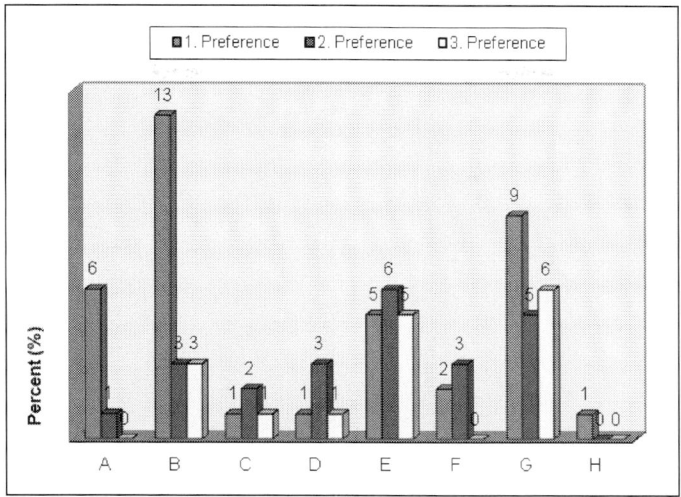

Fig. 2. Primary problems faced by participants with tourism as source of income
(A. Financial difficulties; B. Lack of education in tourism; C. Limitations due to Amasra's
status as a protection site; D. Lack of attention from local administrations; E. Lack of
standards in the tourism sector; F. Lack of branding; G. Lack of local residents' awareness
regarding tourism; H. Other).

The participants indicated that the busiest months of tourism activities in Amasra are July
(74%), August (61%), and June (43%).

According to the participants, the tourist demand for Amasra is due to its coastal location
(31%), the historical and cultural values (30%) and natural beauty of the town and its
surroundings (20%). In addition, the analysis of the ranking of the first, second and third
preferences belonging to the same option revealed that the choice "seafood restaurants" is
of prominence with values close to each other in all the three preferences of the participants
(Figure 3).

86% of the participants stated that the tourism potential of Amasra is not sufficiently
utilized, in constrast with 13% of the participants who thought it was sufficiently used. 1%
of the participants did not answer this question.

In terms of peripheral services developing thanks to tourism, 63% of the participants ranked
guesthouses as the first choice, 28% handicrafts and 19% seafood restaurants (Figure 4).

54% of the participants stated that the impact of tourism in Amasra are positive, while 7%
considered the impact negative and 39% both positive and negative. Among the participants
who indicated that the impact is positive, 44% thought that it is effective in the promotion of
the town. According to the participants, the negative impact of tourism is due to, first, the
degradation of traditional houses because of unplanned development; second, the number
of visitors exceeding the carrying capacity, and third, degradation of natural and cultural
areas, and visual pollution (Table 5).

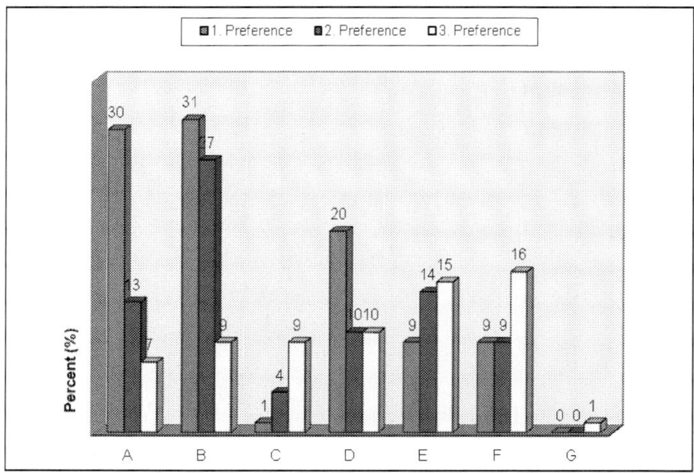

Fig. 3. The aspects of Amasra attractive for tourists (A. Historical and cultural values; B. Coastal location; C. Handicrafts (wood engraving, filigree, etc.); D. Natural beauty of Amasra and its surroundings; E. Seafood restaurants, F. Proximity to Ankara and Safranbolu; G. Being on the routes of tourism agencies).

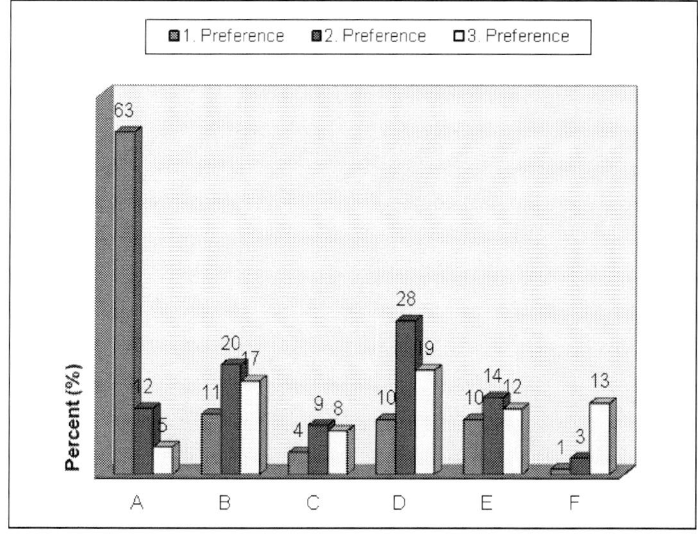

Fig. 4. Peripheral services developing thanks to tourism in Amasra (A. Guesthouses; B. Handicrafts (wood engraving, filigree, etc.); C. Sales of regional organic products; D. Seafood restaurants; E. Hotel-hostel management; F. Tourist ships).

Furthermore, 83% of the participants stated that there are negative environmental factors that could hinder the development of tourism activities in Amasra.

		I. Preference	II. Preference	III. Preference
		Percent (%)		
Positive impact of tourism in Amasra	No response	7	43	67
	Creating comfortable environments with landscape design	7	4	4
	Promotion of Amasra	44	15	5
	Protection of cultural heritage	2	16	9
	Economic contribution to Amasra	40	18	9
	Ensuring the continuity of handicrafts	-	4	6
	Other	-	-	-
Negative impact of tourism in Amasra	No response	52	69	73
	Change in local culture and traditional social structure	8	1	1
	Degradation of traditional houses due to unplanned development	15	5	4
	Degradation of natural and cultural areas due to the number of visitors exceeding the carrying capacity	4	9	1
	Degradation of the peninsula landscape and vegetation	3	4	2
	Air pollution	-	1	-
	Coast and sea pollution	8	8	4
	Soil pollution	-	-	2
	Visual pollution	4	3	8
	Noise pollution	2	-	5
	Other	4	-	-

Table 5. Local residents' perspective on the impact of tourism in Amasra.

The negative environmental factors that hinder the development of tourism activities were listed as some buildings with historical features in a state of neglect (32%), parking problems (23%), and unplanned and dense urbanization (13%) (Figure 5).

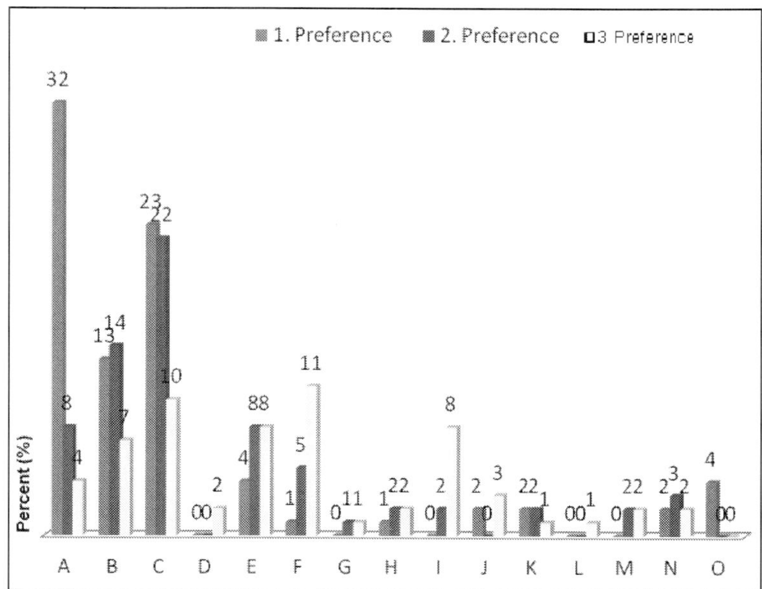

Fig. 5. Negative environmental factors that hinder the development of tourism activities in Amasra. (A. Some buildings with historical features in a state of neglect; B. Unplanned and dense urbanization; C. Parking problems; D. Sewerage; E. Visual pollution; F. Coastal and marine pollution; G. Air pollution; H. Overcrowding and noise; I. Degradation of the nature due to summer houses; J. Degradation of the nature due to road construction works; K. Difficulties in transport; L. Soil pollution; M. Problems due to physical planning and practice; N. Environmental problems caused by Turkish Hard Coal Authority; O. Other).

According to the participants, the forms of tourism that could be developed in Amasta are cultural tourism (61%), sea tourism (17%), and ecotourism (12%) (Table 6).

		I. Preference	II. Preference	III. Preference
		Percent (%)		
	No response	3	29	56
Forms of tourism that could be developed in Amasra	Cultural tourism	61	7	-
	Culinary tourism	4	4	2
	Rural tourism	3	5	5
	Sea tourism	17	23	9
	Nature tourism	12	31	27
	Other	-	1	1

Table 6. Local residents' perspectives on the forms of tourism that could be developed in Amasra.

3.2.3 Local residents' perspectives on the natural and cultural values in Amasra

The majority of the participants stated that urban development is incompatible with the natural and historical pattern of the city and that city walls, historic buildings and artifacts discovered in archaeological excavations are significant contributions to the historical and cultural values of the city. In addition, the state of neglect of the historical buildings of the city was as a prominent opinion. 90% of the participants stated that the natural and cultural values should be protected for sustainable tourism in Amasra (Table 7).

		Percent (%)
Whether new urban development is compatible with the natural and historical pattern of Amasra	No response	1
	Yes	20
	No	79
Perspectives on the city walls, historical buildings and historical artifacts discovered by the archeological excavations in Amasra	No response	2
	They limit land use and building height	12
	They are obstacles to urban development in Amasra	12
	They are important historical and cultural values contributing to Amasra	73
	Other	1
Whether it is necessary to protect the natural and cultural values of Amasra for sustainable tourism	Yes	90
	No	10
Whether the historical buildings and spaces are in a state of neglect in Amasra	Yes	7
	No	93

Table 7. Local residents' perspectives on the natural and cultural values of Amasra.

The participants who indicated that the historical buildings and spaces in Amasra are in a state of neglect stated the reason as lack of local residents' awareness regarding protection (43%), lack of interest by the authorities (23%), and insufficient financial resources (20%) (Table 8).

		I. Preference	II. Preference	III. Preference
		Percent (%)		
Factors for the neglect of historical buildings and spaces in Amasra for the participants who think that they are in a state of neglect	No response	7	41	68
	Lack of local residents' awareness regarding protection	39	21	5
	Lack of interest by the authorities	43	23	7
	Insufficient financial resources	11	13	20
	Other	-	2	-

Table 8. Local residents' perspectives on the cultural values of in Amasra.

3.2.4 Local residents' expectations of and suggestions for new tourism planning and management strategies in Amasra

According to the participants, the activities that could take place in Amasra and its surroundings are visits to the historical and cultural buildings and spaces (54%), hiking (17%) and gastronomical activities (seafood restaurants) (17%). In terms of issues that should be prioritized in urban planning for tourism development in Amasra, 41% of the participants stated that building maintenance and repair should be done, 29% of the participants stated that parking problems should be overcome and 10% indicated the need for maintenance and repair of the city walls. The three most common suggestions in order to extend the duration of stay of daily visitors are as follows: Package programs should be prepared for accommodation facilities (e.g. boat tours, trekking, etc.), the relationship between Kastamonu-Bartın Küre Mountains National Park and Amasra should be established, and trekking routes should be created connection the historical center of Amasra and the natural and cultural assets of its surrounding area. Among the participants' opinions regarding the future of Amasra, the first preference was protection of fishing village identity, the second was protection of the port town identity and the third was qualitative and quantitative improvement in public spaces and green areas (beaches, parks, playgrounds, sports fields, etc.) (Table 9).

According to the participants, the responsibility for the planned development of tourism in Amasra is of local adminstrations (80%), local residents (46%), and civil society organizations (38%).

The results of the Chi-square test indicated that whether new urban development is compatible with the natural and historical pattern of Amasra varies according to the level of education. 33% of the participants whose level of education is primary school stated that new urban development is compatible with the historical and natural structure of the city, while 85% of the participants who have a university degree found it incompatible (Table 10a). It was observed that as the participants' level of education increases, the new development is considered incompatible with the natural and historical pattern in Amasra.

		I. Preference	II. Preference	III. Preference
		Percent (%)		
Activities in Amasra and its surroundings	No response	1	21	31
	Visits to historical and cultural buildings and spaces	54	3	4
	Water sports	9	13	5
	Hiking	11	17	7
	Picnic	1	5	2
	Nature observation	4	7	4
	Gastronomical activities (seafood restaurants)	7	12	17
	Purchase of local food	3	9	4
	Visits to local handicraft workshops / purchase of products	3	9	11
	Boat trips	4	4	15
	Other	3	-	-
Issues of priority in inner city urban planning for tourism development in Amasra	No response	1	20	27
	Maintenance and repair in historical buildings	41	6	7
	Maintenance and repair in the city walls and its environs	10	17	7
	Overcoming parking problem	29	20	18
	Raising standards of accommodation facilities	3	16	14
	Eliminating the factors causing visual pollution in the city	3	9	9
	Natural and historical patterns should be compatible	9	6	7
	Planned development of summer houses should be ensured	2	3	2
	Improvement in public spaces and green areas	-	3	9
	Other	2	-	-
Measures to be taken to increase the duration of stay of daily visitors in Amasra	No response	1	34	45
	Establishing relationship between Amasra and Kastamonu-Bartın Küre Mountains National Park	20	5	9
	Establishing routes to nearby bays by boat tours	8	11	10
	Establishing organic agriculture areas in the environs; visits to these areas and accommodation facilities	6	5	4

	Increasing the attractiveness of river tourism by establishing the relationship between the Bartın River and Amasra	7	9	8
	Preparing package programs in accommodation facilities (e.g. boat trips, trekking, etc.)	33	17	9
	Creating trekking routes between the historical city center and the close-by areas with natural and cultural value	20	19	12
	Other	5	-	3
Opinions on the future of Amasra	No response	1	39	52
	Protection of fishing village identity	30	10	1
	Revitalizing tourism and commerce	28	12	2
	Protection of historical port town identity	19	16	11
	Increase in the hard coal activities	-	3	4
	Protection of archeological and historical assets	11	12	14
	Increase in the number of summer houses	2	2	-
	Qualitative and quantitative improvement in public spaces and green areas (beaches, parks, playgrounds, sports fields, etc.)	8	5	16
	Other	1	1	-

Table 9. Local residents' expectations regarding the new tourism planning and management strategies in Amasra.

In terms of the relationship between the participants' education level and opinions about the city walls, historic buildings and archaeological artifacts discovered in excavations, 33% of the participants with primary school degree considered the historical structure of the city as a factor restricting land use and building height, while 33% of the participants with the same level of education stated that it is a barrier to urban development in Amasra. 80% of the participants with high school degree and 73% with university degree stated that the historical structure of the city is a significant historical and cultural value that contributes to Amasra (Table 10b). Thus, it was found out that the participants' awareness regarding the historical and cultural values increases as education level increases.

In terms of the relationship between the educational background of the participants and opinions on the future of Amasra, it was observed that the participants with primary school degree supported that the fishing village identity of the city should be protected (67%), followed by revitalization of tourism and commerce in the city (33%). While the participants with high school degree (20%) and university degree (20%) stated that the historic port town identity of Amasra should be protected in the future, the participants who graduated from high school (20%) stated the need for qualitative and quantitative improvement in public spaces and green areas. In addition, the participants with university degree indicated that archaeological and historical characteristics of the city should be protected (18%). It was observed that as the participants' educational level increases, the opinions favoring the archeological and historical assets of the city and its identity as a historical port town are more prominent among opinions regarding the future of Amasra. As the participants did not express a significant opinion about the number of summer houses and the increase in the hard coal activities, it could be furthered that these two factors have a negative impact on the city (Table 10c).

The problems faced by the participants with tourism as source of income are lack of training in the field of tourism (35%), lack of local residents' awareness about tourism (22%), lack of standards in the tourism sector in Amasra (14%), and insufficient financial resources (14%) (Table 11).

		Educational background			
Table 10.a. Whether new urban development is compatible with the natural and historical pattern of Amasra		Primary school	Secondary school	High school	University
No response	n	0	-	-	-
	%	-	-	-	-
Yes	n	1	1	9	9
	%	33	17	30	15
No	n	2	5	21	51
	%	67	83	70	85

$\chi2= 103.193^{***}$ p<0.001

Table 10.b. Local residents' perspectives on the city walls, historical buildings and historical artifacts discovered by archeological excavations in Amasra					
No response	n	-	-	-	1
	%	-	-	-	2
They limit land use and building height	n	1	1	4	6
	%	33	17	13	10
They are obstacles to urban development in Amasra	n	1	1	1	9
	%	33	17	3	15
They are important historical and cultural values contributing to Amasra	n	1	4	24	44
	%	33	67	80	73
Other	n	-	-	1	-
	%	-	-	3	-

$\chi2= 58.084^{***}$ p<0.001

Table 10.c. Opinions on the future of Amasra					
No response	n	-	-	-	-
	%	-	-	-	-
Protection of the fishing village identity	n	2	4	10	14
	%	67	67	33	23
Revitalization of tourism and commerce	n	1	-	8	19
	%	33	-	27	32
Protection of the historical port town identity	n	-	1	6	12
	%	-	17	20	20
Increase in the hard coal facilities	n	-	-	-	-
	%	-	-	-	-
Protection of archeological and historical assets	n	-	-	-	11
	%	-	-	-	18
Increase in the number of summer houses	n	-	-	-	2
	%	-	-	-	3
Improvement in public spaces and green areas both qualitatively and quantitatively	n	-	1	6	1
	%	-	17	20	2
Other	n	-	-	-	1
	%	-	-	-	2
X2= 126.081*** p<0.001					

Table 10 a.b.c. Comparison of local residents' perspectives in terms of educational background.

	Primary problems in tourism																
Source of income	No response		Financial difficulties		Lack of education in tourism		Limitations due to Amasra's status as a protection site		Lack of attention from local administration		Lack of standards in tourism sector		Lack of branding		Lack of local residents' awareness regarding tourism		Other
	n	%	n	%	n	%	n	%	n	%	n	%	n	%	n	%	n	%
Tourism	1	3	5	14	13	35	1	3	1	3	5	14	2	5	8	22	1	3

χ2= 88.519*** p<0.001

Table 11. Primary problems indicated by participants with tourism as source of income.

Positive impact of tourism in Amasra varies according to age groups of the participants. The age group 41-60+ stated that tourism contributes to the economy of Amasra, while the age group 21-40 stated tourism contributes to the promotion of tourism in Amasra. The participants showed no awareness with regards to the contributions of tourism to ensuring the continuity of the local handicrafts (Table 12).

Age	Positive effects of tourism in Amasra													
	No response		Creating comfortable environments through landscape design		Promotion of Amasra		Protection of cultural heritage		Contribution to the economy of Amasra		Ensuring the continuity of handicrafts		Other	
	n	%	n	%	n	%	n	%	n	%	n	%	n	%
15-20	1		0		0		0		0		0		0	
21-30	2	6	4	11	18	51	1	3	10	29	0	-	0	-
31-40	4	16	2	8	12	48	0	-	7	28	0	-	0	-
41-50	0	-	1	3	12	46	0	-	13	50	0	-	0	-
51-60	0	-	0	-	2	22	0	-	7	78	0	-	0	-
60 +	0	-	0	-	0	-	1	25	3	75	0	-	0	-

$\chi 2 = 43.612^{***}$ p<0.001

Table 12. Perspectives on the positive effects of tourism in Amasra in terms of age groups.

4. Discussion and conclusions

Together with the rapid economic, political, technological developments and changes in the world, significant changes occur also in tourism. Different forms of tourism as an alternative to mass tourism in Turkey started to develop in the 1990s. These forms of tourism are considered important tools for ensuring sustainable environments as they bear rural and cultural tourism characteristics could be developed in sensitive natural and cultural areas. In this respect, Amasra, an ancient port city located on a peninsula, has rural landscape features. Creating a significant demand for tourism with its natural, cultural and historical values, Amasra was investigated in this study in terms of sustainable tourism planning and management which prioritizes the integrity of ecology and the traditional structure and the participation of local residents. As such, the survey study conducted to determine the local residents' participation in tourism activities, support, expectations and attitudes toward and perceptions of natural and cultural values in terms of sustainable tourism planning and management constituted an important aspect of the present study.

4.1 Tourism development

The results of the survey indicated that the participants were not aware of some of the factors that could hinder tourism activities such as problems related to summerhouses, and environmental problems caused by road construction works and by Turkish Hard Coal Authority.

The analyses demonstrated that the historic coastal town of Amasra has a strong tourism potential. The income of the local residents who are engaged in tourism activities in Amasra is based on running pensions (guesthouses) and seafood restaurants, and handicrafts. The dissemination of pensions and increasing the quality of services is important in terms of tourism development. The other source of income, handicrafts, is based on traditional methods and know-how. The unique aspects of traditional handicrafts should be protected and promoted to international markets through branding. In this respect, efforts should be put in raising the awareness of local residents about tourism and in educational activities.

The demand for Amasra is based mostly on its coastal location, the natural, cultural, historical and gastronomic (especially seafood restaurants) aspects. The continuity of the contributions of these aspects to tourism should be ensured with considering the balance between conservation and use. In addition, the production of local organic products, which are widespread in Amasra, should be promoted and awareness of local residents should be raised. In this respect, tourist tours should be organized to such areas of production.

This study investigated the positive and negative effects of tourism development in Amasra. On the one hand, tourism contributes to the promotion of the city, its economy and to the protection of cultural heritage. On the other hand, it has negative effects on the natural and traditional characteristics due to unplanned land development.

According to the survey results, there is a high tourist demand for Amasra during the summer months (June, July, August) due to the its location on the coast. It is necessary to diversify tourism activities to attract tourists to the city throughout the year and extend the duration of stay. In this respect, particularly cultural and nature tourism are among the types of tourism to be developed in Amasra.

4.2 Natural and cultural environment conservation

Local residents of Amasra are aware of the natural, historic and cultural values of the town and believe that these values should be protected for sustainable tourism. Moreover, they are aware that the new land development is incompatible with the traditional urban pattern.

4.3 The expectations and suggestions of the local residents regarding tourism planning and management strategies in Amasra

First of all, the negative environmental factors (see Figure 6) that could hinder the development of tourism activities in Amasra should be overcome for sustainable tourism planning and management. An effective tourism planning and management in Amasra should be sustainable for the natural and cultural environment, economically efficient and supported by the local community. In addition, public awareness-training seminars on tourism could be organized.

In sustainable tourism planning and management, tourism development should be planned according to the carrying capacity of the city for protecting and maintaining the natural, cultural and historical values and for tourism to be sustainable. In this respect, both natural and social aspects need to be addressed as a whole, and planning should be carried out in accordance with national and international regulations.

According to the United Nations Environment Programme (2009), coordination among destinations within a region usually improves planning. Considering the ties among regional assets and attractions during the planning phase would create opportunities for linking all the regional attractions of a destination. In addition to its own dynamics, Amasra should establish relationships with its hinterland for diversifying tourism activities and making them year-long. In this respect, the natural beaches in the 59-km coastal area of Bartın, one of Europe's major ecosystems, Kastamonu-Bartın Küre Mountains National Park, the Bartın River, the only natural waterway that allows for transport in Turkey, and Safranbolu, which is in the UNESCO World Heritage List, should be integrated into the regional tourism plan as greenway planning.

As a result, tourism planning and management strategies considering the balance between protection and use of natural and cultural values and based on public participation would contribute to sustainable regional development.

5. References

Anonymous, (2001). Bartın-Amasra Meteoroloji İstasyonu 1991-2000 İklim Verileri, Meteoroloji İl Müdürlüğü, Bartın (in Turkish).

Anonymous, (2005). Environmental Status Report Of Bartın-2004. Bartın Governorship Provincial Directorate of Environment Bartın Press, Bartın (in Turkish).

Anonymous, (2007) Mitolojiden Gezginlere Bartın Kültür ve Turizm Envanteri. Bartın İl Kültür ve Turizm Müdürlüğü Yayını. Bartın (in Turkish).

Anonymous, (2012a). Turkey Statistical Institute Population Data for 2008. http://report.tuik.gov.tr/reports.

Anonymous, (2012b). (http://www.bartin.gov.tr/modules.php?name=Tutoriaux&rop=navig&did=39&eid=70).

Aşçıoğlu, E. (2001). Bartın. Bartın Ticaret ve Sanayi Odası Yayını, Bartın. (in Turkish).

Atıcı, E. (2012). Statistically Forecasting Of Some Market Variables Of Florists on Usage Of Wild Plant Species in Istanbul, Turkey. *African Journal of Agricultural Research,* 7(8), 1245-1252, 26 February, 2012.

Cengiz, B., Sabaz M., Bekci, B., Cengiz, C. (2012). Design Strategies for Revitalization on The Filyos Coastline, Zonguldak, Turkey. *Fresenius Environmental Bulletin,* 21(1), 36-47.

Daniel, W.W., Terrell, J.C. (1995). Business Statistics for Management and Economics, Seventh Edition, Houghton Mifflin Company, USA. ISBN: 0-395-71671-3. 972 pages.

Davis, P.H., Mill, R.R., Tan, K. (1988). Flora Of Turkey and The East Aegean Islands. Edinburg Universty Press, Vol. 10. Edinburg, UK.

Dymond, S.J. (1997). Indicators Of Sustainable Tourism in New Zealand: A Local Government Perspective. *Journal of Sustainable Tourism,* 5(4), 279–293.

Ernoul, L. (2009). Residents' Perception Of Tourist Development and The Environment: A Study from Morocco. *International Journal of Sustainable Development & World Ecology,* 16(4), 228-233.

Harrill, R., Potts, T.D. (2003). Tourism Planning in Historic Districts: Attitudes Toward Tourism Development in Charleston. *Journal of the American Planning Association,* 69(3), 233-244.

Israeli, A., Uriely, N., Reichel, A. (2002). Attitudes Of Local Residents Vs. Residents Of Surrounding Areas Towards Tourism Development. *Anatolia,* 13(2), 145-157.

Jackson, L.A. (2008). Residents' Perception Of The Impacts Of Special Event Tourism. *Journal of Place Management and Development,* 1(3), 240–255.

Kapuci, C. (2004) The Effect Of Historical City Structure On The Tourism Potential Case Study: İznik City. Zonguldak Karaelmas University, Graduate School of Natural and Applied Sciences, Department of Landscape Architecture, M.Sc. thesis, Bartın, Turkey. (in Turkish).

Kelkit, A., Özel, A.E., Demirel, Ö. (2005). A study Of The Kazdagi (Mt. Ida) National Park: An Ecological Approach To The Management Of Tourism. *International Journal of Sustainable Development & World Ecology,* 12(2), 141-148.

Kelkit, A., Celik, S., Esbah, H. (2010). Ecotourism Potential Of Gallipoli Peninsula Historical National Park. *Journal of Coastal Research,* 26(3), 562–568.

Kiper, T., Korkut, A., Yılmaz, E. (2011). Determination Of Rural Tourism Strategies By Rapid Rural Asessment Technique: The Case Of Tekirdağ Province, Şarköy County. *Journal of Food, Agriculture & Environment,* 9(1), 491-496.

Lankford, S.V., Howard, D.R. (1994). Developing A Tourism Impacts Attitude Scale. *Annals of Tourism Research,* 21(1), 121-139.

Lee, T.J., Li, J., Kim, H. (2007). Community residents' perceptions and attitudes towards heritage tourism in a historic city. *Tourism and Hospitality Planning & Development,* 4(2), 91-109.

Mason, P., Cheyne, J. (2000). Residents' Attitudes To Proposed Tourism Development. *Annals of Tourism Research,* 27(2), 391–411.

Oztürk Kurtaslan, B., Demirel, Ö. (2011). Pollution Caused By Peoples' Use For Socio-Economic Purposes (Agricultural, Recreation and Tourism) In The Gölcük Plain Settlement at Bozdağ Plateau (Ödemiş-İzmir/Turkey): A Case Study. *Environmental Monitoring and Assessesment,* 175, 419–430.

Pereira, R., Soares, A.M.V.M., Ribeiro, R., Gonçalves, F. (2005). Public Attitudes Towards The Restoration and Management Of Lake Vela (Central Portugal). *Fresenius Environmental Bulletin,* 14(4), 273-281.

Puczkó, L., Rátz, T. (2000). Tourist and Resident Perceptions Of The Physical Impacts Of Tourism at Lake Balaton, Hungary: Issues For Sustainable Tourism Management. *Journal of Sustainable Tourism,* 8(6), 458-478.

Sakaoğlu, N. (1999). Çeşm-i Cihan / Amasra. Bartın (in Turkish).

Sertkaya, Ş. (2001). A Study On The Determination Of Tourism and Recreation Potential Of The Bartın Coastal Zone. Ph.D. thesis. Ankara University, Graduate School of Natural and Applied Sciences, Department of Landscape Architecture, Ankara (in Turkish).

Sethna, R.J. (1980). Social Impact Of Tourism In Selected Caribbean Countries, In D. E. Hawkins, E. L. Shafer, and J. M. Rovelstad (eds.) Tourism Planning and Development Issues, pp. 239–249 (Washington D.C.: George Washington University).

Soutar, G.N., McLeod, P.B. (1993). Residents' Perceptions On The Impact Of The America's Cup. *Annals of Tourism Research,* 20(3), 571–582.

United Nations Environment Programme, (2009). A Three-Year Journey For Sustainable Tourism. Final Report. [accesses 2012 January]. Available from: http://www.unep.fr/scp/tourism/activities/taskforce/pdf/TF%20REPORT_FINAL.pdf

United Nations World Tourism Organization, (2012). Tourism – An Economic and Social Phenomenon; [accesses 2012 January]. Available from: http://unwto.org/en/content/why-tourism

Welford, R., Ytterhus, B. (2004). Sustainable Development and Tourism Destination Management: A Case Study Of The Lillehammer region, Norway. *International Journal of Sustainable Development & World Ecology,* 11(4), 410-422.

Yaslioglu, E., Akaya Aslan, S.T., Kirmikil, M., Gundogdu, K.S., Arici, I., Arici, C.F. (2007). Environmental Protection-Based Village Development: The Case Of Eskikaraagac. *Fresenius Environmental Bulletin,* 16(8), 940-947.

Yatgın, H. (1996). Floristic Compositions Of Amasra. Ms.C thesis. Zonguldak Karaelmas University, Graduate School of Natural and Applied Sciences, Department of Landscape Architecture, Bartın (in Turkish).

Yazgan, M.E., Kapuci, C., Cengiz, B. (2005). Sustainability Of Tourism and Ecotourism in Coastal City Amasra. pp. 259-266. Proceedings of the Seventh International Conference on the Mediterranean Coastal Environment, MEDCOAST 05, E. Özhan (Editor), 25-29 October 2005, Kuşadası, Turkey.

Yazgan, M.E., Kapuci, C. (2007). İznik City with Its Historical Landscape Properties. Environment: Survival and Sustainability- ESS'07. February 19-24, 2007. Vol. 3, pp. 955-962, Cyprus.

9

Woody Plants in Landscape Planning and Landscape Design

Viera Paganová and Zuzana Jureková
Slovak University of Agriculture in Nitra
Slovak Republic

1. Introduction

Vegetation is considered to be a significant element of the landscape. The greenery elements design landscape scenery and have productive and eco-stabilizing functions in the landscape area. Woody plants have a specific position within the greenery elements. They have bigger dimensions in comparison to the other plants, their organs usually have large surfaces and their biomass fills large parts of the overhead space and the soil. Trees and shrubs have great influence on the environment and living conditions of the other organisms. Short-term changes in ecosystems do not significantly impact their lifecycle and survival. Woody plants are long-lived organisms with different adaptability to changes of environmental conditions. The knowledge of the morphological characteristics, as well as biological and ecological qualities of woody plants is very important for their efficient utilization (Paganova, 2006; Paganova et al., 2010; Sjöman, 2012; Valladares et al., 2007).

In urban conditions woody plants have several environmental, economic and social benefits. Trees reduce heat, wind speed and provide shading, increasing the energy efficiency of buildings (Sand, 1994; Simpson, 1998; McPherson & Simpson, 2003). Urban greenery increases the sociological value of the environment, improving the aesthetic and hygienic properties of the particular place. It also increases positive feelings and moods, enjoyment of everyday life and stronger feelings of connection between people and the environment (Dwyer et al., 1992, 2003; Westphal 2003).

However, environmental conditions in urban areas are significantly different to natural habitats. Plants in urban areas are exposed to many negative factors like: water deficit, soil compaction, pollutants, artificial lighting, overheating of the root zone and mechanical injuries. According to Nilsson et al. (2000), street trees are exposed to multiple stresses and their average lifespan is short. Park trees are exposed to moderate stress and compared to street trees their lifespan is relatively high. In urban woodlands the level of stress depends more on climate, soil conditions, recreational patterns and biotic damages rather than on anthropogenic causes (Sæbo et al., 2003).

Proper woody plant selection for specific conditions and establishment of the effective vegetation elements increases the functionality and stability of the landscape area (Paganova, 2004; Paganova, 2006). Like all living organisms, woody plants change in time and space, and knowledge of the nature and dynamics of these changes is essential for their successful utilization in landscape planning.

2. Principles of woody plant selection for landscape planning

The selection of proper plants for specific stand conditions is a very important task that affects success in the landscape planning and landscape design.

In Europe recent studies (Pauleit et al., 2002; Sæbo et al., 2003; Sæbo et al., 2005) documented poor diversity of tree genera and species planted in urban areas. A few genera of woody plants (*Acer, Aesculus, Platanus* and *Tilia*) are used as street trees. The number of species planted in parks, gardens and residential zones is high, however, an increase in the number of species used for urban forestry is essential. Higher diversity of species used in landscape planning and design would increase ornamental value and longevity of the plantings, and decrease costs of establishment and regular maintenance.

Spellenberg & Given (2008) reviewed the general criteria for selection of trees for urban environments. According to their worldwide knowledge, the most important criteria for selecting trees for urban environments are rather pragmatic: suitability of the taxa for local conditions, low maintenance and avoidance of structural problems. Criteria that contribute to landscape design appear to be next in importance. The authors documented a predominance of artificial mixes of woody plant species in urban areas, which possibly increases the detrimental effects of urbanization on nature and habitats.

However, there are many activities which aim at preservation and utilization of indigenous species in landscape planning and landscape design, including urban areas (Breuste, 2004; Dunnet & Hitchmough, 2004; Florgård, 2004). The new planning and design concept allows using parts of nature with specific beauty and amenity for urban spaces as well as using native species and their communities for urban plantings.

Utilization of indigenous species should support increased biodiversity in urban areas with ecologically better balanced plant communities.

3. Scientific tools of woody plant selection for landscape planning

Use of non-traditional woody plant species (species that are not commonly used) in landscape planning and landscape design is quite difficult. In the opinion of many professionals some stereotypes in dealing with urban trees remain with regard to the suitable and unsuitable species for urban environments. Such species are rare on plant markets because their propagation protocols have not yet been elaborated upon and the qualified sources of reproductive material are missing. There are also different opinions about the possibilities of establishment and utilization of tree species on particular stands.

The lack of information and data about growth rate, growth characteristics and maintenance management of non-traditional tree species is another reason why they are not included in landscape and urban plantings. The selection of woody plants established just on their ecological background is not appropriate for specific conditions with cumulative impact of various stresses. Also assessment of growth and physiological parameters directly on trees growing in the field or urban conditions does not give relevant results, because they are influenced by many barely quantified factors.

Fig. 1. Two individuals of *Sorbus domestica* on open stand with significantly different phenotypical traits of the crown architecture (shape, density and angle of branching) (Maceková, 2011).

From the scientific point of view mainly the ecological criteria of selection, the adaptability of trees to local conditions and the functional aspects of trees in the urban environment are studied and evaluated. The ecophysiological characteristics of trees are effective tools for selection. The stress responses of woody plants are considered to be key to the selection criteria and to markers that can be used in tree improvement programmes for specific urban conditions. Although much information can be found on environmental tree physiology (Kozlowski et al., 1991), the stress physiology of urban trees has until now been rather overlooked (Sæbo et al., 2005).

According to Ware (1994), extreme conditions of the original stand can be a good starting point for finding plants adapted to stressful urban environments. Drought is a significant factor influencing the existence and lifecycle of woody plants. However, taxa with similar ecological backgrounds (light-demanding and tolerant to water deficiency) probably use different strategies to overcome water stress impact (Paganova et al., 2010). Explanation of these mechanisms needs more experimental work and larger data files, as well as addition of information on other characteristics of drought impact such as changes of the assimilatory pigments, accumulation of osmoprotectants, water use efficiency (WUE), dynamics of the plant water regime parameters, structural leaf and root adaptations etc.

4. Selection of woody plants for landscape and urban greenery in Slovakia

Service tree (*Sorbus domestica* L.) and wild pear (*Pyrus pyraster* L. Burgsd.) are light-demanding species tolerant to water deficiency during the growing season (Brutsch & Rotach, 1993; Rittershoffer, 1998; Wilhelm, 1998; Paganová, 2003a, 2008; Paganová & Jureková, 2012). According to the ecology background, these woody plants are potentially suitable for extreme conditions of urban and landscape environment.

Fig. 2. Service tree (*Sorbus domestica* L.) with a straight stem, regular crown and thin branching in the late autumn aspect in a vineyard. A "Plus tree" with positive phenotypic characteristics (Maceková, 2011).

Fig. 3. A young wild pear (*Pyrus pyraster* L. Burgsd.) with conical crown and thin branches, growing on abandoned grazing land (Šenšel, 2009).

These woody plants are part of the original flora in Slovakia. For both taxa large seasonal and phenotypic variability has been documented (Pagan & Paganová, 2000; Paganová, 2003b) that represent a good basis for targeted selection of genotypes with the qualities required for landscape planning and design. Species with high phenotypic plasticity it is assumed survive better on stands with changing environmental conditions. They have higher adaptability to changes by modifications of their functions and structures which increases the probability of their survival and reproduction. However, environmental conditions on stands affected by anthropogenic activities change too fast for manifestation of the evolutionary mechanisms of the plant adaptation (Sultan, 2004), so non-hereditary changes of plants – acclimations – take place. This is why detailed evaluation of the adaptability and phenotypic plasticity of non-traditional or rare woody plants is an essential tool when assessing their wider utilization in the urban landscape.

4.1 Assessment of the potential sources of reproductive material of woody plants

Within programme selection focused on woody plant species for landscape and urban greenery the following steps are essential: qualitative assessment of the potential gene pool of the particular taxon, identification of the appropriate sources of reproductive material in the natural conditions and inclusion of the selected components of the gene pool in the selection and breeding programme of woody plants for targeted utilization in landscape vegetation units, as well as in urban areas.

In Slovakia the sources of reproductive material have not yet been identified for *Sorbus domestica* and *Pyrus pyraster*. Therefore, we devote a targeted assessment of phenotypic traits of individuals (growing on the original stands in the landscape), that can be used as a qualified source of the reproductive material (Fig. 2 and 3). The identification of the phenotypes suitable for wider plant production is a significant assumption for systematic planting and use of the mentioned woody plant species.

So far, attention has been paid to a reliable method of phenotypic classification of *Sorbus domestica* on original stands. Options and tools of their interpretation for selection of the sources of reproductive material were also analysed within a field study in the cadastre of the village of Žemberovce (Paganová & Maceková 2011). The scale of quantitative and qualitative parameters was elaborated and confirmed in 2011. For individual trees several parameters were determined which represent tree habit, tree growth and volume production (Fig. 4) (Tab. 1).

Architectural traits, such as patterns of branching or clonal spread and production of terminal versus axial inflorescences, may also vary plastically in certain taxa. These traits provide very important insight into the structural and ultrastructural levels at which phenotypic adjustment takes place in the plant body (WU & Stettler 1998). Therefore, we also apply them within the selection of indigenous woody plant species and their phenotypes suitable for urban landscapes and greenery.

Crown traits, such as branch diameter, branch angle and branch frequency, are also important determinants for the quality of woody plants and timber products (Bowyer et al., 2002). These characteristics also affect the utilization of woody plants in urban areas with limited space for tree growth. According to the literature, variability of the growth habit and other growth characteristics of woody plants has been the object of extensive research and represent a good basis for selection and tree breeding. In contrast, the crown architecture and structural characteristics of trees are less at the centre of interest and only rarely applied in breeding programmes. It should be noted that structural tree characteristics, such as crown shape, crown diameter, crown density, branch diameter, branch angle and leaf area, influence efficiency and magnitude of radiation interception and competitive interactions with other trees (Emhart et al., 2007). These characteristics have significant influence on the use of individual trees on particular stands and its competitiveness and ecological influence. The crown architecture and the active area covered by an individual woody plant have potential influences on the surrounding environment.

Several tree quantitative parameters are measured: tree height (h), stem girth at height 1,3 m ($O_{1,3}$), deployment of living crown (a) – vertical distance between the first living branch (that is a part of the living crown) and horizontal plane of the stem base (URL1). Crown diameter (b) – average horizontal distance between opposite points of the crown projection. This parameter is usually measured in at least four ways (Šmelko, 2000) using a densiometer. The shape and area of the crown projection is determined with GPS (global positioning system). The branch angle is determined in four categories (<30°, 30 - 60°, 60 - 90°, 90°<) according to the deflection of tree branches from the vertical axis (Fig. 4).

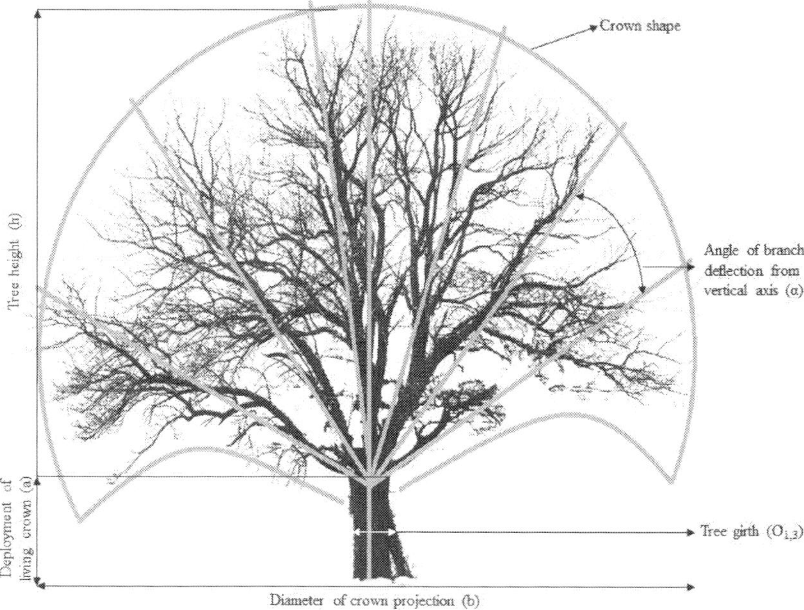

Fig. 4. Selected tree parameters measured and evaluated within the field study of the phenotypic classification of *Sorbus domestica L* (Paganova & Maceková, 2011).

Quantitative parameters		Qualitative parameters
Measured	Calculated	Crown shape
Tree height	Crown density	Trunk development
Sterm girth	Crown volume	Tree trunk shape
Deployment of living crown	Crown projection area	Cross sectional shape of the trunk
Crown diameter	Crown length	Thickness of branches
		Tree branch angle

Table 1. Selected parameters of trees measured and evaluated for *Sorbus domestica* and *Pyrus pyraster* within the study of their phenotypic variability and assessment of the gene pool quality.

In addition to these quantitative parameters, the crown shape, crown density and several other qualitative characteristics (trunk shape, trunk development and cross sectional shape of the trunk) were evaluated within field study. The photo documentation of tree habit was made for each of the evaluated genotypes of *Sorbus domestica* and *Pyrus pyraster*. The records will be used for other graphic processing. Within analytical data processing the other quantitative parameters of the tree habit were calculated: crown length, crown area projection and crown volume.

The relationship between parameters of the tree crown and stem were calculated for both woody plant species based on mentioned quantitative data, as well as the relationship between crown architecture and structural characteristics ⇒ regression analysis; we attempted to define a range of structural parameters of the trees, which are characteristic for high-quality phenotypes (plus trees) of *Sorbus domestica* and *Pyrus pyraster*. The multivariate statistical methods (discriminant analysis etc.) were used for assessment.

Preliminary data obtained within the field study documented distinctive variability of the phenotypic parameters of *Sorbus domestica* (Paganová & Maceková, 2011). Five basic crown shapes were found: conical, ovate, globular, parabolic and semi-globular (Fig. 5) within the population of trees growing in one location. Good phenotypic characteristics were confirmed for 12 individuals among the whole number of 35 individuals of *Sorbus domestica* which were evaluated according to the phenotypic classifications within the field study. These trees can be later (after additional testing and assessment) included under selected sources of the reproductive material of *Sorbus domestica*.

Location of the evaluated individuals on their original stand was determined using GPS, which will facilitate identification of the trees in the future. GPS was used also for determination of the shape and area of the crown projection within the field data collection.

The aim of the tree assessment is establishment of a database of genotypes for both woody plant species (*Sorbus domestica* and *Pyrus pyraster*) from the territory of Slovakia and selection of phenotypes suitable for landscape design and urban greenery. Selection of suitable sources for a reproduction (Drobná & Paganová, 2010) and breeding programme of the mentioned species was identified from the database of genotypes based on the classifications of the phenotypic characteristics (Fig. 6).

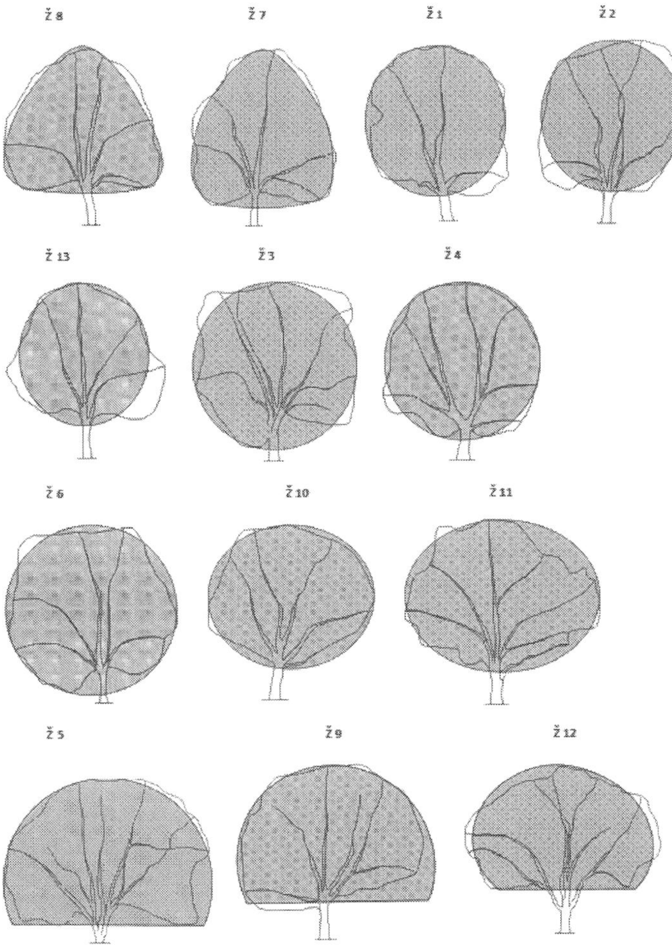

Fig. 5. Models of the crown architecture for *Sorbus domestica* designed according to the data obtained from the field assessment of trees. Quite large phenotypic variability of the specimen growing on one location at Žemberovce is evident. Crown shape – conical (Ž8, Ž7), ovate (Ž1, Ž2, Ž13) globular (Ž 3, Ž4, Ž6), parabolic (Ž10, Ž11) and semi-globular (Ž5, Ž9, Ž12).

4.2 Assessment of the phenotypic plasticity of woody plant species to drought

One-year old seedlings of the wild pear and service tree were analysed. Plant material was produced directly for purposes of the experimental assessment of the physiological parameters of the analysed woody plants. Reproductive material came from the original stands of *Sorbus domestica* (Kosihovce, altitude 250 m) and *Pyrus pyraster* (Trnie, altitude 540 m) in Slovakia. The seeds of both species passed winter pre-sowing treatment in the cold greenhouse with temperatures ranging from +5°C to -5°C and germinated in the boxes with fertilized sowing substrate based on white sphagnum (peat moss). In the phenological stage "expanded cotyledons" (Šenšel & Paganová, 2010) seedlings were placed in the plant boxes (content 1.17 L) filled with fertilized peat substrate (white sphagnum, pH 5.5-6.5; fertilizer 1.0 kg.m^{-3}). Construction of the root boxes enabled the analytical assessment of the root growth and root structures (Fig. 3). The boxes were placed under a polypropylene cover with 60% shading. The plants were regularly watered and maintained on 60% and 40% of the full substrate saturation, two variants according to a differentiated water regime. Variant "stress" was supplied with water at 40% of full substrate saturation and "control" at 60% of full substrate saturation. The model of the differentiated water regime was maintained for 170 days (from April to the end of September).

After exposure to water stress the size of the leaf area (A) was calculated from leaf scans using ImageJ software (URL 2). The dry weight of the roots, shoots and leaves (DW) was determined gravimetrically, additionally leaf water content (LWC) was calculated. For metabolic characteristics, the total chlorophyll and carotenoid content were determined in the leaves according to the methods described by Šesták & Čatský (1966).

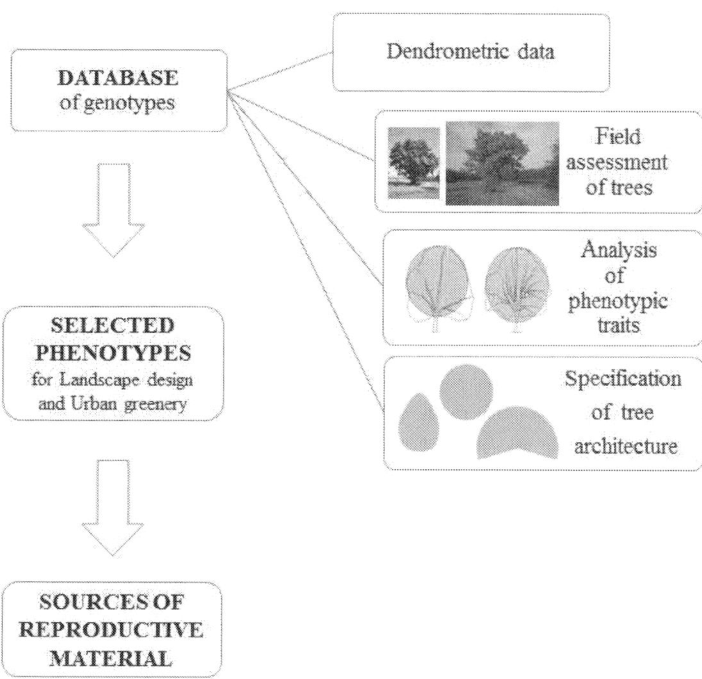

Fig. 6. Assessment and selection of the sources of reproductive material for utilization of indigenous woody plants in landscape planning and urban greenery

Fig. 7. The root box constructed from a metal frame and glass front wall used for the analytical assessment of the root growth and structures (Drobná, 2010).

Data were analysed from two growing seasons in 2010 and 2011 for each taxon under two variations of water regimes (40% and 60% substrate saturation). There was calculated S: R ratio for particular species and both variants of the substrate saturation. The relationship between chlorophyll and carotenoid content of the plants under stress and control conditions, as well as impact of water stress on complex of assimilatory pigments were analysed. A statistical assessment of these parameters was conducted by regression analysis and multivariate analysis of variance of using the statistical software Statgraphics Centurion XV (StatPoint Technologies, USA). A $P < 0.05$ was consisted statistically significant.

The reactions of two woody plant species *Pyrus pyraster* a *Sorbus domestica* on water stress in the juvenile phase of their growth (one-year old seedlings) were analysed in the model experiments with controlled water regime. In the first step of analysis, the size of the leaf area, dry weight of roots, shoots and leaves (DW) were determined, leaf water content calculated and the content of assimilatory pigments was quantified. In the last analysis, the significant interspecific differences in the size of leaf area were documented (Table 2), when *Sorbus domestica* had nearly twice higher leaf area than *Pyrus pyraster*. The difference in the size of the leaf area has not been accompanied by significant differences in accumulation of dry mass of leaves. However, there were differences in the investment of assimilates, *Sorbus domestica* used them for growth of the leaf area and *Pyrus pyraster* for construction of the mesophyll structures. The interspecific differences are documented also by values of the specific leaf area (SLA) for *Sorbus domestica* (SLA = 0.0219 $m^2.g^{-1}$) and for *Pyrus pyraster* (SLA = 0.0157 $m^2.g^{-1}$).

Assessment of the dry mass distribution into particular organs of woody plats confirmed different strategies for analysed species. Comparison of the values of shoot : root ratio (S : R) for analysed taxa is also interesting. There were found statistically significant differences, between wild pear and service tree. According to the obtained data, *Sorbus domestica* preferentially distributes higher ammount of dry mass to roots (S : R = 0.70), while distribution of dry mass is rather balanced between underground and upper organs of *Pyrus pyraster* (S : R = 1.11) (Table 2). The impact of water stress on distribution of dry mass to organs has not been documented. According to the obtained results seedlings under stress and in control conditions had very similar values of the shoot : root ratio (S : R = 0.87 and 0.93). The impact of water stress is manifested in production of dry mass of roots, that represents average value of the parameter for one plant (DW_R=0.59g). While in control conditions the average value of this parameter is nearly twice higher (DW_R=1.22g). Dry mass distribution patterns typical for particular species are determining also under conditions of differentiated water regime. Service tree (*Sorbus domestica*) preferably accumulates dry mass in root system, wild pear (*Pyrus pyraster*) distributes evenly dry mass to upper and undreground organs. Distribution of dry mass in organs of analysed woody plants has not been significantly influenced by drought. Species differentiation in shoot-root ratio (S:R) in favour of the distribution of dry mass in the root was confirmed for young cuttings of poplar Ibrahim et al.(1997).

The content of assimilatory pigments was also evaluated for both species. Significant differences were found in the total chlorophyll content (CC) and the content of carotenoids (CAR) (Table 2). Seedlings of *Sorbus domestica* cumulated a significantly lower content of total chlorophyll and carotenoids in the leaves than did the *Pyrus pyraster* seedlings. The interspecific differences are documented by average values of the total chlorophyll content *Sorbus domestica* (311.67 mg.mm^{-2} *10^{-6}) and *Pyrus pyraster* (490.90 mg.mm^{-2} *10^{-6}) and also by average values of the carotenoid content for *Sorbus domestica* (72.72 mg.mm^{-2} *10^{-6}) and for *Pyrus pyraster* (117.07 mg.mm^{-2} *10^{-6}). The ratio of carotenoid to total chlorophyll content (CAR : CC) had the same value (0.24) for both species. In this context, significant differences in the content of the assimilatory pigments indicate differences in the performance of the assimilatory apparatus of the wild pear and service tree in the juvenile phase of growth.

Content of assimilatory pigments of the seedlings was negatively affected by water stress. Under conditions of water scarcity seedlings produced significantly less chlorophyll (346.02 mg.mm^{-2} *10^{-6}), as well as carotenoids (84.42 mg.mm^{-2} *10^{-6}) than in control conditions (CC=456.55 mg.mm^{-2} *10^{-6}, CAR = 105.37 mg.mm^{-2} *10^{-6}). The ratio of the carotenoid content to total chlorophyll content CAR : CC differs significantly according to level of the substrate

Parameter Source of variance	A (mm²)	DW_R (g)	DW_L (g)	DW_S (g)	S/R	LWC (%)	CC (mg.mm^{-2})* 10^{-6}	CAR (mg.mm^{-2})* 10^{-6}	CAR/CC
Sorbus domestica	12016,5b	1,19b	0,55a	0,20a	0,70a	52,44a	311,67a	72,72a	0,24a
Pyrus pyraster	6286,99a	0,62a	0,40a	0,26a	1,11b	50,09a	490,90b	117,07b	0,24a
Control	12292,3b	1,22b	0,64b	0,34b	0,87a	48,78a	456,55b	105,37b	0,23b
Stress	6011,13a	0,59a	0,31a	0,13a	0,93a	53,75a	346,02a	84,42a	0,25a

Table 2. Average values of the selected physiological parameters of one-year seedlings of *Sorbus domestica* L. and *Pyrus pyraster* L. Burgsd. Seedlings were planted in the root boxes under conditions of differentiated water regime. Statistically significant differences (according to 95% LSD test) between average values of the analysed parameters are documented by different letters.

saturation in variant stress (0.25) and under control conditions (0.23). The content of carotenoids in the leaves increased in conditions of water scarcity. That is evidence of the negative impact of drought on the complex of assimilatory pigments. The results (Table 2) document species differentiation in production of the assimilatory pigments for individuals in the juvenile phase of growth, as well as significant impact of drought on the total content and particular components of the assimilatory pigments.

An interesting view on responses of the analysed woody plants is supplied within comparison of the relationship between parameters leaf dry weight (DW_L) and size of leaf area (A). The relationship between these parameters is highly significant in both variants of the substrate saturation (stress and control) for *Sorbus domestica* (Fig. 8 and 9). The course of values is described by an exponential function in control variant: $Y = (a + b*X)^2$

$$A = (-1.68046E8 + 6.45785E8*DW_L)^2,$$

by contrast, in stress variant is documented with an exponential function: $Y = a + b*X^2$

$$A = 3723.8 + 23848.7*DW_L^2.$$

Under conditions of water scarcity, *Sorbus domestica* accumulated less dry mass in the leaves and significantly reduced the growth of leaf area, that is evident in the course of the functional relationship of these parameters. In the control variant (Fig. 8) the value of leaf dry weight (DW_L= 0.63g) corresponds to leaf area size (A = 14800 mm^2), whilst in the stress variant (Fig. 9) the same value of leaf dry weight (DW_L =0.63 g) corresponds to higher size of the leaf area (A = 11800 mm^2). *Sorbus domestica* significantly reduced growth of the leaf area under conditions of water scarcity and formed thicker leaves, probably for better water management.

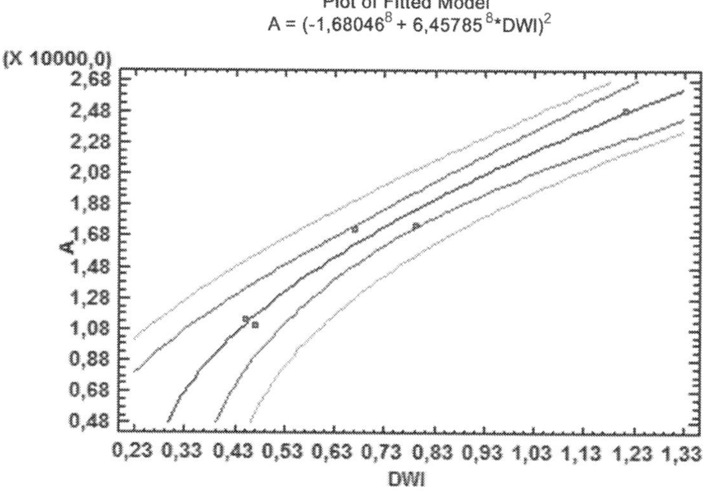

Fig. 8. Simple regression between leaf dry weight (DW_L) and size of the leaf area (A) for seedlings of *Sorbus domestica* growing under control conditions. Correlation coefficient (r) = 0.992349 p value = 0.0008

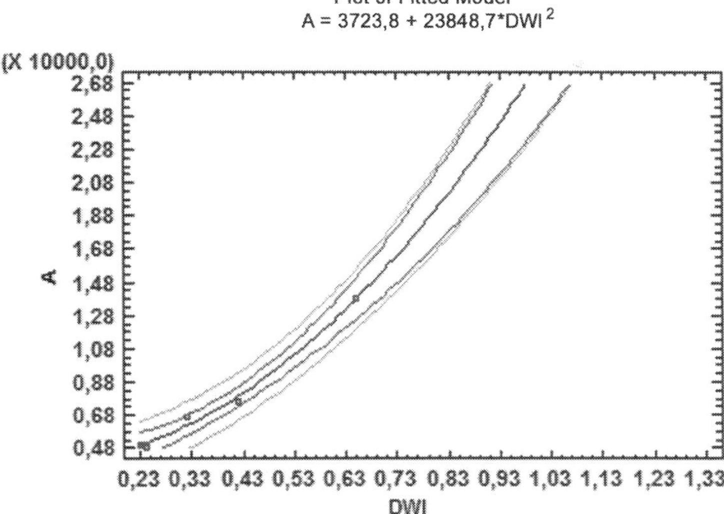

Fig. 9. Simple regression between leaf dry weight (DW$_L$) and size of the leaf area (A) for seedlings of *Sorbus domestica* growing under conditions of water stress. Correlation coefficient (r) = 0.995305 p value = 0.0004

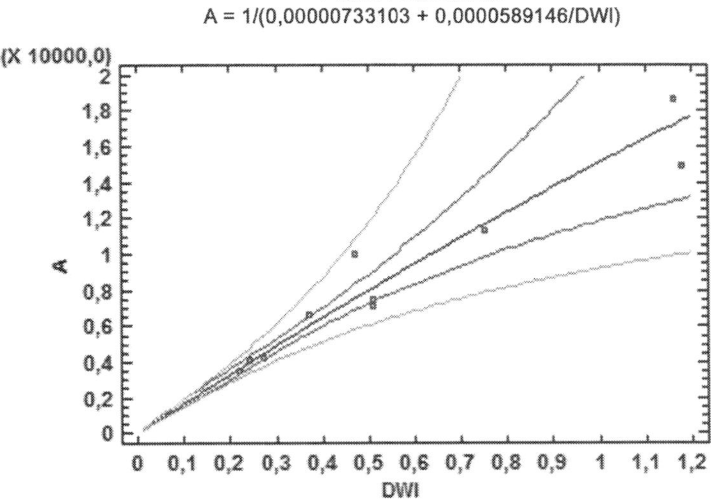

Fig. 10. Simple regression between leaf dry weight (DW$_L$) and size of the leaf area (A) for seedlings of *Pyrus pyraster* growing under control conditions. Correlation coefficient (r) = 0.980787, p value = 0.0000

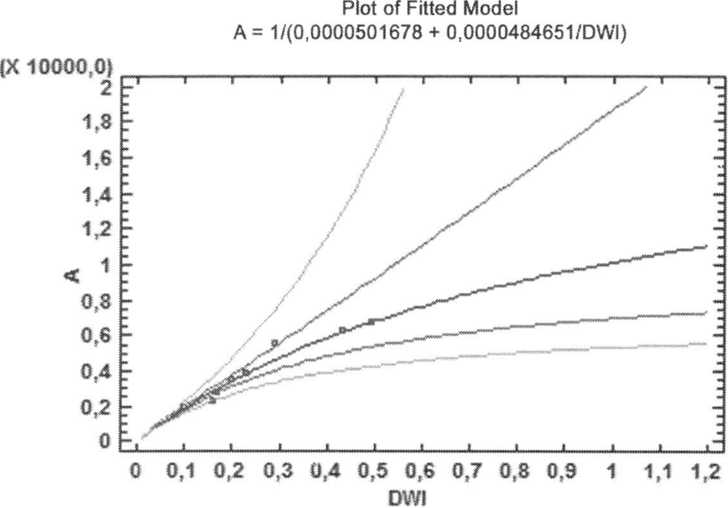

Fig. 11. Simple regression between leaf dry weight (DW_L) and size of the leaf area (A) for seedlings of *Pyrus pyraster* growing under conditions of water stress. Correlation coefficient (r) = 0.984167, p value = 0.0000

The relationship between size of the leaf area (A) and leaf dry weight (DW_L) is highly significant and closely correlated (r = 0.98) for the wild pear (*Pyrus pyraster*). It is described by double reciprocal function: $Y = 1/(a + b/X)$ in both variants of the substrate saturation (stress and control).

The reduction of growth of the leaf area under water stress conditions has been documented also for *Pyrus pyraster* according the course of the functional relationship between the analysed parameters. In the control variant the leaf area 9000 mm^2 corresponds to leaf dry weight DW_L= 0.60g, whilst in the stress variant (Fig. 11) the same value of leaf dry weight corresponds to lower leaf area to about 7000 mm^2. Wild pear also demonstrated a reduction of the leaf area in conditions of water scarcity. However, wild pear forms a lower leaf area in comparison to the service tree generally.

In this context it is interesting to note that between the wild pear and the service tree seedlings analysed in the juvenile phase of growth, significant differences in leaf water content (LWC) were not confirmed (Table 2). The average value of the leaf water content for the wild pear (*Pyrus pyraster*) was 50.2% and for the service tree (*Sorbus domestica*) the average value of the same parameter was 52.4 %. Non-significant differences of LWC were found also for seedlings growing in conditions of the different level of substrate saturation: control (48.8%), stress (53.7%).

Both species are able to maintain balanced water content in leaves, even in conditions of water scarcity. However, they probably use different mechanisms of adaptability to water stress.

5. Conclusions

This study describes criteria and tools for woody plant selection for landscape planning, where there is potential to use a wider range of species. A limiting factor for selection of other woody plant species is the lack of information about their eco-physiological characteristics. These are key characteristics of the adaptability of woody plants to changed environmental conditions in the cultural landscape and urban areas.

The new planning and design concept allows for taking parts of nature with its specific beauty and amenity for urban spaces, as well as using native species and their communities for urban plantings. Utilization of indigenous species should support an increase of the biodiversity in urban areas with ecologically better balanced plant communities.

The advantage of using native woody plant species in landscape planning and design is the broad base of their genetic resources in the landscape and this represents a sufficient basis for the selection of the most suitable phenotypes and individuals.

Efficient use of indigenous species of plants and trees in landscape planning requires well-defined selective criteria within natural populations of plants, thus enabling distinguishing from within the native populations genotypes with the appropriate traits for specific environmental conditions.

The evaluation of woody plants and their responses to specific conditions and stress factors in the urban environment and cultural landscape requires exact assessment methods and techniques.

The results presented synthesize information about woody plants obtained from field research on original stands in the landscape, as well as findings obtained from experimental research held under controlled conditions (study aimed at assessing the impact of drought on some physiological parameters).

Sorbus domestica and *Pyrus pyraster* are considered to be light-demanding woody plants with similar ecological requirements on environmental conditions. However, their adaptability and response to water scarcity are different.

Data about the phenotypic structure and properties of the natural genetic resources were collected within a programme of selection of woody plant species for landscapes and urban environments. According to the ensemble of the quantitative and qualitative phenotypic traits the most valuable architectural individuals can be selected. The selected genotypes are recommended for further testing under controlled and regulated conditions in order to determine their adaptability to negative (stress) factors. These selection principles were used within assessment of the indigenous woody plant species *Sorbus domestica* and *Pyrus pyraster* in Slovakia.

Reactions of woody plants on water stress were evaluated in the juvenile phase of their growth. Significant interspecific differences in the size of leaf area were documented, with *Sorbus domestica* having nearly double the leaf area of *Pyrus pyraster*. The difference in the size of the leaf area has not been accompanied by significant differences in accumulation of dry mass of leaves. However, there were differences in the investment of assimilates, *Sorbus domestica* used them for growth of the leaf area and *Pyrus pyraster* for construction of the mesophyll structures.

Dry mass distribution patterns characteristic for particular species were also determined under conditions of differentiated water regimes. The service tree (*Sorbus domestica*) preferably accumulates dry mass in the root system, while the wild pear (*Pyrus pyraster*) distributes dry mass evenly to upper and underground organs. Distribution of dry mass in the organs of the analysed woody plants was not significantly influenced by drought.

Content of assimilatory pigments of the seedlings was negatively affected by water stress. Under conditions of water scarcity, seedlings produced significantly less chlorophyll and carotenoids than in control conditions. The carotenoids content to total chlorophyll content ratio (CAR : CC) values differed significantly according to level of the substrate saturation. The content of carotenoids in the leaves increased in conditions of water scarcity. That is evidence of the negative impact of drought on the complex of assimilatory pigments. The results document species differentiation in production of the assimilatory pigments for individuals in the juvenile phase of growth, as well as the significant impact of drought on the total content and particular components of the assimilatory pigments.

Between the evaluated species significant differences of the leaf water content (LWC) were not confirmed in the juvenile phase of growth. Both species are able to maintain balanced water content in the leaves, even in conditions of water scarcity. However, they probably use different adaptability mechanisms to water stress.

6. Acknowledgment

This research was supported by research grant projects VEGA 1/0426/09 "Plant adaptability and vitality as criteria of their utilization in urban environment and VEGA 1/1170/11 „Anthropogenic impacts on production and quality of surface runoff from small basins in conditions as climate change" from Slovak Grant Agency for Science.

7. References

Bowyer, J.L., Shmulsky, R. & Haygreen, J.G. (2002). Forest products and wood science: An introduction. Iowa State Press, Ames, Iowa, 554 p.

Breuste, J. H. (2004). Decision making, planning and design for the conservation of indigenous vegetation within urban development. *Landscape and Urban Planning*, 68 (2004) 439-452.

Brütsch, U. & Rotach, P. (1993). Der Speierling (*Sorbus domestica* L.) in der Schweiz: verbreitung, Ökologie, Standortsansprüche, Konkurrenykraft und waldbauliche Eignung. Schweiz. Z.Forstwes., vol. 144, No. 12 :967-991.

Drobná, L. & Paganová, V. (2010). Germination rate and qualitative characteristics of true service tree (*Sorbus domestica* L.) seeds from different stands. Acta Horticulturae et Regio Tecturae. *Acta horticulturae et regiotecturae*. 2010. Vol. 13, spec. issue, p. 17--20. ISSN 1335-2563.

Dunnett, N. & Hitchmough, J. (2004). The dynamic landscape: Design, Ecology and Management of Naturalistic Urban Planting. 2004, Spon press, Taxylor & Francis group: London and New York. 484 pp ISBN 0-415-25620-8

Dwyer, J.F., McPherson, E.G., Schroeder, H.W. & Rowntree, R.A. (1992) Assessing the benefits and costs of the urban forest. *Journal of Arboriculture* 18(5)227-234.

Dwyer, J.F., Nowad, D.J. & Noble, M.H. (2003). Sustaining urban forests, *Journal of Arboriculture* 29:49-55.

Emhart, V. I., Martin T. A., White, T.L. & Huber, D.A. (2007). Clonal variation in crown structure, absorbed photosynthetically active radiation and growth of lobolly pine and slash pine. *Tree Physiology* 27, 421-430

Florgård, C. (2004). Preservation of indigenous vegetation in Urban areas – an introduction. *Landscape and Urban Planning* 68(2004) 343-345.

Ibrahim, L., Proe, M.F. , Cameron, A.D. (1998). Interactive effects of nitrogen and water availabilities on gas exchange and whole-plant carbon allocation in poplar. Tree physiol.,18(7):481-4872002 Heron Publishing-Victoria, Canada

Kozlowski, T. Kramer , P.J. & Pallardy, S.G. (1991) The Physiological Ecology of Woody Plants. New York, Academic Press, 1991, ISBN 978-0124241602, 657p.

McPherson, E.G. & Simpson, J.R. (2003). Potential energy savings in buildings by an urban tree planting programme in California, Urban Forestry& Urban Greening 2:73-86.

Nilsson, K. Randrup T.B & Wandall B.M. (2000). Trees in the urban environment. In: The Forest Handbook (Ed. Evans J), Volume 1:347-361. Blackwell Science, Oxford.

Pagan, J. & Paganová, V. (2000). Variability of service tree in Slovakia (Premenlivosť jarabiny oskorušovej (*Sorbus domestica* L.) na Slovensku). Acta Facultatis Forestalis, vol. 42., 2000, p. 51-67. ISBN 80-228-0982-9

Pauleit, S., Jones, N., Garcia-Martin G., Garcia-Valdecantos J.L., Riviere L.M., Vidal-Beaudet L., Bodson, M. & Randrup T.B. (2002). Tree establishment practice in towns and cities – results from a European survey. Urban Forestry & Urban Greening 1(2):83-96.

Paganová, V. (2003a). Wild pear *Pyrus pyraster* (L) Burgsd. requirements on environmental conditions. In: Ekológia, Bratislava, vol. 23, 2003, No.3, p. 255-241.

Paganová, V. (2003b). Habit and dimensions of wild pear (*Pyrus pyraster* /L./ Burgsd.) in Slovakia. Folia Oecologica, vol. 30., 2003, no. 1, p. 7-19. ISBN 80-967238-2-0

Paganova, V. (2004). Planning arrangement of vegetation in the agricultural landscape. In: Demo, M. & Látečka, M. (eds). Designing of the sustainable farming system in the landscape. (Projektovanie trvalo udržateľných poľnohospodárskych systémov v krajine). Nitra : Slovenská poľnohospodárska univerzita, 2004. 603 - 629. ISBN 80-8069-391

Paganová, V. (2006). Biodiversity research of the rare woody plants in Slovakia as a condition of their conservation and utilization in the landscape / In Biotechnology 2006 [elektronický zdroj] : sborník Zemědělské fakulty JU, 15.- 16.2.2006 České Budějovice. - České Budějovice : Jihočeská univerzita, 2006. - ISBN 8085645-53-X. - S. 1121-1124.

Paganová, V. (2008). Ecology and distribution of service tree *Sorbus domestica* (L.) in Slovakia. In: Ekológia, Bratislava, 2008, 27(2):152-168.

Paganová, V., Jureková,Z., Dragúňová, M. & Lichtnerová, H. (2010). Monitoring of physiological characteristics for assortment selection of woody plants in urban areas. *Acta horticulturae et regiotecturae*. 2010. Roč. 13, spec. issue, s. 7--11. ISSN 1335-2563.

Paganová, V. (2011). Design of the vegetation modifications in agricultural landscape. In: Demo, M. (ed.) Sustainable farming system's designing. Projektovanie udržateľných poľnohospodárskych systémov v krajinnom priestore. Vyd. Nitra : Slovenská poľnohospodárska univerzita, 2011. 487- 524 . ISBN 978-80-552-0547-2

Paganová, V. & Jureková, Z. (2012). Adaptabilty of woody plants in aridic conditions. In: Evapotranspiration - Remote Sensing and Modeling. -- 1st. ed.. -- 514 p.. -- 978-953-307-808-3 Rjeka : InTech, 2011. -- p. 493-514.

Paganová, V. & Maceková M. (2011). Significance of the phenotypic classification of the Service tree (*Sorbus domestica* L.) in Slovakia. Význam fenotypovej klasifikácie jarabiny oskorušovej (*Sorbus domestica* L.) v podmienkach SR. In: Aktuálne otázky štúdia introdukovaných drevín : Proceedings of the scientific conference. Arborétum Mlyňany SAV. p. 136-142. ISBN 978-80-970849-8-1.

Rittershoffer, B. (1998). Förderung seltener Baumarten im Wald. Auf den Spuren der Wildbirne. InAllgemeine Fosrtzeitschrift /Der Wald, 16, p. 860-862.

Sæbo, A., Benedikz, T. & Randrup, T. (2003). Selection of trees for urban forestry in the Nordic countries. Urban Forestry and Urban Greening 2 (2003): 101-114

Sæbo, A., Borzan Ž., Ducatilion, C., Hatizstathis A., Lagerström, T., Supuka J., García-Valdecantos, L., Rego, F. & Van Slycken, J. (2005). The Selection of Plant Materials for Street Trees, Park Trees and Urban Woodland. In: Konindejik, C., Nilsson, K., Randrup, T., Schpperijn, J. (eds.) Urban Forests and Trees: A Reference Book. 2005. Springer Verlad. Berlin Heidelberg. 257-280 ISBN 354025126X

Sand, M. (1994). Design and species selection to reduce urban heat island and conserve energy. In: Proceedings from the Sixth National Urban Forest Conference: Growing Green Communities. Minneapolis, Minn. (pp. 148-151). Washington D.C.: American forests.

Simpson, J.R. (1998). Urban forest impacts on regional space conditioning energy use: Sacramento and county case study. Journal of Arboriculture. 24(4):201-214.

Sjöman, H. (2012). Trees for Tough Urban Sites. Doctoral thesis, Swedish University of Agricultural Sciences, Alnarp 2012

Spellenberg, I. & Given, D. (2008). Trees in urban and city environments: a review of the selection criteria with particular reference to nature conservation in New Zealand cities. In : Landscape Review, 2008, 12 (2) 19-31.

Sultan, S.E. (2004). Promising directions in plant phenotypic plasticity. In: Perspectives in Plant Ecology, Evolution and Systematics. Elsevier GmbH, 2004, Vol. 6/4, pp. 227-233

Šenšel, R. & Paganová, V. (2010) Identification key of the wild pear (*Pyrus pyraster* L. Burgsd.) - phenological stages and potential for application in research and practice. Acta horticulturae et regiotecturae. 2010. Vol. 13, spec. issue, s. 21--23. ISSN 1335-2563.

Šesták, J. & Čatský J. (1966). Metody studia fotosyntetické produkce rostlin, Akademia, ČSAV Praha, 393 p.

Šmelko, Š. (2000). Dendrometria. Vysokoškolská učebnica. Vydavateľstvo TU vo Zvolene. ISBN 80-228-0962-4, 399s.

URL1: *Inventarizace lesů, Metodika venkovního sběru dat* [cit. 2011-09-09]. Dostupné na internete: http://www.uhul.cz/il/metodika/metodika6/kap_3_6_0.pdf

URL2: ImageJ. Image Processing and Analysis in Java. http://rsbweb.nih.gov/ij/

Valladares, F., Gianoli, E. & Gómez, J., M. (2007). Ecological limits to plant phenotypic plasticity. New Phytologist 176: 749-763

Westphal, L.M. (2003). Urban greening and social benefits: A study of empowerment outcomes. Journal of Arboriculture 29(3): 137-147.

Wilhelm, G., J. (1998). Im Vergleich mit Elsbeere und Speierling Beobachtungen zur Wildbirne. AFZ. Der Wald, *16*: 856–859.

Wu R. & Stettler, R.F. (1998). Quantitative genetics of growth and development in *Populus*. III. Phenotypic plasticity of crown structure and function. *Heredity* 81: 299–310.

10

Integration of Infrastructures in Landscape – An Opportunity to Landscape Planning Improvement

Maria José Curado and Teresa Portela Marques
CIBIO - Centre in Biodiversity and Genetic Resources, University of Porto
Portugal

1. Introduction

This paper presents a work that was developed during three years, in a partnership between the company EDP Distribuição, SA (Grupo EDP- Energias de Portugal SA) and the Research Centre in Biodiversity and Genetic Resources of the University of Porto (CIBIO/ UP), financially supported by the PPDA - Promotion of Environmental Performance Plan - approved by the Energy Services Regulatory Authority (ERSE). The objective of this work was the production of a Manual of Good Integration Landscape Practices of the Infrastructure of the Distribution Network. This manual aims to present a set of strategies, guidelines and practices for landscaping integration, in mainland Portugal, of electrical infrastructures, namely: Substations, Lines of High, Medium and Low Voltage, Transformer Stations and Urban Cabinets.

The work had two major challenges: on the one hand, the development of technically sound solutions and proposals from the standpoint of landscape integration of these infrastructures, minimizing their impact; on the other hand, ensure ease of use and the application of this technical manual for non-specialists in the field of landscape. Thus, it was essential to know how the planning and design of these infrastructures in the company was carried out, developing a compatible methodology for landscape integration.

2. Conceptual process for the development of the manual

The principle from which the methodological and conceptual process began, developed for the construction of this Manual of Good Integration Landscape Practices of Electrical Infrastructures (figure 1) was that electrical infrastructures cause always an impact in the landscape. On the majority of cases, this impact is negative since it causes an intrusion and a change in landscape character. Therefore, it is essential to analyze and study these two variables - the landscapes and the electrical infrastructures.

Concerning landscape, it is important to understand its character that reflects the interaction of the various components of landscape, namely physical, biological, social, cultural, economic and visual. The identification of the character of the landscape allows the identification of the types of landscape with homogeneous characteristics, which requires

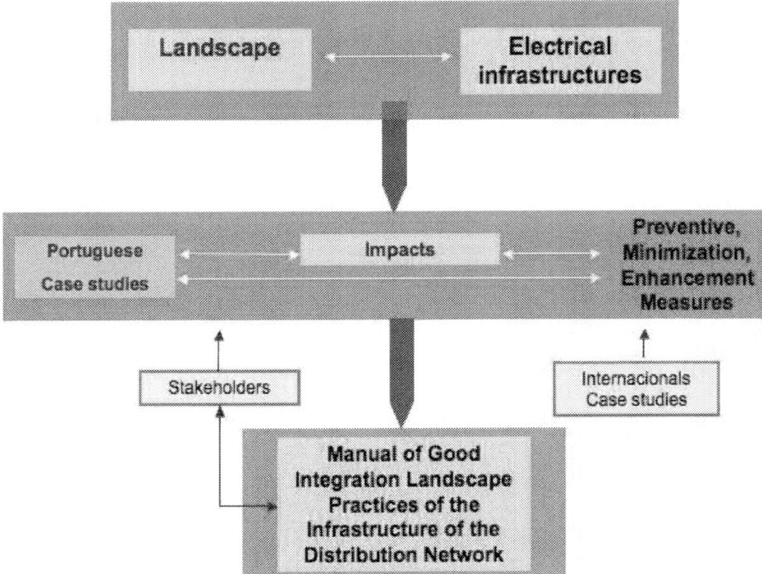

Fig. 1. Conceptual process for the development of the Manual

similar interventions, i.e., it allows the assessment of the capability of each landscape in receiving (from a spatial and visual standpoint) elements foreign to its nature, such as electrical infrastructures.

In the study of the electrical infrastructures, on the one hand, legal and technical aspects, constraints to the implementation of each infrastructure, were taken into account. On the other hand, aspects of shape and size, essential for the determination of its visual and landscape impact, were considered.

In parallel, it was carried out a study of practical cases - international case studies already implemented and cases presented by EDP Distribution for which landscape integration proposals were developed. The process carried out in these studies and the results obtained in terms of proposals, contributed to the definition of measures that constitute the core of this Manual. For the compiling of these measures, in addition to the bibliography research, the practical application in specific cases of new or conversion layouts of infrastructures was essential. It was a joint work of the University of Porto and the technicians of the company, which has proved to be very important for the validation of the proposed measures. On the other hand, it was also essential as a means of raising awareness and to introduce new approaches to be applied by the company technicians.

Finally the input from stakeholders heard during this project was taken into account.

3. Identification of landscape types

Landscape is something very complex and variable that, in the Portuguese context, is heavily accentuated by the geographic location, the orographic variety of the country, and by the interaction of various ecological factors leading to a really rich and diverse landscape. This makes it advisable to attempt to identify homogeneous areas from the point of view of their character which is reflected in different types of landscape. The method developed for the identification of the types of landscapes, within this Manual, refers to a process of sequential selection of the main variables of the landscape, which ought to be considered as more relevant to landscape integration of electrical infrastructures: degree of urbanization, orography, and vegetation cover (figure 2).

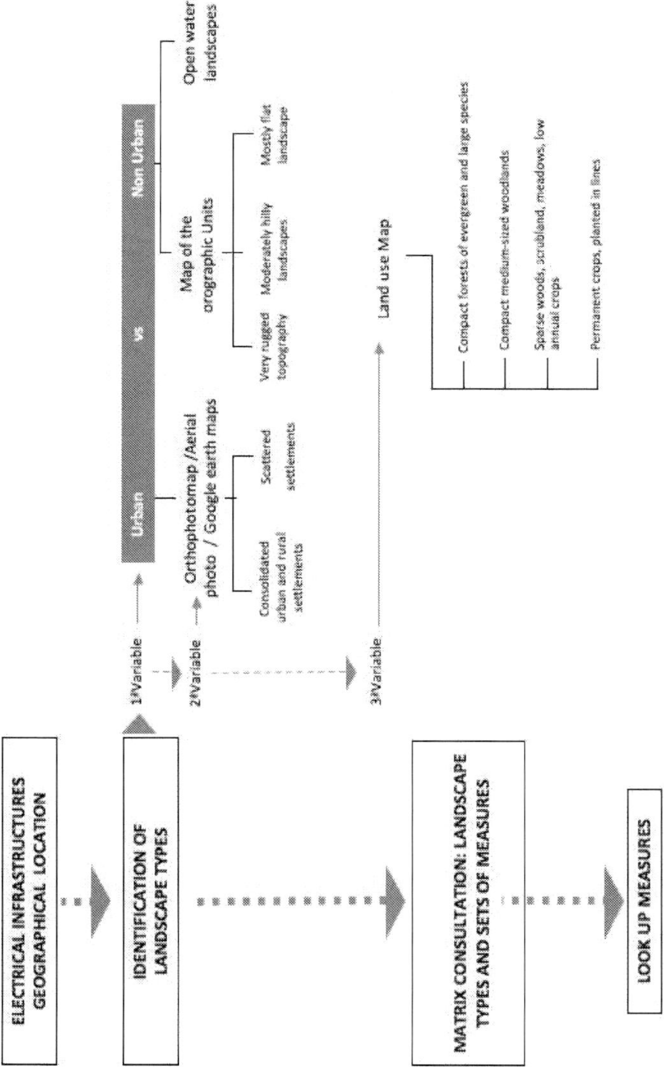

Fig. 2. Conceptual Framework for the consultation of the Manual

At first, two situations were distinguished which, by their nature, bring very different conditions: 'urban areas' and 'non-urban areas', i.e. landscapes dominated or not dominated, respectively, by the territory construction and infrastructure. Indeed, the urbanization of the territory and the consequent infrastructures introduce a significant artificiality that results in landscapes with a distinct character that encompass a very specific approach. 'Urban areas' are also landscapes where people stay longer, which means that social issues about the quality of the places where they live, work and enjoy themselves, must be taken into account.

Consequently, for what was named 'urban areas' two types of landscape were identified - (1) *consolidated urban and rural settlements* and (2) *scattered settlements*. This classification was defined taking into account the degree of urban consolidation, which is reflected into a higher or lower unity/density/continuity of built space. The first case presents a significant degree of consolidation while in the second case the level is low.

Since one of the aims of this process is to ensure ease of use and the application of this technical manual for non-specialists in the field of landscape, 'identification guides' were developed, using either illustrative processes or descriptive processes, to facilitate the procedures of identifying the type of landscape where an infrastructure will be placed.

In the 'identification guide for urban areas', *Google earth* was used as an universal tool of easy and simple use in order to find examples of *consolidated urban and rural settlements* (high/medium or medium/small density of building fabric inserted in the rural landscape) - figures 3 and 4 – as well as examples of scattered settlements (dispersed/diffused building fabric contiguous and directly related to consolidated urban settlements and dispersed/diffused building fabric not contiguous to consolidated settlements) - figures 4 and 5.

Source: Google Earth

Fig. 3. Example of a consolidated urban settlement: Bragança.

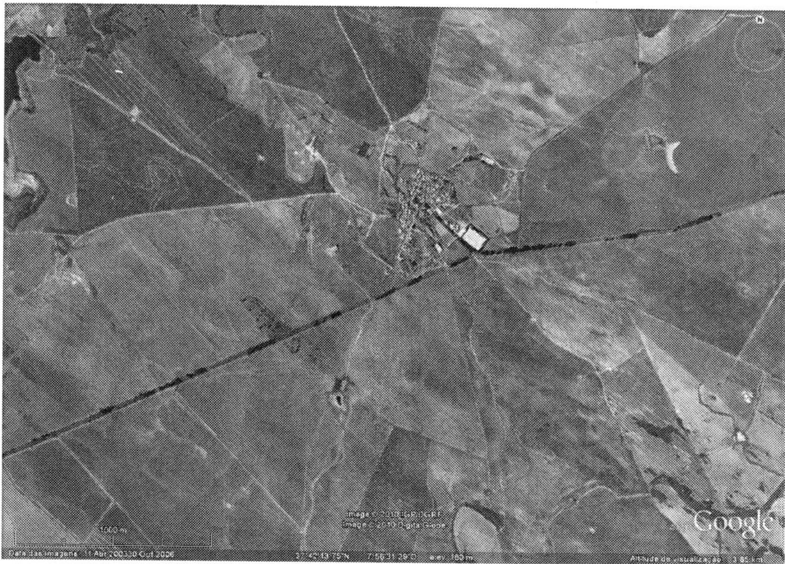

Source: Google Earth

Fig. 4. Example of a consolidated rural settlement: São Marcos de Ataboeira.

Source: Google Earth

Fig. 5. Example of a scattered settlement (dispersed/diffused building fabric) contiguous and directly related to consolidated urban settlements: Joane – Guimarães

Source: Google Earth

Fig. 6. Example of a scattered settlement (dispersed/diffused building fabric) not contiguous to consolidated settlements: Avintes.

In what concerns 'identification guide for non-urban areas', the first variable taking into account was orography - on a national scale, the main distinguishing factor to consider is the terrain. Four macro units were identified:

Unit 1 - Prevalence of landscapes with very rugged topography, i.e. valleys and hills with slopes and significant variation in height
Unit 2 - Prevalence of moderately hilly landscapes, i.e. valleys and hills with moderate slopes and variation in height
Unit 3 - Prevalence of mostly flat landscape
Unit 4 - Landscape with very diverse orography, being present moderately rough and flat reliefs, with a strong component of urbanization and infrastructure.

The objective of mapping these units (figure 7) is to provide guidance for the identification of each of these types of orography, identifying the prevalence of each one of them in the different regions of the country. It should be mentioned that the Portuguese landscape is very diverse, with large variations in orography within a short geographical area - therefore it was decided to choose to identify large patches where such a geographical predominance is found.

The second variable taking into account to 'identification guide for non-urban areas' is vegetation, considering its higher or lower capacity of visual absorption. In this case a land use map should be employed and the following types are considered:

1. Compact forests of evergreen and large species (e.g. maritime pine and eucalyptus).
2. Compact medium-sized woodlands (evergreen and/or deciduous, ex. native woods of oak, umbrella pine...).

3. Sparse woods, scrubland, meadows, low annual crops (e.g. cork oak and holm oak open woodland (*montado*), arable crops).
4. Permanent crops, planted in line (e.g., vineyard, orchards).
5. Open water landscapes (e.g., lagoons, estuaries, reservoirs and large rivers)

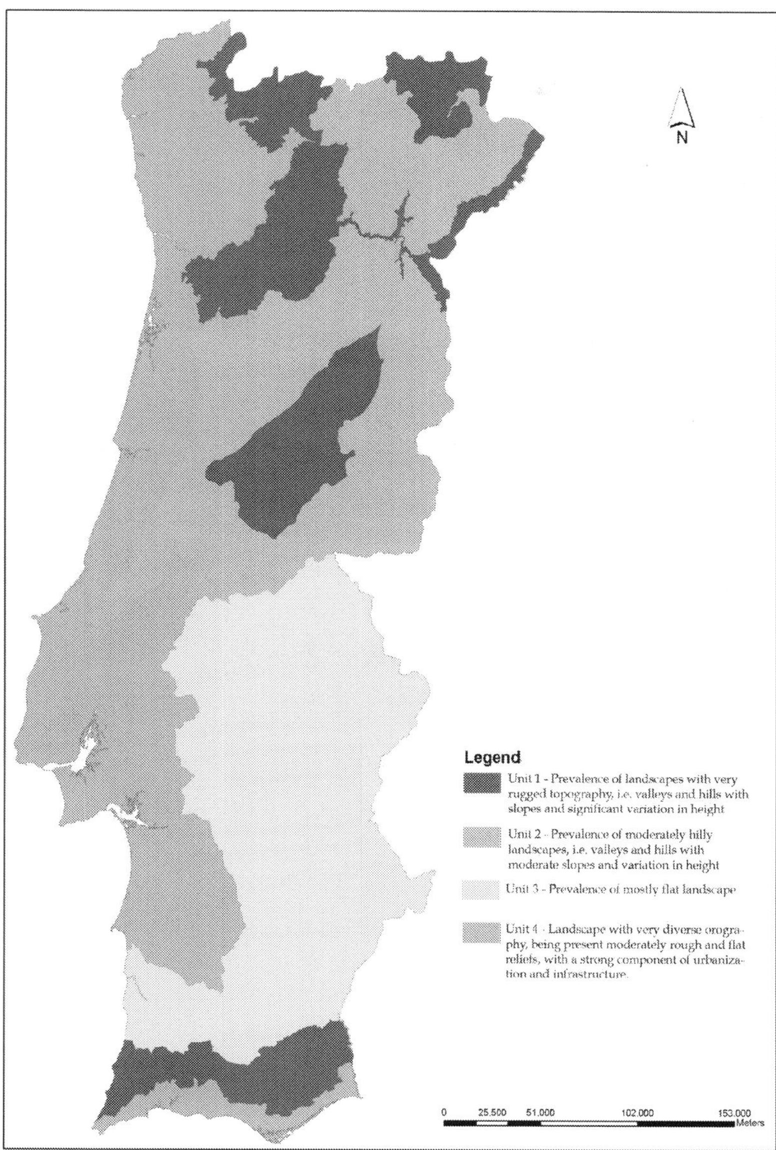

Fig. 7. Map of the orographic units for landscape integration of electrical infrastructures

4. Measures of landscape integration of electrical infrastructures

Having identified the type of landscape, it is possible to identify, through a guideline, the landscaping integration measures to consider in every situation. These measures are organized taking into account technical reasons, but also the process of planning and design of the company. Consequently, groups of Preventive Measures, Minimization Measures and Enhancement Measures were organized (figure 8).

Landscape types			SET OF MEASURES																	
			Preventive															Minimization		Enhan.
			1	2	3	4	5	6	7	8	9	10	11	12	13	14	15	16	17	18
Urban Areas	Consolidated urban and rural settlement		X	X	X									X	X	X	X	X	X	X
	Scattered settlements		X	X		X								X	X	X	X	X	X	X
Non Urban Areas	Prevalence of landscape with very rugged topography	Compact forests of evergreen and large species (e.g. maritime pine and eucalyptus)	X	X			X	X		X	X			X		X		X	X	X
		Compact medium-sized woodlands (evergreen and/or deciduous, ex. native woods of oak, umbrella pineE)	X	X			X	X		X		X		X		X		X	X	X
		Sparse woods, thickets, meadows, low annual crops	X	X			X	X						X		X		X	X	X
		Permanent crops, planted in line (e.g., vineyard, orchards)	X	X			X	X					X	X		X		X	X	X
	Prevalence of moderate hilly landscapes	Compact forests of evergreen and large species (e.g. maritime pine rnd eucalyptus)	X	X				X		X	X			X		X		X	X	X
		Compact medium-sized woodlands (evergreen and/or deciduous, ex. native woods of oak, umbrella pineE)	X	X				X		X		X		X		X		X	X	X
		Sparse woods, thickets, meadows, low annual crops	X	X				X						X		X		X	X	X
		Permanent crops, planted in line (e.g., vineyard, orchards)	X	X				X					X	X		X		X	X	X
	Prevalence of mostly flat landscape	Compact forests of evergreen and large species (e.g. maritime pine and eucalyptus)	X	X				X	X	X	X			X		X		X	X	X
		Compact medium-sized woodlands (evergreen and/or deciduous, ex. native woods of oak, umbrella pineE)	X	X				X	X	X		X		X		X		X	X	X
		Sparse woods, thickets, meadows, low annual crops	X	X				X	X					X		X		X	X	X
		Permanent crops, planted in line (e.g., vineyard, orchards)	X	X				X	X				X	X		X		X	X	X
	Open water landscapes (e.g. lagoons, estuaries, reservoirs and large river)		X	X				X	X					X		X		X	X	X

Fig. 8. Matrix for identification of landscape integration measures in relation to landscape

Preventive Measures are used primarily to support the development of new projects, namely the layout of transmission and distribution overhead lines and the siting and implementation of substations, transformer stations and distribution cabinets. These strategic measures focus on large scale questions with a wider but integrated eye on the landscape. They are based on a macro view of the landscape, focusing on its organization and its biophysical components.

Minimization Measures apply both in existing situations and in new situations, after the consideration of the preventive measures. They aim to mitigate the impacts that infrastructures can cause, nevertheless, on the landscape. They are based on a micro view of the landscape and they focus on the physical and cultural components of the landscape. They aim the very specific integration on the ground and, as such, actions can impact both in terms of infrastructures and in terms of landscape through earth modelling, planting schemes, physical treatment of infrastructures, among others.

Enhancement Measures reflect the added value that the implementation of an infrastructure can bring, in terms of landscape. In other words, the infrastructure should be regarded as something useful or interesting to users of that landscape, through the inclusion of the social, environmental and visual purposes.

4.1 Preventive measures: Planning and design of the layout of infrastructures

Fifteen sets of Preventive Measures were considered, which are shortly described as follows:

- Set 1 - General measures for all infrastructures: they refer to issues related to preliminary studies of landscape, namely the analysis of the planning instruments of the territory, the detailed survey of the intervention area (terrain, type of land use, land cover, natural and cultural values, roads, existing overhead lines and other infrastructures), the production of studies on view-sheds and 3D simulation projects, among others (fig.9).

Fig. 9. 3D simulations for different line designs

- Set 2 - General Measures for transmission and distribution lines: it includes examination and evaluation of the best solutions, based on studies of visibility from the main access points since landscapes with greater capacity for visual absorption should be considered as preferential for the implementation of lines. At the same time, landscapes with high scenic and natural/cultural values, namely those with a high degree of integrity and conservation, should be kept free of overhead lines.
  It is also considered using the same pole as support for multiple distribution lines or for various infrastructures, namely telecommunications.
- Set 3 - Transmission and distribution lines in urban centers and rural consolidated settlements: consider placing the line underground as a priority solution, particularly in heritage areas. When this is not possible, consider to associate the layout of the line with other linear infrastructures and preserve, free from infrastructures, the zone of visual influence of natural and cultural heritage elements.
- Set 4 - Transmission and distribution lines in scattered urban zones and peripheries of the consolidated urban areas: select the most infrastructure areas (predominance of roads, industrial areas...) for the layout of the line.
- Set 5 - Transmission and distribution lines in hilly areas: focus on the possible deployment in less illuminated slopes, avoiding the ridge lines.
- Set 6 - Transmission and distribution lines in valley areas: prioritize the layout in areas where the valley is more engaged, in other words at points of lower visibility, following the natural depressions so that they are concealed.
- Set 7 - Transmission and distribution lines in flat zones: take in account land use - if the line finds a forest patch, focus the layout on the inside of the patch; if the line finds a

permanent crop area in line focus on the straight layout, following the linear array of the landscape; in the water landscapes, consider placing the line underground or using existing infrastructures over the water plan, e.g. bridges.

- Set 8 - Transmission and distribution lines in forest patches: focus on the crossing by stands of lower landscape quality, in particular, eucalyptus and acacia tree species, at the expense of forest patches of higher landscape value (visually and ecologically).
- Set 9 - Transmission and distribution lines in compact forests of evergreen and large species: the layout of the lines should follow existing forest roads or forest clearings; and also consider the use of off line easement areas; in the slopes steeper than 1:3, not considering paths perpendicular to the line of greatest slope (fig.10).

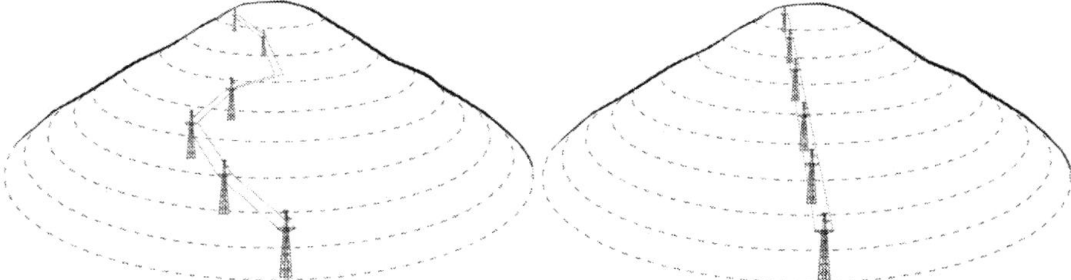

Fig. 10. Scheme explaining the preferable layout (left) of a distribution line in hilly landscapes – avoid perpendicular lines to the steepest gradient.

- Set 10 - Transmission and distribution lines in compact medium-sized woodlands: safeguard the permanence of the existing vegetation in the 'right-of-way' areas so that the height of the pole should be as small as possible so that, together with the line, they have reduced visibility above the trees canopy.
- Set 11 – Transmission and distribution lines in agricultural areas of permanent crops: give preference to placement of poles along the lines of culture and paths since line design should follow the geometry of the fields.
- Set 12 - General measures for substations: prioritize areas with land use with less visual and ecological value and in accordance with the topography of the place to minimize the areas of excavation and landfill; develop appropriate slopes to allow their landscaping with the use of vegetation; consider, in the purchase of land to implement a substation, a marginal area to develop a landscape project aiming its integration (fig.11).
- Set 13 - Substations in urban areas: consider the integration of substations in buildings and the construction of compact substations or underground substations; consider the possibility of creating a rooftop as an opportunity for providing space of interest for public use.
- Set 14 - General measures for transformer stations: consider the integration of transformer stations in existing built structures or attached to them, the incorporation in slopes or the combination with other elements of terrain modeling; consider the transformer stations as equipment subject to architectural design (fig.12).
- Set 15 - General measures for distribution cabinets: consider the establishment of cabinets on the walls of urban properties or on other structures, to avoid cluttering the sidewalks; consider the distribution cabinets as equipment subject to architectural design.

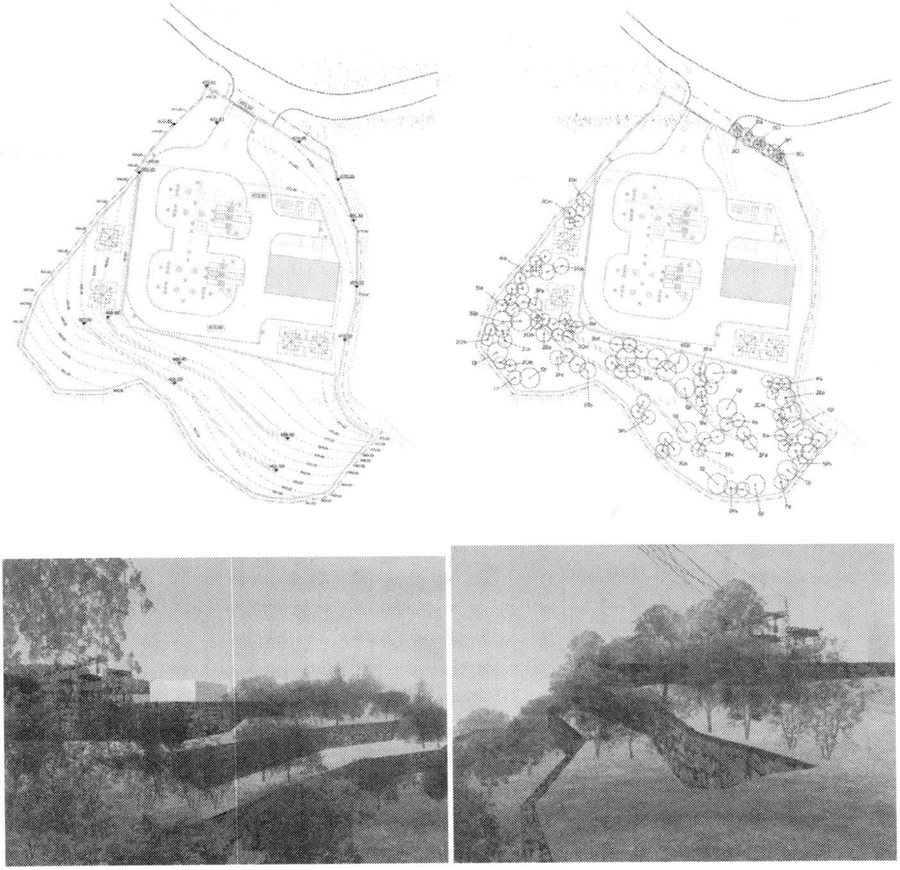

Fig. 11. Proposed integration of a substation in northern Portugal through terracing and planting schemes according to the character of the prevailing landscape

(a)   (b)

(c)   (d)

Fig. 12. Landscape integration of transformer stations in relation to structures (a) transformer station built-in a wall; (b) transformer station attached to an existing construction; (c) and (d) transformer station built by extending an existing building.

4.2 Minimization measures: Integration project and landscape treatment

Two types were considered: measures applicable to infrastructures and measures applicable to landscapes:

- Set 16 - At the infrastructure level, minimization measures include actions such as the physical treatment of the infrastructure – poles and towers, substations, transformer stations, distribution cabinets – as far as form, scale, colour, texture, and pattern are concerned, taking into account the formal features of the landscape in which they are placed.
- Set 17 – At the landscape level, minimization measures include landscape restoration treatment that mitigate the physical impact of the infrastructures in the landscape and actions that reduce the visual impact of the infrastructures themselves, namely: consolidation and planting of slopes, platforms and edges of new road access created by the installation of infrastructures; planting clumps along the 'right-of-way' areas to minimize its linear and disruptive effect; planting in the vicinity of the infrastructure in order to visually absorb it, on a scale of proximity, or next to the main points of visibility in order to frame the views to the infrastructure.

4.3 Enhancement measures: Intervention strategies for the benefit of the landscape and people, offered by the installation of infrastructures

- Set 18 - Consider the improvement of areas surrounding infrastructures or 'right-of-way' areas for social and recreational uses or to nature enhancement (green corridors, urban agriculture, bike paths, habitat restoration programs, water retention basins, among others); consider poles with particular shapes, involving in its design artistic considerations – infrasculpture or camouflaged poles; consider distribution cabinets and transformer stations as urban equipment, namely through its conversion (camouflage, coverage…) into elements of great visual interest (fig.13).

Fig. 13. Enhancement of a landscape, affected by the placement of electrical infrastructures, aiming to accommodate social and recreational functions - case study in Lisbon region.

5. Conclusion

As conclusion, it may be stated that this process - the design and development of the Manual and the interaction between landscape specialists and company technicians - was the beginning of a journey where issues concerning the value of the landscape played a major role, independently of the scale of intervention. Actually, this was the chief focus of this work – to convey the idea that landscape is a major asset that has to be respected and protected according to its quality, sensibility and character. As specialists in landscape, the authors consider that it is important not only to develop worthy landscape integration projects, but also to share the perception on landscape issues and principles with non-experts who also act in the landscape. This view implied the formulation of guidelines that could be well understood and applied to the formulation of proposals by non-specialists. Some joint projects carried out, under this work, between the authors and technicians of the Electricity Company has evidenced that the proposed guidelines can not only help to design better layouts, as far as landscape is concerned, but also to provide comprehensive guidance, i.e., serve as a framework for well-balanced designs, in general terms and after taking into account the necessary technical issues. The Manual was only published last December and the authors expect to get more reactions from technicians and then assess the results of its implementation in practice. This monitoring and evaluation will allow future revisions and the refinement of the Manual.

6. References

Arriaza, M; Cañas-Ortega, JF; Cañas-Madueño, JA; Ruiz-Aviles, P. (2004). Assessing the visual quality of rural landscapes. *Landscape and Urban Planning* 69, pp.115-125. Elsevier B.V.

Bell, Simon (1993). *Elements of visual design in the Landscape.* United Kingdom: Spon Press.

Bell, Simon (1999). *Landscape: pattern, perception, and processs.* London : E & FN Spon.

Betz, Mary. (1984). *Methodology for visual resource evaluation in transmission line environmental impact assessment.* Paper presented in Facility siting and routing 1984 – Energy and Enrironment – Banff, Alberta.

Bishop, I. (2003). Assessment of visual qualities, impacts and behaviours, in landscape, by using measures of visibility. *Environment and Planning,* vol 30, pp677-688.

Blair, William. (1986). Visual Impact Assessment in Urban Environments. *Foundations for visual Project analysis,* edited by Smardon,R; Palmer, James; Felleman,J. wiley-interscience publication.

Brown, Thomas; Daniel, Terry. (1987). Context effects in perceived environmental quality assessement: scene selection and Landscape quality ratings. *Journal of environmental Psychology.* N°7. pp.233-250. Academic Press Limited.

Busquets i Fàbregas, Jaume dir. (2007). *Buenas Práticas de Paisaje: líneas de guia, Generalitat de Catalunya.* Departament de Política Territorial i Obras Plúbliques – Direcciò General d'Architectura i Paisatge, Barcelona, Dez. 2007

Craik, Kenneth; Zube, Ervin (1975). *Issues in Perceived Environmental Quality Research.* U.S.A.: Institute for Man and Environment; University of Massachussets (Amherst).

Crowe, Sylvia. (1958). *The Landscape of Power.* London: The Architectural Press.

Daniel, T.C.; Vinning. (1974). Methological issues in the assessement of Landscape quality, *Behaviour and Natural Environment.*

Daniel, Terry. (2001). Whither scenic beauty? Visual landscape quality assessment in the 21st century. *Landscape and Urban Planning* 54. pp.267-281.Elsevier Science B.V.

Delgado Mateo, Santiago (2003). *Metodología para la realización de los estudios de impacto paisajístico en líneas eléctricas de transporte.* Tese de doutoramento: E.T.S.I. Agrónomos (UPM).

EDP. *Recomendações para a construção de novas linhas em áreas classificadas com estatuto de protecção   de* natureza..http://www.edp.pt/EDPI/Internet/PT/Group/Sustainability/Environ ment/ImpactEvaluation/RecommendationsNewLines.htm , accessed Fev. 2009

Electrobras. (1988*). Manual de construção de Redes.* Vol.6. Rio de Janeiro: Editora Campus.

Entidade Reguladora dos Serviços Energéticos. (2007). *Ligações às Redes de transporte e distribuição de Energia Eléctrica: Resumo de disposições aplicáveis.* Lisboa.

Escribano, Ascensión. (2000). *Propuesta de una metodologia de análisis del paisaje para la integración visual de actuaciones forestales: de la planificación al diseño.* Tese de Doutoramento. Universidad Politécnica de Madrid: Escuela Técnica Superior de Ingenieros de Montes. Madrid.

Fabos, Julius (1973). *Model for Landscape Resource Assessment. Amherst.* Water Resources Research Center University of Massachusetts at Amherst.

Fry ,G., Tveit , M.S., Ode, Å., Velarde, M.D. (2009). The ecology of visual landscapes: Exploring the conceptual common ground of visual and ecological landscape indicators, *Ecological Indicators*, Volume 9, Issue 5, pp. 933-947

Ghosn, Rania (ed.) (2010). *New Geographies 2: Landscapes of Energy.* Cambridge Mass: Harvard University Graduate School of Design.

Gill, R. (2005). *Electric transmission line routing using a decision landscape based methodology.* Tese de mestrado. College of engineering and the faculty of the graduate school of Wichita state university.

Gill, R; Jewel, W; Grossardt, T.; Bailey, K. (2006). *Landscape features in routing transmission line routing.* IEEE xplore.

Higuchi, Tadahiko (1988). *The visual and spatial structure of landscapes.* Cambridge, Mass.: MIT Press

Hull, R.B.; Bishop, I. (1988). Scenic impacts of electricity transmission towers: the influence of landscape type and observer distance, *Journal of Environment Management*, 27, pp.99-108.

Jensen, Ron. (1996). *Artists and the New infrastructure.* Places: Place debate: phoenix public art plan. p.59.

Jorgensen ,A. (2011). Beyond the view: Future directions in landscape aesthetics research, *Landscape and Urban Planning*, Volume 100, Issue 4, pp. 353-355.

Krause, Christian. (2001). Our visual landscape Managing the landscape under special consideration of visual aspects. *Landscape and Urban planning* n°54. pp.239-254. Elsevier Science B.V.

Kroloff, R. *From Infrastructure to identity*, Places.

Marshall, R; Baxter, R. (2002). Strategic Routering and Environmental Impact Assessment for Overhead Electrical Transmission Lines, *Journal of Environmental Planning and Management.* 45(5), pp. 747-764.

National Grid. *A Sense of Place: Design Guidelines for development near high voltage overhead lines.*< www.nationalgrid.com> accessed June 2010

Ode, Å., Sundli Tveit, M., Fry, G. (2010). Advantages of using different data sources in assessment of landscape change and its effect on visual scale, *Ecological Indicators*, Volume 10, Issue 1, pp. 24-31

Rodrigues, M., Montañés, C., Fueyo, N. (2010). A method for the assessment of the visual impact caused by the large-scale deployment of renewable-energy facilities, *Environmental Impact Assessment* Review, Volume 30, Issue 4, pp. 240-246

Ross, Robert. (1979). The Bureau of Land Management and Visual Resource Management – An Overview. Our National Landscape: A Conference on Applied Techniques for Analysis and Management of the Visual Resource, pp 666-670. Berkley: United States Department of Agriculture: Forest Service

Sanoff, Henry (1991). *Visual research methods in design.* New York : Van Nostrand Reinhold.

Smardon, R.C.; Palmer, J.F.; Felleman, J.P. (1986). *Foundations for Visual Project Analysis.* New York: John wiley and Sons.

Smardon, Richard. (1979). *Protype Visual Impact Assessment Manual.* School of landscape architecture, University of New York

Soini, K. Pouta, E; Salmiovirta,M; Uusitalo,M; Kivinen, T. *Perceptions of power transmission lines among local residents: A case study from Finland.*

Soini, K., Pouta , E., Salmiovirta, M., Uusitalo , M., Kivinen, T. (2011). Local residents' perceptions of energy landscape: the case of transmission lines, *Land Use Policy*, Volume 28, Issue 1, pp. 294-305

Taylor, J; Zube, E; Sell, J. (1987) *Landscape Assessement and perception research methods. Methods in Environmental and Behavioral research.* In: Bechtel, R.B., Marans, R.W., Michelson, W., (Eds.), Chapter 12.pp. 361-393. New York: Van Nostrand Reinhold Company.

The Landscape Institute; Institute of Environmental Management & Assessment. (2003). *Guidelines for Landscape and Visual Impact Assessment.* Second Edition. London: Spon Press

Watson, David. *Zone of Visual Impact Analysis.* < www.davidwatson/zvi.php> accessed March 2009.

Yeomans, W.C. (1979). *A Proposed Biophysical Approach to Visual Absorption Capability.*

Zube, Ervin (1976). *Studies in Landscape Perception.* U.S.A.: Institute for Man and Environment; University of Massachusetts (Amherst).

Zube, Ervin; Pitt, David; Anderson, Thomas (1974). *Perception and Measurement of Scenic Resources in the Southern Connecticut River Valley.* U.S.A.: Institute for Man and His Environment; University of Massachusetts (Amherst).

11

Ecological Landscape Planning, with a Focus on the Coastal Zone

Canan Cengiz

Bartın University, Faculty of Forestry, Department of Landscape Architecture
Turkey

1. Introduction

How people utilize the land and their socio-economic activity taking place are the principal causes of the changes occurring in land cover, and thus, affecting greatly environment at a global level. However, the land use and management decisions taken often fail to consider this influence on ecology.

Human activity reflected particularly in urban and agricultural land use alters the capacity of the Earth, through its impacts on the physical material and ecological systems. This in turn adversely influences the basic resources that humans need and together with the gradual population growth lead to significant changes in land use (Dale et al., 2001). Despite this significance, however, land use decisions often neglect these impacts. Land use and land management should seek to establish a balance between different and often-conflicting interests regarding the use of the land, such as resource extraction, agriculture, industry, urban development, and complex ecological systems (ESA Committee on Land Use, 2000).

As some species and resources need structural and functional integrity of landscapes, landscape conservation approaches should be present in decision-making processes. Moreover, the multi-faceted characteristics of environmental problems necessitate the incorporation of ecological, socio-cultural and economic approaches in planning and the cooperation of individuals from different disciplines. Ecological landscape plans present a significant opportunity for implementing landscape conservation approaches and for contributing to the sustainability of landscapes. Ecological landscape planning has five stages: division into landscapes, inventory of nature conservation value and socio-cultural factors, landscape analysis, landscape plan, regeneration of biotopes (SCA Skog, 2011). The following section presents an overview of landscape ecology and details its principles and their use in landscape planning. The third section discusses the integration of ecological planning approaches in landscape planning, the fourth section coastal zone planning.

2. Landscape ecology principles and landscape planning

What landscape planning signifies today used to be considered within the concept land use planning until three decades ago. Landscape planning, as a concept, emerged due to the growing awareness and concerns about problems and the developments that took place in the society (Marsh, 2005). Although similar at first sight with land use planning, as both of them deal with the macro environment, landscape planning focuses on the resources and systems of landscape in the planning and management decisions.

Coined in the late 1930s and developed thanks to aerial photography, landscape ecology originally focused on the spatial patterns created by the environment and vegetation. Ecology studies the interactions of organisms with their environment, and a landscape is a mosaic with ecosystems and land uses. Landscape ecology focuses on heterogeneous land mosaics, where the distribution, movement and flow of living beings and materials could be easily observed and foreseen. The principles of landscape ecology, particularly taking the landscape as the unit of study, later gained prominence in landscape planning. Several authors, like McHarg (1969) and Steiner (1991), sought to bridge the gap between landscape ecology and planning and gave way to the development of ecological approaches of landscape planning. More recently, the concept 'ecological landscape planning' has gained prominence (Cook & Lier, 1994). Whereas it is commonplace in landscape planning to use administrative boundaries or watersheds (Cook & Lier, 1994), the methodology of ecological landscape planning is based on landscape ecology. In addition, landscape ecology is related to land evaluation. The focus of land evaluation has changed considerably since the 1960s, from classification and potentiality, to feasibility and lastly sustainable land use in the 1990s (Peng et al., 2006). As both concepts share a common emphasis on social, economic and ecological values, landscape ecology could be utilized in relation with sustainable land use evaluation (Peng et al., 2006; Turner, 1989).

Ecological approaches of landscape planning constitute guidelines that shed light on various steps of planning processes such as data collection and analysis, participation and eventual monitoring (Langevelde, 1994). Ecological principles are functional in maintaining the integrity of landscape by increasing connectivity and minimizing fragmentation and land degradation. Below, four landscape ecological principles are presented: patches, edges and boundaries, corridors and connectivity, and mosaics (Dramstad et al., 1996).

2.1 Patches

Patches can have both positive and negative impacts on landscape. While forest patches between agricultural areas may prove beneficial for the ecological health, a landfill next to a sensitive wetland may have an adverse effect (Dramstad et al., 1996). Below, patches are categorized according to size, number, and location (Table 1).

Patch Size	Patch Number	Patch Location
Edge habitat and species		
Interior habitat and species	Habitat loss	
Local extinction probability	Metapopulation	Extinction
Extinction	dynamics	Recolonization
Habitat diversity	Number of large patches	Patch selection for
Barrier to disturbance	Grouped patches as	conservation
Large patch benefits	habitat	
Small patch benefits		

Table 1. Categorized of patch

2.2 Edges and boundaries

An edge is the outer section of a patch displaying different characteristics than the interior conditions of a patch, in terms of vertical and horizontal structure, width, and species composition and abundance. These differences constitute the edge effect and the edge acting as a transition zone between habitats presents opportunities for landscape planners to facilitate the achievement of an ecological goal. While the shapes of patches can be natural, i.e. due to their boundaries, they can as well be artificial, i.e. administrative, and thus, differ to a varying extent from natural edges (Dramstad et al., 1996). (Table 2).

Edge	Boundaries	Shapes of patches
• Edge structural diversity • Edge width • Administrative and natural ecological boundary • Edge as filter • Edge abruptness	• Natural and human edges • Straight and curvilinear boundaries • Hard and soft boundaries • Edge curvilinearity and width • Coves and lobes	• Edge and interior species • Interaction with surroundings • Ecologically "optimum" patch shape • Shape and orientation

Table 2. Edge, boundaries, and shapes of patches

2.3 Corridors and connectivity

Habitat loss and isolation, results of spatial processes such as fragmentation, dissection, perforation, shrinkage and attrition, necessitate the establishment of connections within the landscape. In the face of these challenges, it is ever more fundamental to preserve the integrity of landscape corridors, such as wildlife corridors and river systems can as well be thought as barriers to wildlife movement, as in the example of roadways, railroad and canals (Dramstad et al., 1996). Pattern and scale can be used to assess the integrity of a landscape (Table 3).

Corridors	Barriers	Stream and River Corridors
For species movement	**Road and windreak barriers**	
Controls on corridors functions	Roads and other "trough" corridors	Stream corridor and dissolved substances
Corridor gap effectiveness		
Structural versus floristic similarity	Wind erosion and its control	Corridor width for main stream
Stepping Stones		Corridor width for a river
		Connectivity of a stream corridor
Stepping stone connectivity		
Distance between stepping stones		
Loss of a stepping stone		
Cluster of stepping stones		

Table 3. Catagorize of corridors and connectivity

2.4 Mosaics

As abovementioned, the connectivity of the corridors within a landscape is an indicator of its ecological condition. These corridors often form networks of connectivity, circuitry, and mesh size and are useful for planners to assist movements across a land mosaic (Dramstad et al., 1996) (Table 4).

Networks	Fragmentation and Pattern	Scale
• Network connectivity and circuitry • Loops and alternatives • Corridor density and mesh size • Intersection effect • Species in a small connected patch • Dispersal and small connected patch	• Loss of total versus interior habitat • Fractal patches • Suburbanization, exotics, and protected areas	• Grain size of mosaics • Animal perceptions of scale of fragmentation • Specialists and generalists • Mosaic patterns for multihabitat species

Table 4. Networks, fragmentation and pattern, and scale

3. Ecological landscape planning

The conservation movement emerged in as early as the 19th century, as a response to the negative effects of development on the land. Despite its long history, it was not until the 1960s when landscape planner Ian L. McHarg started to advocate for an 'integrated landscape planning' approach indicating the application of the concept in planning to establish a balance between human activity on land cover and the environment (Marsh, 2005). The influence of McHarg was significant in the adoption of an ecological perspective in land use planning, which started to take into consideration the carrying capacity of the environment.

Ecological planning developed in the mid-19th century as part of landscape planning, which seeks to safeguard the land and looks for the optimum development of ecologic-biological diversity, structural and visual diversity of the landscape (Ayaşlıgil, 1997, as cited in Tozar & Ayaşlıgil, 2008). It creates and preserves an optimum landscape pattern in terms of ecology, structure and visual aspects by protecting natural resources and seeks to balance and reduce the adverse effects of different land uses by different sectors to a minimum level (Tozar & Ayaşlıgil, 2008). In doing so, ecological landscape planning prioritizes the complex biophysical and sociocultural relationships taking place within a bioregion (Smart Communities Network, 2004). Moreover, another aspect of ecology-based planning is the emphasis it places on not only natural factors but also social and cultural processes that should be involved in the decisions taken about the land use (Cengiz, 2009; Dale & Haeuber, 2001; Markhzoumi & Pungetti 1999).

A planning methodology should consider the injuring activities in an area, their impact on ecology, and types of use affected by these activities (Bierhals et al., 1974, as cited in Altan, 1982). The Integrated Ecological Assessment (IEA) is a useful method in understanding factors related to ecosystems, factors related to human activity on land cover, and the synthesis of these factors (Bourgeron et al., 2001). Considering the scope of the IEA, its use in an assessment study should follow three steps: integration of data (i.e. biophysical, biological, land use, and socio-economic data) (Fig. 1.), relationships between different analytical and planning activities, and depiction of spatial boundaries.

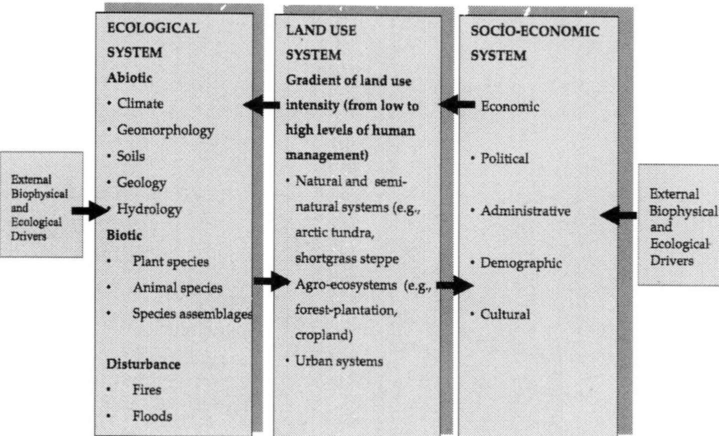

Fig. 1. A simple model of the relationships among regional biophysical, biological, land-use and socio-economic systems

Planning, particularly environmental planning, is a profession primarily related with the use of resources. Its prominence has increased within the last three decades due to the growing awareness about scarce natural resources and the ecological and cultural values at risk, in the face of wide-ranging problems, such as desertification, impacts of sectors such as industry and tourism, and filling wetlands (Marsh, 2005). Thus, key ecological principles are diversified, too, and they constitute a series of guidelines that demonstrate how these ecological principles are used in the decisions about land use and management (Dale et al., 2001) and consider various contexts in fulfilling multi-faceted goals (Table 5).

Examine the impacts of local decisions in a regional context,
Plan for long term change and unexpected events,
Preserve rare landscape elements, critical habitats, ans associated species,
Avoid land uses that deplete natural resources over a broad area,
Retain large contiguous or connected areas tht contain critical habitats,
Minimize the introduction and spread of nonnative species,
Avoid or compensate for effects of development on ecological processes,
Implement land use and land management practices that are compatible with the natural potential of the area.

Table 5. Checklist of factors to be considered in making a land-use decision.

Landscape planning is undergoing change due to new requirements. Its previous main task of controlling spatial uses and the development of nature and the landscape has extended. Implementation of the European requirements for the Natura 2000 network, for the Water Framework Directive (WFD), the Floods Directive as well as the Strategic Environmental Assessment (SEA) can be made considerably easier and can be coordinated with the help of landscape planning.

"As general, coordinating planning, within the scope of landscape planning, existing nature conservation concepts are merged and the nature conservation sub-objectives are coordinated with each other and possible alternative objectives and measures are named. The landscape planning plans are therefore also the suitable instrument for cross-sectionally oriented coordination of nature conservation and landscape management issues with other interests and claims "(Fig. 2) (Haaren et al., 2008).

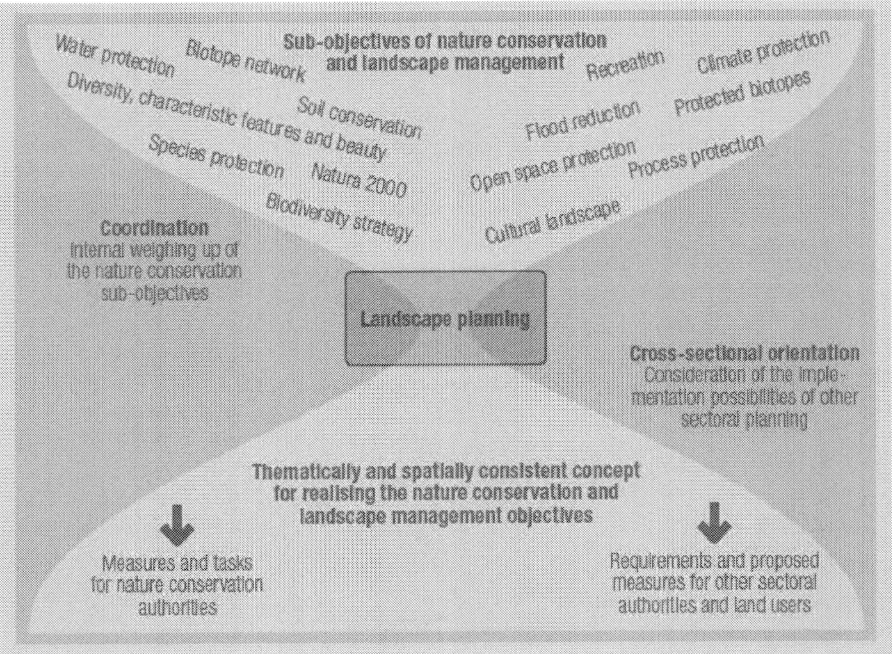

Fig. 2. Coordination and cross-sectional orientation in landscape planning in Germany (Haaren et al., 2008).

3.1 Ecological landscape planning methods

Ecological planning methods are examined under two main headings: Landscape Suitability Approach I (LSA I) and Landscape Suitability Approach II (LSA II). The five methods of the LSA-I that uses natural landscape characteristics in determining the suitability of a piece of land for a certain land use are as follows: the Gestalt method, the Natural Resources

Conservation Service capability system, the physiographic-unit method, the resource-pattern method, and the suitability method. Landscape suitability approach II brought about some refinements and new approaches in both theory and method. The suitability of the landscape is determined by the dialectical balance between the economic, social and biophysical factors. The methods of LSA II are landscape classification method, landscape-resource survey method, allocation-evaluation method, strategic suitability method, Australian approach to regional land use planning, Steiner method and Golany method (Tozar & Ayaşlıgil, 2008).

In the last three decades, the growing social consciousness about the negative effects of human activity on the nature and the increasing number of environmental laws worldwide necessitated the development of nature protection methods. Consequently, significant theoretical developments took place in LSA. The variations of LSA are among the most common methods used in ecologic planning. The LSA methods are applied in two basic stages (Tozar & Ayaşlıgil, 2008). In these stages,

- The area is divided into identical cells (similar features and same dimensions), and
- The suitability of each cell could be analysed according to different criteria and techniques for each type of land use.

3.2 GIS and remote sensing usage in planning

Lenz & Stary (1995) point out the need to use clear concepts in planning processes. The Geographic Information System (GIS) techniques are useful in this respect as they allow for multi-disciplinary approaches and complex data to be used in landscape planning in a comprehensible way (Cengiz, 2009; Cengiz et al., 2011; Ozyavuz & Yazgan, 2010; Tixerant et al., 2010). While generic classifications that lacked in depth and detail were used widely in forming landscape typologies (Antrop, 2000, as cited in Van Eetvelde & Antrop, 2007), recent practices make it possible to obtain detailed landscape typologies, such as zoning which demonstrates different uses in the selected areas (Clark, 1996) and the typologies based on GIS techniques like overlay and spatial analysis. A landscape classification depends on aspects such as the aim of classification, the way of defining the landscape through typology or chorology, the method chosen between holistic and parametric methods, the data quality and the hierarchical scales (Van Eetvelde & Antrop, 2007).

Use of GIS in landscape planning brings substantial advantages. Content and technical requirements and standards are necessary which enable reciprocal data exchange so that the data of other sectoral administrations can be used for landscape planning and so that the results of landscape planning can be incorporated in the information systems and planning of other technical disciplines. Consideration of the relevant existing standards (e.g. ISO standard 19115 for documentation using metadata) in GIS-assisted landscape planning makes it easier, or indeed makes it possible in the first place, to forward and make use of the data and information acquired within the scope of landscape planning in a relatively uncomplicated way. In addition, it would be important to increase the use of standard methods, classifications and structuring in order to merge information from different landscape planning, e.g. within the scope of an SEA or Environmental Impact Assessment (EIA). The use of GIS supports the integration of landscape planning content during the planning process is made easier (Haaren et al., 2008):

- The plan produced is no longer a comprehensive data packet which remains unchanged until it is updated. The use of GIS enables the plans to be updated as needed with little effort. Independent of this, the nature conservation concept must be evaluated at suitable intervals and changed if necessary.

- The data on which landscape planning is based can be directly evaluated for pending planning tasks and if necessary linked with other information. This makes it easier to use landscape planning for other planning, because the planning authorities can specifically retrieve the contents of landscape planning according to their requirements.

"Remote sensing – by satellites such as LANDSAT (USA) and SPOT (France) – is very helpful in planning process. The high spatial resolution of multiband radiometers on LANDSAT and SPOT, well proved for land survey, also works moderately well for shallow-water survey (where waters are clear and cloud cover is low). Remote data have their best use in coastal zone planning and management when coupled to digital mapping and GIS technology (Salm, et al, 2000)."

"Remote Sensing Platforms: Light reflectance-based remote sensing technologies can generally be grouped according to the resolution (pixel size) of the resulting data. This resolution is affected by both the altitude of the platform from which data are collected and the design of the instrument or camera. Low-resolution satellite platforms such as NASA's SeaWIFS (Sea Viewing Wide Field-of-View Sensor) and NOAA's AVHRR (Advanced Very High Resolution Radiometer) produce images where each pixel represents an area of 1 to 10 sq. km. Moderate-resolution satellite platforms such as LandSat, SPOT, and human-occupied spacecraft (e.g., the Space Shuttle or International Space Station) produce images where each pixel represents an area of 10 - 30 sq. meters. Instruments mounted on fixed wing aircraft and helicopter platforms produce images where each pixel represents an area of 1 - 5 sq. meters. Classified remote sensing platforms from the National Technical Means (NTM) Programme produce images where each pixel represents an area of less than 1 sq. meter (Salm, et al, 2000)."

4. Coastal zone planning

4.1 Coastal zone

Unlike lakesides and river sides, sea coasts enjoy a special status due to being a zone of transition between the land and sea, two great compositions in the world. This status has turned sea coasts into a resource that should be utilized in terms of certain uses and functions. These uses or functions include (Arslan, 1988):

- Providing industrial or urban settlement in ports through a connection between sea transport and land transport,
- Providing settlement areas to make use of marine products,
- Discharging urban and industrial waste in a cost-effective way,
- Providing as many ideal settlement areas as possible in order to benefit from the positive influences of the sea and beach on climate,
- Establishing favorable environments for agricultural uses,
- Providing favorable areas for tourism activities with their visual features and natural resources.

Ecological features and natural resources found in coastal zones, which penetrate from the coastline into the land to a certain degree, have an influence on human life and make it possible for them to benefit from coastal zones in different ways. Among the natural resources in coastal zones are (Arslan, 1988):

- Wetted areas and outfall bays,
- Alluvial pools, which resemble lagoons in shape, formed in coastal areas following tidal currents,
- Natural resources that must be protected for future scientific research, and educational, instructional and social activities,
- Arable areas and those areas that are suitable for forestry,
- Reserve areas,
- Mineral deposits,
- Beaches and dunes,
- Areas and waters that can be used for recreation,
- Visual features.

In addition to their natural structures and biological diversities, coasts are an ecosystem in which nature is connected to cultural texture and different types of flora/fauna communities with different characteristics are enabled to live, reproduce and grow. Establishing a strong link between land and sea resources, coastal ecosystems play a key role in regulating the life quality of living creatures. The circle in coastal ecosystems is closely intertwined with and depends heavily on the natural structure of coastal zones, their geological features, their micro-climatic impacts, their hydrologic features, their flora and fauna, their soil structure, human activities, cultural structures and the way human beings use water. Coastal zones are dynamic compositions that can be different depending on the quality and intensity of human activities on them. One of the most significant factors in landscape, population growth has rapidly caused human beings to diversify their demands on coastal resources, which, in turn, has led to an increase in use pressure on unit area.

Three fundamental living environments (water, air and land) today are irreversibly polluted and destroyed as a result of rapid structuring and industrialization, population growth, insufficient awareness of environment, uncontrolled human activities and poor land use planning decisions. Ecological balance is becoming more and more upset, which leads to environmental pressure on natural ecosystems about the continuance of biological environments and sustainability of resources. Their distinctive ecological properties and cultural values make coastal areas delicate landscapes (Lindgren, 2010; Scialabba, 1998; Cicin-Sain & Knetch, 1998). Besides, their high landscape value enables multi-faceted spatial solutions to be developed. An example of these solutions would be cultural use for touristic and recreational purposes (Marin et al., 2009; Scialabba, 1998). Coastal destruction is the most common form of natural deterioration. Fill areas that are formed in these regions cause destruction of the fauna through filling of the sea. Therefore, after the construction of fortifications, designs should be developed to be reestablish the balance between the sea and the flora, considering the natural species of that region (Cengiz et al., 2012). It is essential that ecological planning approaches based on the balance between protection and use should be developed in order to minimize environmental pressures on coastal zones and to sustain these delicate areas.

Accordingly, the guidelines to be considered in coastal planning should focus on (Clark, 1996):

- A high-quality protection of coastal zones,
- Defining high-quality zones to be protected,
- Defining and protecting delicate coastal habitats,
- Defining special areas and habitats that are suitable for development,
- Determining and controlling the level of pollution from point sources through surface flows,
- Determining the economic structure and environmental pressures that have an influence on the protection and development of coastal zones,
- Raising public awareness.

4.2 Boundaries and zoning in coastal zone

"One of the major problems in attempts to conserve coastal and marine ecosystems is determining their boundaries to use them in the protected area design. Protected area boundaries used to be dependent mainly on three variables, namely geological features, political districts or costs. If ecological boundaries cannot be identified in an appropriate manner, the result will be inappropriate boundaries and zoning of the protected area. There is no consensus on the ideal size and design of Marine Protected Area (MPA), some being in favor of "disaggregation" whereas others favoring "aggregation". Although the method of disaggregation is suitable for terrestrial protected areas, they are not equally effective when it comes to underwater areas. The best approach to the latter group of areas seems to be "aggregation" coupled with an effective use zoning scheme (Salm et al., 2000)."

"The requirements of local residents, tourism development and the conservation values and needs within an MPA often conflict with each other. It is possible to make tourism in MPAs harmonious with conversation of most areas. Even so, the construction of tourist facilities around places bordering the MPA might lead to certain damage. MPAs are often designed so that they allow for controlled and sustainable uses within their boundaries. However, the MPA should have certain zones allocated for certain appropriate uses. The method of zoning is commonly used to make sure that the most sensitive and ecologically valuable areas are free of people and the impact of visitors is limited (Salm et al., 2000)."

"One activity might be more suitable for a habitat than others. Therefore, areas should be zoned in a way that i) damaging activities are kept out of sensitive habitats, ii) intensive use is permitted only in certain sites, and iii) conflicts are prevented through a separation of incompatible activities (Salm et al., 2000)."

Zoning methodology

Activities in management zones are designed in reference to the objectives of the reserve. The intensity of management changes from one zone to another.

"Defining the core zones, or sanctuaries. "Core zones" are defined as habitats with high conversation values vulnerable to disturbances. Such areas can be used by humans only at a

minimum level and managed for a high level of protection. In accordance with both conservation objectives and replenishing depleted stocks, areas should be allocated for a breeding population of the key species and their support systems. Core zones should be designed so that they will contain as many diverse habitats as possible (Salm et al., 2000)."

"Defining the use zones. Dedicated zones in a protected area are sites with special conservation value and can tolerate different types of uses by human beings. It is useful to map different neighboring habitats and to ensure that the protected area boundary has as many of these as possible. There should be harmony between the types and locations of required zones and the range of activities. Areas among and around these zones can be considered as general conservation zones (Salm et al., 2000)."

"Defining buffer zones. It is sometimes necessary to have a buffer zone which allows for a more liberal but still controlled range of uses. Such zones are set up in order to protect the area from encroachment and other activities that might have an impact on ecosystems. Buffer zones are a significant way of keeping external influences out of MPAs, for currents can carry nutrients, pollutants and sediments over great distances. The way an external buffer and the MPA is managed is different, the latter requiring cooperation of authorities outside of the MPA (Salm et al., 2000)."

4.3 Emergence of coastal zone planning in the U.S.A and Europe

Ecological planning in European countries today is supported by laws and regulations and integrated into the hierarchy of planning. In The Federal Republic of Germany between 1950 and 1970, for instance, attempts to rapid industrialization for developmental purposes led to irreversible damage to natural environment. In response to this, the Nature Conservation Act and relevant regulations were accepted in 1970s so that natural resources would be enabled to continue their ecological and biological functions. These laws and regulations included such precautions as (Kiemstedt, 1998):

- Ensuring functionality of current natural resources,
- Protecting natural resources and making them usable,
- Ensuring the continuance of landscapes formed by natural creatures.

Those who are responsible for making and implementing decisions with physical plans are governed by these laws and regulations. These plans consist of such stages as:

- Landscape framework plan,
- Landscape program,
- Land use plan.

The United Nations Conference on Environment and Development (1992) discussed, among other things, the protection and rational use of coastal zones and emphasized the importance of holistic coastal management. Similarly, it was emphasized, in the 787th meeting of the Organization for Economic Cooperation and Development in July 23, 1992, that strategic planning and holistic approach to coastal issues should be developed and a reasonable balance should be established between carrying capacity and tourism development in coastal zones (Anonymous, 1993).

In the U.S.A, where coastal areas were dominated by certain land uses such as business, industry, housing, tourism and recreation, coastal protection and planning attempts started in California in 1967. In 1972, the government accepted the Coastal Zone Management Act in order to protect, develop and utilize resources in coastal zones. According to popular opinion in the U.S.A, the state where the coastal zone project was first put into effect was California. The coastal zone of California is 1770 km in length, from the Mexican boundary to the southern-western boundary of Oregon. The propositions included in the project were (Arslan, 1988):

- **Coastal Waters:** With the aim of maintaining the quality of coastal waters, to launch a fund for refining all dangerous pollutants and effluent water before the discharge, to compensate for the damage caused by oil spills in a quick manner and to clean the waters as soon as possible,
- **Soil:** With the aim of protecting the flora in coastal zones, to conduct meticulous studies, to monitor sandpits and stone pits, to protect areas where salmons and whitefish lay their eggs, to open fertile plains and valleys only to agricultural activities,
- **Coastal View:** To ensure that such large facilities as industry centers and shopping malls are founded into the land and that urban development is in consistent with natural structure,
- **Development:** To ensure that new developments are concentrated on places where infrastructure is favorable, that rural area planning does not change natural character, that no power plant is included unless they are really necessary and that special precautions are taken for protecting coastal features in all respects,
- **Energy:** To ensure that a policy of energy conservation is adopted in coastal zones and that there are incentives to undertake projects on obtaining energy from the sun, wind, geothermal resources and methane gases,
- **Transportation:** To use air and sea transportation services at full capacity,
- **Coastal Accessibility:** To ensure that the public can access to the coast,
- **Recreation:** To identify the zones that are used for recreational purposes (intensive, medium, low), to determine the carrying capacity of the zone and to compose management plans,
- **Educational and Scientific Use:** To ensure that special precautions are taken in order to protect areas of historical, archeological, educational or scientific value,
- **Restoration of Coastal Resources:** To develop landscape restoration techniques in order to protect and rehabilitate the coasts whose ecological features have been disrupted,
- **Expenditures:** To launch a fund for long-term expenditures of the project.

Cancún (Mexico) is a good example of threats brought about by coastal tourism and precautions taken. Cancún is located on the Yucatan peninsula, eastern Mexico, and a worldwide famous holiday spot. In 1970, the city was defined as a significant tourist attraction because of its climatic conditions, a 19 km-lenght white beach, population of coconut palms, clean water, rich water products, a 50 km²-lenght well-protected lagoon system, being the second largest coral reef barrier of the world that starts 30 m off the coast, having rich underground water resources and being an important archeological area. The distinguishing natural feature of Cancún is the Nichupte Lagoon System (Clark, 1996).

The basic problem here was the increase in the number of holiday resorts established in the city, which had been developing as a holiday spot. The increases in the areas of employment, the number of annual tourists and the number of accommodations were regarded as an accomplishment; however, environmental quality was neglected. Among the indicators of the change in environmental quality were (Clark, 1996):

- Disruptions in the Nichupte Lagoon System,
- Impacts on underground water resources,
- Openings in the flora along the highways,
- Decreases in the number of beaches and dunes,
- Problems associated with accessing or using beaches and dunes, and
- Visual pollution. The economy was based on tourism, which led to increases in the amount of pressure.

Planning and rehabilitation activities for the City of Cancún were focused on:

- Urban development based on protecting the nature,
- Protecting the environment, ways of using natural resources,
- Revision of nature-human relationship,
- Aesthetic and functional regulations in the area of tourism. The scope of the planning was mainly defined as the harmony between the natural environment and urban components and the integration of homogenous, dynamic and proper complexes. The areas of tourism were reorganized through precautions that imposed a limit on the use of natural resources with a consideration into aesthetics and functionality.

5. Conclusion

"A major problem in land use and management is achieving a reconciliation among such conflicting goals as resource-extractive activities, infrastructure of human settlement, recreational activities, services provided by ecological systems, support of aesthetic, cultural and religious values and maintaining the compositional and structural complexity of ecological systems. The more different goals stakeholders have, the more difficult land use decisions can be. The process of planning and decision-making should consider the ecological, socio-cultural and economic values of the landscape so as to achieve a new spirit of reconciliation between landscape conversation and changing demands (Dale et al., 2001)."

"It seems that the problems in ecological planning are brought about by disagreements between land use and environment. These disagreements are usually of five origins: i) initially poor land use decisions ii) environmental change iii) social change as well as technological change iv) violations of human values about the mistreatment of the environment v) geographic or spatial scale (Marsh, 2005)."

The leading reasons for disruptions in the natural structure of the land in coastal zones are irrational settlement processes, irregular and unplanned recreational and touristic facilities, industrial uses, agricultural activities, fill areas and highways that are too near to coasts. The increase in the number of ecosystems ruined by planning decisions in planning attempts of several scales without a consideration into natural resources makes ecological planning obligatory. Ecological planning approaches are defined as reflections of practices on environmental values within the framework of principles of sensitive area use in the relationship between human beings and the nature (Langevelde, 1994). Ecology-based planning should take natural and social processes into account (Markhzoumi & Pungetti, 1999). Instead of short-term gains, coastal zones should be used sustainably and consciously with a balance between protection and use and they should be maintained as natural heritages. They have the potential to make contributions to the economy of a country or region for long years. Therefore, it is evident that the main objective in future practices in coastal zones should be using them in a planned way and handing them down to future generations without disruptions in their current ecological values.

6. References

Altan, T. 1982. A computer aided investigation on the application of ecological landscape planning in regional scale and the determination of the land use suggestions in Cukurova. *PhD thesis* (unpublished). Cukurova University, Institute of Natural Sciences, Department of Landscape Architecture, Adana.

Anonymous, (1993). Coastal zone management, Integrated Policies, *OECD*, Paris.

Arslan, M. (1988). Recreational Planning on the Coastal Areas. *Text book of Ankara University, Graduate School of Natural and Applied Sciences, Department of Landscape Architecture*, Ankara. (Unpublished).

Bourgeron, S.P., Humphries, H.C., Jensen, M.E. and Brown, B.A. (2001). *Integrated ecological assessment and land use planning*. Springer Verlag New York, Inc. USA.

Cengiz, C. (2009). Ecological Planning in Coastal Areas: Yalova-Armutlu Case Study. Ankara, Turkey: Ankara University, Graduate School of Natural and Applied Sciences, Department of Landscape Architecture, *Ph.D. thesis.* (in Turkish).

Cengiz, B., Smardon, R. C. and Memlük, Y. (2011). Assessment of river landscapes in terms of preservation and usage balance: a case study of the Bartın River floodplain corridor (Western Black Sea Region, Turkey), *Fresenius Environmental Bulletin,* 20 (7). pp.1673-1684.

Cengiz, B., Sabaz, M., Bekci, B. and Cengiz, C. (2012). Design strategies for revitalization on The Filyos Coastline, Zonguldak, Turkey. Fresenius Environmental Bulletin, 21(1); 36-47.

Cicin-Sain B; Knetch R.W. (1998). *Integrated Coastal and Ocean Manegement, Concepts and Practices.* Island, Washington, 517p.

Clark, J.R. (1996). *Coastal zone management handbook.* Lewis Publishers, CRC Press LLC, Boca Raton, Florida USA.

Cook, E. A., and Lier, H. V. (1994). Landscape planning and ecological networks: an introduction. *Landscape planning and ecological networks.* Editors: Edward A. Cook and Hubert N. Van Lier. Elsevier Science B. V.Amsterdam, The Netherlands.

Dale, V.H., Brown, S., Haeuber, R.A., Hobbs, N. T., Huntly, N. J., Naiman, R. J., Riebsame, W. E., Turner, M. G., and Valone, T. J. (2001). Ecological Guidlines for Land Use and Management. *Applying ecological principles to land management.* Editors: Virginia H. Dale and Richard A. Haeuber. ISBN 0-387-95099-0 Springer – Verlag New York, Inc. USA.

Dramstad, W. E., Olson, J. D., and Forman, R. T. T. (1996). *Landscape Ecology Principles in Landscape Architecture and Land use planning.* Harvard University, Graduate School of Design. American Society of Landscape Architects, USA.

Eetvelde, V. V. & Antrop, M. 2007. Landscape Identification and Assessment, Examples of Belgium. *Turkey on the way implementation of the European Landscape Act, International Participation Meeting*, May, 17-20. Ankara

ESA Committee on Land Use, (2000). *Ecological Principles for Managing Land Use*, The Ecological Society of America's Committee on Land Use. Available from: (http://www.epa.gov/owow/watershed/wacademy/acad2000/step8esa.html   ). Date of Access: 05.01.2012

Haaren, C. v., Galler, C., Ott, S. (2008). *Landscape Planning, The basis of sustainable landscape development*. Bundesamt für Naturschutz / Federal Agency for Nature Conservation.

Kiemstedt, H. (1998). Requirements of the ecological movement toward Country and regional planning- concrete examples. *Ecological basis International Symposium Proceedings*, Editor: Semra ATABAY. ISBN 9754610932, 9789754610932. January, 18-19. Printing-Publishing Center of Yıldız Technical University, Istanbul.

Langevelde, F. (1994). Conceptual integration of landscape planning and landscape ecology, with a focus on the Netherlands. *Landscape planning and ecological networks*. Editors: Edward A. Cook and Hubert N. Van Lier. Elsevier Science B. V.Amsterdam, The Netherlands.

Lenz, R. J. M. and Stary, R. (1995). Landscape Diversity and Land Use Planning: A Case Study in Bavaria. *Landscape and Urban Planning*. Elsevier Science.Vol. 31. Page, 387-398

Lindgren, D. (2010). Assessment of Ecological Value of Coastal Areas Using Morphometry and Secchi Depth: A Case Study with Data from the Swedish Coast, *Journal of Coastal Research*, Vol 26 (3), 429-435

Marin, V., Palmisani, F., Ivaldi, R., Dursi, R., and Fabiano, M. (2009). Users' perception analysis for sustainable beach management in Italy, *Ocean & Coastal Management*, 52 (2009) 268-277.

Markhzoumi, J. and Pungetti, G. (1999). *Ecological landscape design and planning the Mediterranean context*. New York.

Marsh, W. M. (2005). *Landscape Planning Environmental Applications*, Fourth Edition, John Wiley & Sons, Inc.

Ozyavuz, M. and Yazgan, M. E. (2010). Planning of Igneada Longos (Flooded) Forests as a Biosphere Reserve, Journal *of Coastal Research*, Vol. 26, No. 6.

Peng, J., Wang, Y., Li, W., Yue, J., Wu, J., and Zang, Y. (2006). Evaluation for sustainable land use in coastal areas: A landscape ecological prospect. *International Journal of Sustainable Development & World Ecology* 13 (2006) 25-36

Salm, R. V., Clark, J. R., and Siirila, E. (2000). *Marine and Coastal Protected Areas. A Guide for Planners and Managers*. Third Edition. International Union for Conservation of Nature and Natural Resources Gland, Switzerland .

SCA skog, (2011). *Ecological landscape planning*, Svenska Cellulosa Aktiebolaget SCA (publ). Available from: (http://www.sca.com/en/skog/environmentnature/ecological-landscape-planning-/). Date of Access: 17.01.2012

Scialabba, N. (1998). *Integrated coastal area management and agriculture, forestry and fisheries*. Food and Agriculture organization of the united nations. Rome 1998 ISBN . 92-5-104132-6

Smart Communities Network, (2004). *Key Planning Principles, Ecological Landscape Planning.* Available from: (http://www.smartcommunities.ncat.org/landuse/luecldsc.shtml). Date of Access: 20.01.2012

Tixerant, M. L., Gourmelon, F., Tissot, C., Brosset, D. (2010). Moddeling of Human Activity Development in Coastal Sea Areas, *Journal Coastal Conservation,* DOI 10.1007/s11852-010-0093-4

Tozar, T. & T. Ayaşlıgil, (2008). Ecological planning methods for sustainable management of natural resources. *Journal of the Faculty of Forestry, Istanbul University,* Seri A, Cilt:58. Sayı 1. Istanbul.

Turner M.C. 1989. Landscape ecology, the effect of pattern on process. *Annual Review of Ecology Systematics;* 20:171-9.

Section 2

Landscape Design

12

Landscape Perception

Isil Cakci Kaymaz
Ankara University
Turkey

1. Introduction

"... landscape is composed of not only of what lies before our eyes but what lies within our heads."

D.W. Meinig (1979)

Landscape, as a term, has been subject to a wide range of disciplines, such as art, history, geography, ecology, politics, planning and design. Although it has been associated with mainly physical features of an environment, today the term landscape refers to much more than just scenery. Landscape is a complex phenomenon which evolves continuously through time and space. It is a reflection of both natural processes and cultural changes throughout time. Landscapes can be a product of either only natural processes (natural landscapes) or human intervention on natural ecosystems (cultural landscapes). Nowadays, it is almost impossible to encounter with a natural landscape in our daily lives. Most of the natural landscapes have been modified by human activities. Hence, they are embedded with symbolic meanings of our societies' cultural diversity and identity. On the other hand, the deterioration of natural ecosystems has become an important issue in sustainable development, since we depend on natural resources to survive. Thus, as natural and cultural heritages, landscapes need to be protected and managed in the context of sustainability. In 2000, Council of Europe adopted the European Landscape Convention (ELC) to promote sustainable planning, protection and management of European landscapes. ELC defines landscape as:

"...an area, as perceived by people, whose character is the result of the action and interaction of natural and/or human factors".

The definition of ELC puts an emphasis on the perceptual dimension of the landscape. Since landscape involves a subjective experience, it encompasses a perceptive, artistic and existential meaning (Antrop, 2005). Figure 1 shows the components of a landscape, which hence influence perception of the landscape. There is a mutual relationship between individual and the surrounding environment. People are intrinsically involved with their living environments to survive. They use and shape the physical environment to meet their physical and social needs. While environments are shaped by people, people are inspired and shaped by their environments as well. Thus, perception of the environment or the landscape has become an area of concern of various disciplines in order to understand and explain this interaction between people and their physical settings.

Perception is the process in which information is derived through senses, organized and interpreted. It is an active process which takes place between the organism and environment (Hilgard, 1951 in R. Kaplan & S. Kaplan, 1978). S. Kaplan (1975) states that information is central to organism's survival and essential in making sense out of the environment, to which perception is assumed to be oriented. Perception of our environment helps us to understand and react to our environment. Environmental perception is different to object perception in many ways (Forster, 2010; Ungar, 1999);

- The components of the environment are diverse and complex. Therefore perception of the environment is not immediate and it takes time.
- Scale affects perception of the environment. Environments are larger and, hence more complex systems.
- Environment surrounds people. Thus it is perceived and experienced from inside.
- Navigation skills are needed in environmental perception.
- People usually interact with their environment for a purpose. As a result, we select spatial information related to our purpose.

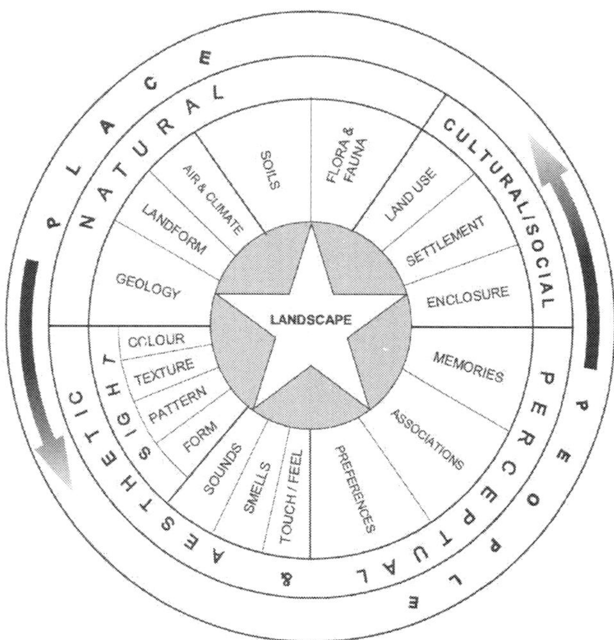

Fig. 1. What is landscape (Swanwick, 2002).

Porteous (1996) discusses that there are two basic modes of perception; autocentric, which is subject centered, and allocentric, which is object centered. He explains that sensory quality and pleasure are involved in autocentric senses, while allocentric senses involve attention and directionality. He states that vision (except color perception) is mostly autocentric, and most sounds (except speech sounds) are autocentric.

The perception of the physical environment is not merely a physiological phenomenon. It is also influenced by the individual's experiences, and both social and cultural factors. Knox and Marston (2003) points out that *"different cultural identities and status categories influence the ways in which people experience and understand their environments"*. Thus, perception of our surrounding environment is learnt, selective, dynamic, interactive and individual (Lee, 1973).

Theories of perception provide foundation for research in psychology. Environmental psychology is the branch of psychology which deals with relationships between physical environment and human behavior. It is a multidisciplinary field where perception of the environment is a fundamental subject. Environmental perception research includes topics such as cognitive mapping, landscape (environmental) preferences, way finding, restorative environments, all which should be considered in landscape planning and design. Landscape architecture aims to create livable, pleasant and sustainable outdoor environments. Although the findings of environmental psychology research can enlighten and influence landscape architects in context of research and practice, it is hard to say that a firm link has been established between two disciplines so far. There is a mutual relationship between people and their physical environments which influences each other. Thus, landscape architects must acknowledge that perception of the environment plays an essential role in comprehension of this relationship.

This chapter presents an overview to landscape (environmental) perception research in context of landscape planning and design. It discusses perception of the landscape based on two fundamental senses; sight and hearing. Firstly, theories and research methodology on visual perception and aesthetics will be presented in order to provide guidance for visual landscape design and planning. Secondly, the concept of soundscape will be briefly introduced and discussed to promote awareness on the importance of sound as a landscape element in design and planning.

2. Visual landscape perception

In landscape planning and environmental impact assessment studies, evaluation of visual landscape character is often based on assessment of physical characteristics of landscapes (such as topography, land cover etc.) and is done by experts. On the other hand public or user preferences are generally neglected. This section aims to present and provide understanding of psychophysical and cognitive dimensions of visual landscape perception for landscape designers and planners.

Although we receive spatial information through many of our senses (seeing, hearing, smelling and feeling) sight is assumed to be the most valued sense. More than 80% of our sensory input is through sight (Porteous, 1996). Hence, most of the environmental perception, and likewise landscape assessment studies, focus on visual dimension of the perception process. Assessment of landscape character is fundamental to decision making process in landscape planning. Landscape assessment is a tool for determination of landscape quality and provides a systematic analysis and classification for sustainable management of landscapes. Within this context, the criteria for landscape perception studies are mostly scenic beauty or preference (Palmer, 2003).

Landscapes are aesthetic objects. There isn't a universally accepted theory for landscape aesthetics. According to Maulan *et al.* (2006) neglect of scenic or preferred landscapes during development stage is one of the problems. Bourassa (1990) argues that landscape aesthetics is beyond the traditional theories of aesthetics. Based on Scruton's approach, he states that people experience and respond to the whole scene, therefore *"it is not relevant to speak of the aesthetics of individual objects in the landscape (e.g. buildings) without asking how those objects contribute to the wholes (landscapes) of which they are only parts"*. For Bourassa (1988) there are two principles for landscape aesthetics, namely biological and cultural. The biological principle states that *"aesthetic pleasure in landscape derives from the dialectic of refuge and prospect"*. On the other hand, *"aesthetic pleasure derives from a landscape that contributes to cultural identity and stability"*. As a product of either natural processes or human intervention, natural and cultural landscapes involve intrinsic (objective) and artistic (subjective) aesthetic values. Thus, theories of aesthetics may provide a basis for landscape scenic beauty assessments. Brief history of aesthetics in philosophical context is given below.

2.1 A brief overview of history of aesthetics

Scenic beauty of the landscape or in a broader sense environmental aesthetics has been an area of concern for assessing visual quality of landscapes and landscape preferences. Although the involvement of aesthetics in environmental psychology and landscape assessment studies does not date back very far, it has been a subject for philosophy since ancient times. The word "aesthetic" is derived from *aisthanesthai*, Greek word for "to perceive" and *aistheta*, which means "perceptible objects" in Greek. The term "aesthetics" was first coined by Alexander Baumgarten, a German philosopher, in 1735. Before that, "beauty" was the focus of the aesthetical debates of philosophers.

The question of "what is beauty" has been central to theories of aesthetics since classical Greek times (Porteous, 1996). According to Socrates, (469-399 B.C.) there is a mutual connection between beauty, truth and symmetry (Hofstadter, 1979 in Barak-Erez & Shapira, 1999). He believed that beauty was desirable for youth and he linked beauty to being good and morality (Lothian, 1999). For Plato (427-347 B.C.), there is an "essential universe", the perfect universe; and there is the "perceived universe" where we perceive the reality through our senses as imperfect copies. Plato believed the beauty was an "idea" and the beauty we perceived in the "perceived universe" was not the real, original beauty, but just an imperfect copy. On the other hand, Aristotle (384-322 B.C.) discusses beauty in context of mathematics. He believed that beauty was associated with size and order, and there were three components of beauty; integrity (*integras*), consonance (*consonantia*), and clarity (*claritas*). Beauty was accepted as a sign of God's existence after Christianity emerged and during medieval times.

With Renaissance, approaches towards aesthetics in ancient Roman and Greek times returned back with the movement *Classicism*. In this period, beauty was associated with order, symmetry, proportion and balance. In the end of 17th century, modern aesthetics emerged in Britain and Germany. For John Locke (1632-1704); *"beauty consists of a certain composition of color and figure causing delight in the beholder"* (Carson, 2002) and therefore, it was a subjective quality. Likewise, British philosophers David Hume and Edmund Burke believed that aesthetics was a subjective concept. According to Hume (1711-1776), people decide whether an object was beautiful or not by their feelings. Burke (1729-1797) identified beauty as a *"social quality"* and linked beauty with the feeling of affection, particularly toward the other sex. According to him, the feeling of the beautiful is grounded in our social nature (Vandenabeele, 2012). On the contrary, German philosopher Immanuel Kant's (1724-2804) approach to aesthetic judgment was based on logic and deduction (Lothian, 1999). He believed aesthetic judgments were based on the feeling of pleasure and they were disinterested. Daniels (2008) explains disinterestedness as *"... a genuine aesthetic judgment does not include any extrinsic considerations toward the object of judgment itself, such as political or utilitarian concerns"*. Therefore, Kant claimed that aesthetic judgments were both subjective and universal. However, German philosophers Friedrich Schiller and Wilhelm Hegel rejected Kant's subjective approach on aesthetics (Lothian, 1999). Schiller (1759- 1805) claimed that beauty was the property of the object, thus aesthetic experience was rather objective. On the other hand, Hegel (1770-1831) believed that aesthetics was concerned with the beauty of art and beauty of art is higher than the beauty of nature. Like Schiller, for Hegel beauty was the property of the object. According to Baumgarten (1714-1762), who coined the term aesthetics, beauty is not connected to the feeling of pleasure or delight, indeed beauty is an intellectual category and perfection of sensitive cognition is a precondition for beauty (Gross, 2002). In 19th century, romanticism focused on nature as an aesthetic resource. In this period, landscape was viewed in objectivist terms and considered as having intrinsic qualities (Lothian, 1999). However, nature lost its importance as an aesthetic object by the end of 19th century and during the 20th century art has become the main concern for aesthetic debates.

George Santayana, Benedetto Croce, John Dewey and Susanne Langer are amongst the modern era philosophers on aesthetics. Spanish-American philosopher George Santayana (1863-1952) believed that beauty was a subjective concept, rather than objective. He defined beauty as the pleasure derived from perception of an object (Lothian, 1999). Croce (1866-1952) interprets aesthetics as an experience. For Croce, intuition is basis for the sense of beauty. Dewey's (1859-1952) aesthetics is based on experience as well. In contrast to Kant's disinterestedness principle, Dewey's aesthetics require involvement and engagement (Lothian, 1999). While Dewey suggested that aesthetic experience was a biological response, Langer (1895-1985) strongly rejected this idea (Bourassa, 1988). Langer's aesthetics is based on the concept of semblance. According to Langer, semblance of a thing is an aesthetic symbolic form which constitutes its direct aesthetic quality (Kruse, 2007).

Although philosophical theories of aesthetics may seem relatively relevant to landscape assessment, landscape planners and designers need to understand the fundamentals of aesthetic theories of art and nature in order to develop valid and efficient approaches towards evaluation of landscape aesthetics in context of landscape planning and design. According to Berleant (1992), the idea of environment possesses deep philosophical assumptions about our world and ourselves, thus the study of aesthetics and environment can provide mutual benefit in this changing world.

2.2 Theories on perception and preferences

2.2.1 The biophilia hypothesis

The biophilia hypothesis was developed by Edward O. Wilson, biologist in Harvard University, in 1984. The biophilia hypothesis proclaims that human beings have an inherent need for affiliation with natural environments and other forms of life. Wilson suggests that preferences for natural environments have a biological foundation as a result of human's evolutionary process. Since human beings spent most of their evolutionary history in natural environments as hunters and gatherers, they have a hereditary inclination towards establishing an emotional bond with nature and other livings. Ulrich (1993) explains the proposition for biophilia as that during evolution certain rewards or advantages associated with natural settings were crucial for survival and humans acquired, and then retained, positive responses to unthreatening natural settings. He states that human's positive responses to natural settings in terms of such as liking, restoration and enhanced cognitive functioning might be influenced by biologically prepared learning. On the other hand, McVay (1993) questions whether biophilia hypothesis can influence our attitudes towards our world in a more environmental friendly manner. He emphasizes the need for realization of our evolutionary based need for affiliation with nature by everyone who shares the responsibility of human future.

2.2.2 Prospect-refuge theory

British geographer Appleton's prospect-refuge theory stems from his habitat theory which proposes that human beings experience pleasure and satisfaction with landscapes that responds to their biological needs (Porteous, 1996). Appleton's habitat theory basically depends on Darwin's habitat theory, but with an aesthetical dimension. For Appleton, aesthetic satisfaction is "a spontaneous reaction to landscape as a habitat" (Porteous, 1996). On the other hand, prospect-refuge theory is about preferences for landscapes which provide "prospect" and "refuge" opportunities. Prospect-refuge theory is based on human's urge to feel safe and to survive. During our evolutionary past as hunters and gatherers, a broader sight of view and opportunities to hide when in danger were essential for survival. Thus, Appleton believes that we intrinsically tend to prefer environments where we can observe and hide. However, ironically, the places with prospect and refuge opportunities are also favorable for potential offender (Fisher & Nasar, 1992). The offender may hide from, wait for and attack to his victim in environments which offer prospect and refuge. Fisher and Nasar (1992) suggested that places with low prospect and high refuge lead to feelings of fear and unsafety. Although Appleton's theory is concerned with natural environments, physical organization of a space is clearly linked to the feeling of safety. Therefore, same principles can be adapted to design in urban environments.

2.2.3 Berlyne's and Wohlwill's approaches to environmental aesthetics

Exploratory behavior, physiological arousal and experimental aesthetics were amongst the main interest areas of psychologist David E. Berlyne (1924-1976). He developed a psychobiological approach towards aesthetics. According to Berlyne, environmental perception is a process of exploratory behavior and information transmission which are triggered by the amount of conflict or uncertainty in the environment (Chang, 2009). Berlyne's theoretical framework involves two main concepts; arousal potential and hedonic response. He identified four factors, which he called "collative properties" that determined the arousal potential of a stimulus; (i) complexity (diversity of the elements in the environment), (ii) novelty (presence of novel elements), (iii) incongruity (extent of any apparent 'mis-match' between elements), and (iv)surprisingness (presence of unexpected elements) (Ungar, 1999). The arousal potential of the stimulus results in hedonic response in the observer. Berlyne (1972) hypothesized that there is an inverted U-shaped relation between collative properties and hedonic response; increase in arousal also increases pleasure up to a point, however beyond a certain point hedonic response will lessen (Galanter, 2010; Nasar, 1988a). Thus medium degree of arousal potential has a positive effect on preference, while low or high degrees of arousal potential cause negative response (Martindale, 1996).

Wohlwill's studies on environmental aesthetics are based on Berlyne's theory. Both Berlyne and Wohwill regarded arousal and hedonic value as an important aspect of aesthetic response (Nasar, 1988b) Similar to Berlyne, he proposed that there was an optimal level of information in a landscape and too much information was stressful while too little information was boring (Mok *et al.*, 2006). He also extended Berlyne's arousal theory and hypothesized (1974) that there is an adaptation level where environmental stimulation is at optimal degree for an observer and larger changes in the adaptation level produce negative response (Bell *et al.*, 2001; Ungar, 1999). Adaptation level depends on an individual's past experiences, thus it differs from person to person and furthermore changes in time if exposed to a different level of stimulation (Bell *et al.*, 2001).

2.2.4 Information processing theory

Rachel and Stephen Kaplan of University of Michigan are leading researchers in the field of environmental psychology. They have many published works on human-environment relationship. Kaplans' information processing theory (1979) is amongst the most influential and well-known theories on landscape preferences. Information is the fundamental concept of their approach. Information has been central to human experience and survival throughout the evolution of human being (Kaplan *et al.*, 1998). Not only we need to gain information to make sense out of the environment, but an individual also values environments with promising information for exploration (Kaplan *et al.*, 1998; S. Kaplan, 1975). Understanding of an environment aids an individual to know what is going around and feel secure. On the other hand, people want to explore by seeking more information and look for new challenges (Kaplan *et al.*, 1998). Furthermore information is important to people's ability to function well in the environment (Maulan *et al.*, 2006). Aesthetics reflects the functional potential of things and spaces (S. Kaplan, 1988a).

We gather information from our environment through our senses, mostly through visual sense. Kaplans' theory suggests that information is derived through the contents and the organization of the environment. Organization of an environment is an important variable in perception since it affects the degree of making sense. S. Kaplan (1975) states that acquisition of knowledge should be related to environmental preference. Results of their studies show that scenes with large expanses of undifferentiated land covers, dense vegetation and obstructed views are low in preference (Kaplan *et al.*, 1998). They suggest that if visual organization of spaces is homogenous within an environment, then it suggests that nothing is going on. Besides, there is little to focus on and sameness causes difficulty in keeping interest in the environment. On the other hand although dense vegetation has a rich content, it lacks of clear focus which confuses one. People also are discomforted when the view is blocked, they feel insecure because it is hard to tell what to expect. On the contrary scenes with spaced trees and smooth ground have been found to be high in preference. They explain that in contrast to large expanses and obstructed views; such combinations of settings provide a clear focus and invite entry.

Based on their results, the Kaplans developed a preference matrix which comprises of four informational factors which affect preferences of landscape (Figure 2). These factors are; coherence, complexity, legibility and mystery. Coherence and complexity of a setting can be understood as soon as when one enters or views the setting, thus they happen in the picture plane (2D) and they are perceived immediately. In contrast, to perceive legibility and mystery degrees of a setting requires time, an involvement with the environment. Hence, they are inferred factors and this inference about the third dimension occurs in longer (a few milliseconds longer) and unconsciously.

PREFERENCE MATRIX

	Understanding	**Exploration**
2-D	Coherence	Complexity
3-D	Legibility	Mystery

Table 1. Kaplans' preference matrix (Kaplan *et al.*, 1998).

Coherence: Coherence of a setting is about the order and organization of its elements. If a place is coherent, then people can easily make sense out of the setting. Kaplan *et al.* (1998) suggest that coherence can be achieved through repeat of themes and unifying textures; however limited degree of contrast is also helpful. Coherence is similar to gestalt principles of organization that states elements are perceived in groups rather than parts (S. Kaplan, 1975).

Complexity: Complexity refers to the degree of diversity of landscape elements. The more complex an environment is, the more information it involves. According to Kaplans' theory, greater variety in a setting would encourage exploration. They argue that coherence and complexity shouldn't be confused since a highly coherent setting can still also be very complex.

Legibility: The concept of legibility is about orientation. Way-finding is important for an individual in terms of feeling secure and safe. It is about reading the environment and making sense out of it. Distinctiveness contributes to legibility of an environment. Hence, landmarks or focal points may increase the legibility of a setting. However, one has to experience the setting first, in order to realize what is distinctive and what is not. Spaciousness also supports legibility by increasing the individual's range of vision (S. Kaplan, 1975). S. Kaplan, (1975) points out that fine texture is also a legibility component; the finer the texture, the easier to distinguish figures from ground.

Mystery: Mystery is the component of preference related to exploration. It is about the setting's potential of promising information. Mystery requires an inferential process (S. Kaplan, 1975). Mystery motivates people for exploration in order to gain new information. There are various ways to create mystery in a landscape. Kaplan *et al.* (1998) suggest that a curved path or vegetation that partially obstructs the view can add mystery to an environment.

The Kaplans suggest that we prefer environments that involve all of the four components explained above. They also emphasize that information needs to be central in environmental design and management. However, handling and managing information can also be stressful for people. According to Kaplan *et al.* (1998), our capacity for directed attention is limited, and mental fatigue occurs if one is forced to receive and manage information above his capacity. Mental fatigue may cause difficulties in or loss of concentration, impulsive actions, anger and irritability. Hence, the designers should be aware of the risks of creating settings that offers too many information.

2.2.5 Gibson's Theory of Affordances

Psychologist James J. Gibson has developed his "Theory of Affordances" based on an ecological approach towards visual perception. In his work "The Theory of Affordances" (originally published in 1979, 1986) he describes the environment *as the surfaces that separate substances from medium in which the animal lives*. He continues that the environment offers and provides affordances to the animal. The term "affordances" has been first coined by Gibson, himself. An affordance can be described as a possible action which properties of an object allow or suggest for the observer. For example, a bench affords sitting. Affordances are perceived directly and they are relative to the observer. Gibson states that although the needs of observer can change, the affordance of an object does not change. Gibson's theory is rather different from the conventional perception theories. His theory has received criticism, mainly for being unclear and underestimating the complexity of perception process.

2.2.6 Gestalt principles of visual perception

Gestalt theory was developed by German psychologists Max Wertheimer, Kurt Koffka and Wolfgang Köhler in the early 20th century. The German word *die Gestalt* means "form" or "shape" and Gestalt theory of perception can be summarized as that people tend to perceive things as wholes rather than separate parts. It proposes "laws of organization in perceptual forms" (Wertheimer, 1938) which have been applied by various design disciplines. Basically, people perceive visual stimuli as organized or grouped patterns. Gestalt principles related to spatial design are briefly explained below.

Figure-ground relationship: As Köhler (1938) states *"figure perception is represented in the optic field by differences of potential along the entire outline or border of the figure"*. Thus, contrast plays an important role in distinguishing figure from the ground. The most famous example that demonstrates figure-ground relationship is probably the Danish psychologist Edgar Rubin's "Rubin's vase" (Figure 2). The figure-ground relationship is related to legibility in spatial design.

Fig. 2. Figure-ground relationship in Rubin's vase (Baluch & Itti, 2011).

Proximity: Objects located close to each other tend to be perceived as groups. For example; the number *"3012"* is perceived as two different numbers when a space inserted in the middle: *30 12*.

Similarity: Objects that have similar visual characteristics such as color, shape, direction etc. are perceived in groups (Figure 3).

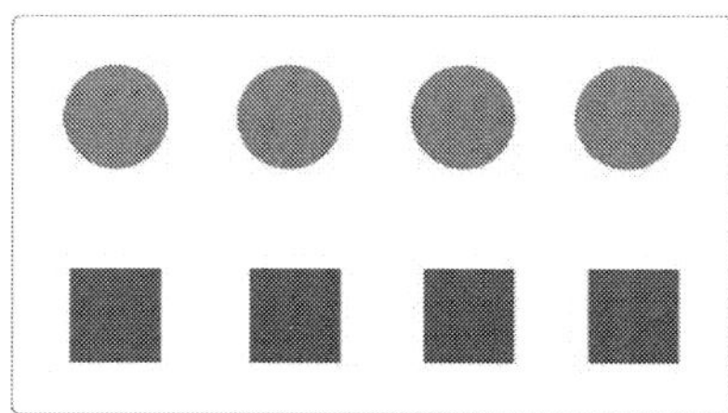

Fig. 3. Gestalt factor of similarity.

Continuation: Graham (2008) explains continuation as *"continuation occurs when the eye follows along a line, curve, or a sequence of shapes, even when it crosses over negative and positive shapes"* (Figure 4).

Fig. 4. Factor of continuation (Graham, 2008).

Closure: There is a tendency to close and mentally complete the missing parts of an image which is visually incomplete (Figure 5).

Fig. 5. Factor of closure (Graham, 2008).

2.3 Overview of research methodology

Basically, there are two approaches in visual landscape assessment; objective and subjective. Objective approach to visual landscape assessment assumes that visual quality of the landscape is an inherent characteristic and physical attributes of the environment determine its aesthetic value. On the contrary, subjective approach assumes that visual quality is in the eye of beholder and aesthetic value of an environment can be determined through subjective evaluation. There are also studies which have integrated both objective and subjective approaches.

Zube *et al.* (1982) identified four research paradigms on landscape assessment and perception which are; expert, psychophysical, cognitive and experiential paradigms (Taylor *et al.*, 1987).

The expert paradigm: this paradigm is based on expert judgments of visual quality of landscapes. Evaluation of landscape quality depends on formal characteristics of the landscape such as landform, vegetation, color, texture etc. Another assumption of this paradigm is that natural ecosystems have the greatest aesthetic value. This paradigm is criticized for the lack of user environment and being incompatible with users' perceptions (Lekagul, 2002). Furthermore, S. Kaplan (1988a) points out that experts perceive visual environment different to other people, and expert judgments are "a dubious source of objective judgment" about what other people really care about in the landscape.

The psychophysical paradigm: In psychophysical paradigm, in contrast to expert paradigm visual quality of the landscape is evaluated by the general public or special interest groups. The main assumption of this paradigm is landscapes have a stimulus property which is external to the observer who perceives the landscape without conscious thinking. Ranking and sorting are widely used techniques in visual assessments within this paradigm.

The cognitive paradigm: The cognitive paradigm focuses on why people prefer particular landscapes. The research is directed mostly towards developing a theoretical basis. In contrast to psychophysical paradigm, cognitive paradigm assumes that cognitive processes influence aesthetic judgments. Mostly verbal evaluation techniques, such as semantic differential analysis and adjective checklists, have been used to evaluate preferences and meanings. Most of the evolutionary theories on environmental perception (e.g. prospect-refuge theory and information processing theory) form a basis for this paradigm. However, this paradigm neglects the physical environment and rather focuses on meanings associated with landscapes (Taylor *et al.*, 1987).

The experiential paradigm: This paradigm focuses on human-environment interaction. Human experiences affect the landscape's perceived value. This approach is commonly used in "sense of place" studies and mainly by geographers. However, experiential approach is more subjective than cognitive and psychophysical paradigms; therefore reliability and validity of the results are hard to be measured (Taylor *et al.*, 1987).

Although paradigms explained above may seem completely different from each other, each contributes to overall comprehension of environmental perception. In terms of design and planning, the expert paradigm has been the most used approach in visual landscape assessment. However, there is a certain need for involvement of public or users in order to create enjoyable places for people. Fenton & Reser (1988) criticize that human geographers and landscape architects tend to use *atheoretical* and *apsychological* methods while psychologists use mainly theoretically derived psychometric methods. Professional differences might make it difficult to find a common basis for theoretical and methodological research. Nevertheless, collaboration of disciplines involved in environmental perception studies is essential to resolve some of the conflicts.

2.4 Landscape preferences

Assessment of landscape preferences is widely studied in environmental perception research. Landscape preference studies aim to investigate how and why people prefer some environments to others. People judge and interpret their environments and they respond to environments in terms of affective responses. Environmental preference is not luxury for people but essential and tied to basic concerns (R. Kaplan & S. Kaplan, 1989). Kaplan sees preference as an indicator of aesthetic judgment (1988b) and as a complex process which involves perception of things and space and reacting to them in terms of their potential usefulness and supportiveness (1988a). According to Charlesworth (1976), species has to be able to both recognize and prefer environments in which it functions well (S. Kaplan, 1988a). Preference for specific landscapes is about the organization of the space, rather than the individual elements (R. Kaplan & S. Kaplan, 1989), hence designers should focus on the integrity of different landscape elements.

The bio-evolutionary perspective on landscape preferences were explained in the previous section (Section 2.2): long history of human evolution is believed to be the reason for why we prefer some environments to others. One consistent finding of environmental preference research is that people prefer naturalness or natural environments to human-modified environments (e.g. R. Kaplan & S. Kaplan, 1989; van den Berg *et al.*, 2003). Presence of water also increases the preference ratings (Hull & Stewart, 1995; Yang & Brown, 1992). Natural scenes are also assumed to contribute to well-being by reducing stress levels, and to have positive influence on functioning and behavior (Ulrich *et al.*, 1991). It is assumed that preferences for savanna-like landscapes are linked to human evolutionary history, as an adaptation to East Africa savannas for survival (Falk & Balling, 2010). Ulrich (1979) found that homogenous ground texture, medium to high levels of depth, presence of a focal point, and moderate levels of mystery leads to high level of preferences in natural scenes (Porteous, 1996).

Complexity has been one of the central concepts in environmental preference research. Although R. Kaplan & S. Kaplan (1989) have found that coherence is more significant in explaining preferences, Ode & Miller (2011) suggest that landscape preferences have a relationship between measurements of complexity. Their study on rural landscapes showed that *"a landscape with an unequal distribution of land cover, a moderate amount of land cover, and a low level of aggregation is more likely to be preferred over a landscape with many land-cover classes, equal distribution, and strong aggregation"*. Complexity is also found to have a positive influence on urban landscape preference (Falk & Balling, 2010).

Environmental preference research generally focuses on natural or rural environments and there is little research on urban landscape preferences. This might be due to the fact that urban environments are highly complex structured; there are too many kinds of elements (both natural and cultural) that form urban structure. Moreover, social dynamics have important influence on shaping urban environments. Hence, it is rather difficult to measure and to assess landscape preference determinants in urban landscapes. One of the preference studies in urban environment was conducted by Nasar and his colleagues (1988a). They investigated the visual preferences for urban street scenes. Nasar used bipolar adjectives to describe the environments; closed-open, simple-diverse, chaotic-orderly, dilapidated-well-kept, vehicles prominent-vehicles not in sight, and nature (greenery) not in sight- nature (greenery) prominent. He found (just like he expected) that people preferred ordered, natural, well-kept, and open scenes with vehicles not prominent. However, Nasar was cautious about the interdependence of the variables; he expressed the need for further research for explanation of the relationship between these variables. Nevertheless Nasar suggested that moderate novelty, increase diversity, increased contrast among buildings, good maintenance, order, more vegetation and reduced vehicle prominence might produce highly preferred urban environments.

According to Bourassa (1990), aesthetic response occurs at both biological and cultural levels. Falk & Balling (2010) also state that *"that human landscape preferences is best understood as a continuous progression of aesthetic ideals, tempered by social convention, passed on from one generation to the next through human culture"*. But do culture and socio-demographic factors really affect preferences? There are several cross-cultural studies that investigate preferences for landscapes and landscape elements. Generally, the results show that despite cultural differences, people seem to have similar preferences for specific landscapes; however the concepts of novelty and familiarity can affect preferences for people from different cultures. Familiarity plays an important role in feeling secure and safe. People feel comfortable and relaxed in environments which they are familiar to (Kaplan *et al.*, 1998). On the other hand too much familiarity may become boring and people seek for novelty. For instance, Yang & Brown (1992) found that traditional Japanese style landscapes and water presence were highly preferred by people from both Korean and Western cultures. However they also found that while Koreans preferred Western style landscapes, Western tourists preferred Korean style landscapes. A similar result was found by Nasar (1988a). His study results showed that although there were consensus on preferences for ordered, natural, open and well-kept scenes; Japanese subjects highly preferred the American scenes and vice versa. His findings supported Berlyne's assumption that people prefer novelty to familiarity. He also pointed out that the results would have been different if subjects had been chosen from older population since Sonnenfeld (1966) claims that younger people prefer novelty and others familiarity (Nasar, 1988a). In their study Yang &Kaplan (1990) investigated landscape style preferences of Korean and Western individuals. They found a cross-cultural similarity in preferences in favor of landscapes with natural styles. Landscapes with rectangular or formal designs were less preferred by both groups.

Lyons (1983) showed that there is a strong relationship between age, gender, residential experience and landscape preferences. She found that preference levels changed in different age groups, adolescent male and females had different preferences, urban and rural residents had different preferences, familiar vegetational biomes were preferred highest, and there was no evidence that landscape preferences were shaped by innate or evolutionary factors. Yu (1995) also reported that people from different living environments (rural vs. urban) had different preferences; rural residents had high preference for novelty and modernity. He also indicated that landscape preferences were strongly influenced by education levels. However, his findings did not show any significant relation between gender and preferences.

Landscape preference studies are generally based on public or user (non-expert) evaluations. Ranking, rating or sorting of visual stimuli and verbal instruments are popular tools in determination of landscape preferences. Participants are asked to rank, rate or sort visual stimuli according to their preferences. The outcomes can be evaluated in terms of most and least preferred scenes, preference predictors (e.g. coherence, diversity, naturalness etc.), correlations between preference and predictors, content analysis of preferred environments or comparison of different landscape characteristics.

Although photographs and slides have been widely used as visual stimuli in preference research, there has been a constant debate on the representational validity of them. While some researchers have found that photographs can be adequate and valid resources to use (Dunn, 1976; Shuttleworth, 1980; Stewart *et al.*, 1984), some others do not agree with this idea (Kroh & Gimblett, 1992; Scott & Canter, 1997). R. Kaplan (1985) points out that use of photographs is less in cost and easy to administer, however sampling of the environments and selection of photographs require careful attention. In-situ assessments are time consuming, expensive and not practical. Besides, other variables of the landscape, (such as air condition and brightness) may vary during assessments and that might affect visual preference judgments of observers. On the other hand, a landscape is definitely more than just a scene and it is dynamic, however photographs and slides reflect landscapes as more static. Sevenant & Antrop (2011) state that depending on the character of the landscapes, some vistas are better presented by panoramic photographs, while some by normal photographs; thus, horizontal angle of view should be considered while selecting photographs. Palmer & Hoffman (2001) also support using panoramic images to increase validity. They also suggest that comparing the ratings of representations and actual field conditions from several individuals would help to establish validity of representations.

Current technology allows further visualization techniques such as computer graphics, 3D-modelling, virtual reality, GIS-based photorealistic visualisation, etc. (Sevenant & Antrop, 2011). However, validity issues remain the same. In their study, Bishop & Rohrmann (2003) concluded that "computer simulations do not necessarily generate the same responses as the corresponding real environment". On the other hand, detailing seems to be an important aspect in computer visualizations; higher detail levels are believed to increase the validity (Bishop & Rohrmann, 2003; Daniel & Meitner, 2001). Nevertheless photographs still seem to be the most popular tools as surrogates for actual landscapes. However as concern about validity increases, researchers will need to prove reliability of their results and we'll see much more debate on this issue.

Alternatively, sometimes verbal instruments such as verbal descriptions and bipolar adjective scales are used for assessment of landscape preferences. People can explain their preferences better by using words rather than rating or ranking visual stimuli. Although verbal assessments are quick and low-cost, analysis of the data may not be easy. Different people may use different adjectives or descriptions for the same preference judgment. Therefore content analysis of verbal descriptions should be done by experts or trained individuals in order to improve accuracy of results. On the other hand bipolar adjective lists, or semantic differentials, have been criticized for presenting adjectives selected by the researcher and therefore limiting people. However, Echelberger (1979) states that semantic differential may contribute to landscape preference assessment. On the contrary R. Kaplan (1985) claims that using adjectives does not tell much about preferences.

2.5 Environmental images and cognitive maps

Cognition involves perception, thinking, problem solving and organization of information and ideas (Downs & Stea, 1973). Hence, environmental cognition can be defined as perception, understanding, organization and retrieval of spatial information. Through cognition, we construct images of our environment which help us to find our way in our daily lives. These constructed environmental images form mental representations which are unique to the individual. This process is called cognitive mapping. Memory plays a crucial role in cognitive mapping. As S. Kaplan (1978) states *a cognitive map is based on familiar objects and events*. Hence, cognitive maps can change or improve depending on the individual's experiences.

People derive information from their environments through neurophysiological processes, but they also rely on personality and cultural factors to produce cognitive images (Knox & Marston, 2003). Thus, cognitive maps are highly personal constructs. A cognitive map of an individual can be quite different from an actual physical map in terms of accurate distance and structural organization. Simplification and distortion are two most important attributes of cognitive maps (Knox & Marston, 2003). The images might be incomplete or have inaccurate distance estimates. Nevertheless, cognitive images reflect how we see our environments and how we connect places to each other. Consequently, people's orientation and navigation through space can affect their quality of life. Sense of orientation helps people to feel confident and less anxious (Kaplan *et al.*, 1998). Cognitive maps help people to establish their routes and find their way, no matter how incomplete or distorted they are.

The term of "cognitive map" was first introduced by Tolman (1948) in his study where he investigated the spatial behavior of rats in a maze (Göregenli, 2010). However it was Kevin Lynch (1960) who pioneered cognitive mapping studies in urban design and planning with his famous work "The Image of the City". Lynch puts an emphasis on the concept of legibility for structuring and identifying the environment. Legibility plays an important role in way-finding and environmental images are fundamental for way-finding. An environmental image is a product of both immediate sensation and the memory of past experience (Lynch, 1960). Clarity of environmental images, thus the degree of legibility facilitates one's way-finding. Lynch identified five key elements of urban form which determine the legibility of an urban environment; paths, edges, districts, nodes and landmarks. Although paths were found to be the dominant elements of environmental images, Lynch emphasizes that all of the elements operate together and interrelation of these elements are important in creating legible urban environments.

Lynch's work has been mainly criticized for its small sample size and research technique; his five elements of legibility had already been established before interviewing the subjects. Later, he (1984) also criticized his own work for not being practical but being academically interesting (Pacione, 2005). His work also neglected the importance of symbolic meanings associated with places. Lynch was aware of the influence of meaning attached to a place on one's environmental images, however his work focused on urban form and he stated that form should be used to strengthen the meaning in urban design. Still, his legibility framework is still considered as fundamental and influencing in cognitive mapping studies in urban environments.

Today's fast paced urban life-styles urge us doing our daily tasks in a limited time. Hence, difficulties in getting to the desired destination may cause people to feel stressed out. As a primary component of cognitive mapping, legibility should be considered as an essential objective in place-making. Cognitive maps can be used in landscape architecture to investigate the relationship between characteristics of outdoor environments and perceived legibility. Evaluation of existing structure and organization of the environments will provide landscape architects to improve their place-making strategies in terms of design and planning.

3. Perception of soundscapes

Sound, as a landscape element, has not received much interest in landscape design and planning compared to vision. Listening to an environment is generally not the primary activity or interest of a person (Jennings & Cain, 2012); however information provided by the visual landscape play a great role in realizing our daily activities. The concept of soundscape has recently gained attention of planning and design disciplines where focus is generally on the visual aspect, rather than the acoustic. One of the reasons might be that most of the time designers' and planners' lack of scientific knowledge on acoustics. Concepts like "weighted sound levels", "absorption coefficient", and logarithmic measurements may seem unfamiliar and intimidating. Although noise mapping is quite a popular tool in environmental assessment studies, sound is rarely considered as a design element in landscape. Sound does not literally mean "noise". While some sounds can be disturbing, some sounds can give pleasure to an individual. However, sound as a sensory experience is rather different from vision. Acoustic space does not have obvious boundaries and is less precise in terms of orientation and localization (Porteous, 1996). Therefore assessment of sound as a design element is much complex than the visual dimension. Sound is an important element of a place which affects individual's perception and understanding of an environment. People derive information from sounds, just like visual environment. Sound can act as a guide for way finding or a cautionary signal for alert. In the context of space, soundscape can be defined as the acoustic character of an environment.

Urban environments are diverse and complex acoustic environments. They include different kinds of sound resources. Therefore, outdoor acoustic environment studies are mostly concerned with urban soundscapes. Evaluation of urban soundscapes is crucial not only for noise mitigation but also to assessment of acoustic comfort which is integral to overall environmental quality.

The term of soundscape was first coined by R. Murray Schafer. In his book "The Tuning of the World" (1977), he describes the soundscape as any acoustic field of the study; it may be a musical composition or a radio program or an acoustic environment. Influenced by Gestalt figure-background relationship (see section 2.2.6), Schafer identified three elements of a soundscape; (i) keynote sounds, (ii) sound signals, and (iii) soundmarks. Keynote sounds are background sounds and can be perceived subconsciously. Schafer suggests keynote sounds might have an effect on our behavior and moods since they are permanently there, whether we hear them consciously or not. Traffic sound is often given as a keynote example for contemporary urban environments. Sound signals, are foreground sounds and are listened to consciously (e.g. sirens). Finally, soundmarks (derived from landmark) are unique to that environment or to people in the community, thus they need to be protected. These elements have established a foundation for many soundscape studies so far.

The most noticeable study on the relationship between landscape architecture and soundscape is Hedfors's (2008) book "Site Soundscapes: landscape architecture in the light of sound". In his book, he analyzes sound in context of landscape architecture. He suggests a hypothetical model, named "the model of prominence" as a starting point for landscape architects. The model is also grounded on Gestalt figure-background relationship, like Schafer's work. It is based on description of the sounds. In Hedford's model figure-ground relationship is combined with two other dimensions; intensity and clarity (Figure 6). According to Hedford, a soundscape can be described as clear if prominent sounds are strongly experienced against a weak background. However, if prominent sounds are weaker than the background, then the soundscape becomes crowded. If both prominent sounds and the background are experienced equally strong, the soundscape can be described as powerful. On the contrary, if both are experienced weak, the soundscape becomes mild.

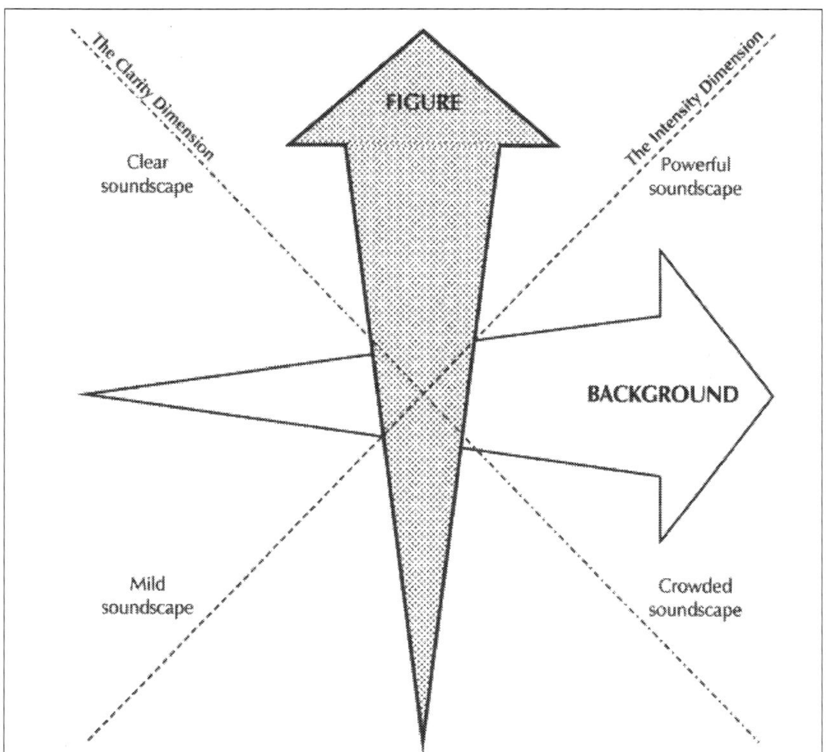

Fig. 6. The model of prominence (Hedfords, 2008).

Similar to natural landscapes, natural sounds such as bird sound and water sound are highly preferred by people. This may be explained through evolutionary perspective on landscape perception as well as therapeutic effects of natural landscapes (please refer to section 2.4). In fact, relaxation is found to be an important factor for urban open soundscapes (Yang & Kang, 2005). It is known that natural sounds such as bird and water sound can help people feel relaxed (Carles *et al.*, 1999).

Since sound perception is assumed to be a personal and therefore unique phenomenon, most researchers believe that perception of the acoustic environment is affected by personal factors such as demographics and culture. Yu & Kang (2010) found that people preferred natural sounds with increasing age and education level. Their results showed no significant correlation between preferences and occupation, and residence status. They found that gender influenced preference only for some sound types (e.g. bird sound).

Anderson *et al.* (1983) emphasize the importance of expectations in people's sound evaluations (Hedfords, 2008). People might tolerate or appreciate undesirable sounds if they expect to hear them in an environment. For instance some traffic sounds were found to be appreciated in urban environments. Thus, cultural and life-style differences might play role in evaluations of environments with different sound levels.

Sound types have also been found to be related to acoustic comfort evaluations; pleasant sounds, with either high or low sound levels, are perceived to improve the acoustic comfort (Yang & Kang, 2005). The source of the sound type can also affect preference. Zhang & Kang (2007) found that while "music on the street" was rated as favorite by 46% of the participants, 15% rated for music from stores, and only 2% rated for music from cars. Perception of the soundscape is also influenced by the activity involved and hence listening situation (Jennings & Cain, 2012).

Although sound level measurements (e.g. A-weighted levels) are widely used in soundscape research, it is also indicated that perception of the acoustic environment is independent from sound levels (Jennings & Cain, 2012; Szeremeta & Zanin, 2009). Reducing sound levels do not always improve perceived quality of acoustic environment (Yang & Kang, 2005). Furthermore, elimination of negative sounds from the environment does not necessarily make the acoustic environment more positive, may even generate anxiety (Cain *et al.*, 2011). However, Yang & Kang (2005) have found that background sound level is an important factor in evaluating soundscape in urban open public spaces; they suggest that reduced background sound level can help to create comfortable acoustic environments.

Visual perception also affects sound perception; Faburel & Gourlot (2009) found that visual images can reduce the negative effect of a sound, equivalent of up to 10dB decrease in the sound pressure level (SPL) (Solène, M., 2011). Yang and Kang (2005) also concluded that visual factors affect acoustic comfort evaluations and they suggested that interaction of visual and auditory perception work together *"as an aesthetic comfort factor"*. Carles *et al.*'s (1999) study supports this idea. They presented varying combinations of visual and auditory stimuli and participants were asked to rate each image, each sound, and finally each combination. It is found that sounds in the scenes containing vegetation or abundant water were rated higher; hence they concluded that visual and acoustic information can reinforce or interfere with each other. Furthermore, people are less annoyed by the sounds when the source is not visible (Solène, M., 2011). Zhang & Kang (2007) proposed some suggestions for creation of soundscapes in urban environments. They state that if SPL is higher than 65-70 dBA, then people will feel annoyed. Figure 7 shows their design suggestions.

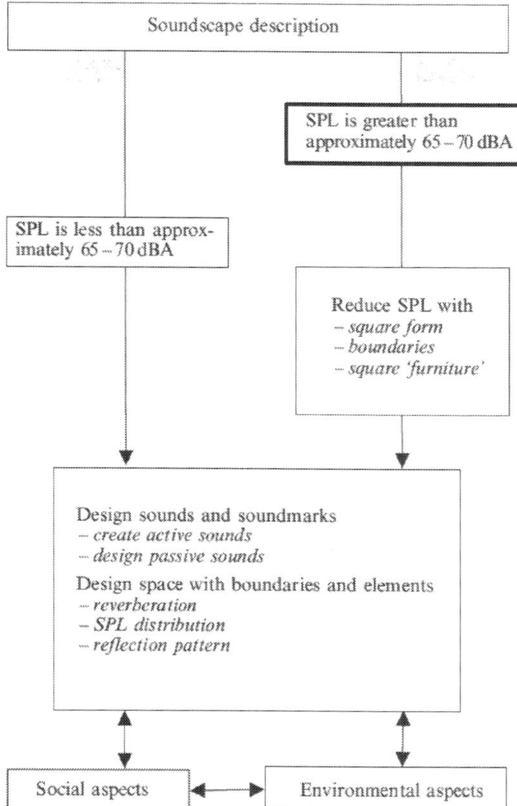

Fig. 7. Soundscape design strategies (Zhang & Kang, 2007).

Research methodology

Besides sound level measurements, perceived acoustic quality and preferences are often evaluated through interviews, questionnaires and soundwalks. During sound level measurements A-weighted equivalent continuous sound level (LA_{eq}) is measured and statistical sound levels (L_{eq90}, L_{eq50}, and L_{eq10}) are calculated. Equivalent continuous sound level presents the average level of noise over a time period, while statistical sound levels are used to define maximum, intrusive, median and background sound levels (Kang, 2007). Identification of the sounds which are perceived by the user is generally the first step of soundscape preference research. This procedure provides researchers an insight to perceived foreground sounds, background sounds and soundmarks. Rating scales are frequently employed in preference research. Participants are asked to rate the acoustic environment or a particular sound. The rating scale generally involves bipolar verbal descriptions such as like-dislike, noisy-quite, favorite-unfavorite etc. Semantic differential analysis is another tool which is commonly used to investigate people's emotional responses towards sound types and acoustic environments. These techniques can easily be applied any time to anyone without a hearing deficit or disorder. However, the soundwalk technique depends on more conscious listening state. During soundwalks, participants observe and listen to their environments very carefully while walking along a path and make judgments on the acoustic environment.

Previous studies have been carried out either in laboratories or in-situ. Binaural recordings are preferred for laboratory studies. Laboratory conditions are also suitable for assessment of the interaction between visual and auditory stimuli with different characteristics. However, if the soundscape of a particular place is to be assessed, considering primarily that visual perception affects perception of acoustic environment. it is best to carry out the study in-situ. Furthermore, for researchers from design and planning disciplines it might be difficult to establish and maintain laboratory conditions for acoustic evaluations due to lack of technical and scientific knowledge. Therefore, a multidisciplinary approach might be helpful in designing and conducting more effective research on soundscapes.

Zhang & Kang (2007) proposed a detailed system for investigation of urban open space soundscapes. They identified four facets; (i)characteristics of each sound, (ii)acoustic effects of the space, (iii)characteristics of users and (iv)other aspects of physical and environmental conditions. Description of each facet is summarized below.

Sound: For each sound, it is recommended that both steady state and the statistical SPL, spectrum, temporal conditions, source location, source movement and the psychological and social characteristics should be taken into account. Temporal conditions include variation (hour, day, season), duration and impulsive characteristics. Meaning, natural or artificial sounds, relation to activities, soundmarks and listening state (descriptive or holistic) should be considered in context of psychological and social characteristics.

Space: The shape of the space, boundary materials, street and square furniture, landscape elements, reverberation, reflection pattern and/or echogram, general background sound and sounds around the space are the characteristics to be considered related to the space.

People: Social, demographic, cultural characteristics of users and acoustic condition at users' home, work etc. should be assessed.

Environment: Microclimate conditions, lighting, visual, landscape and architectural characteristics are among the environmental characteristics that need to be taken into consideration.

Jennings & Cain (2012) propose a framework which uses Kano model in order to provide designers and planners a tool for predicting impacts of design interventions on soundscape. Kano model is generally used in product development for determination of customer needs and satisfaction. Jennings & Cain (2012) suggest that Kano model can help to clarify thinking, since perception of the soundscapes is a complex process. To summarize their proposal, there are three attributes to be assessed in Kano model; basic requirements, performance requirements and excitement (or attractive) requirements. The first step is to satisfy basic requirements, such as fulfilling legislative requirements for noise control. Then performance requirements need to be assessed in order to find out user needs and expectations. Use of emotional perceptual dimensions, interactive simulations and soundwalks might be helpful in this step. Finally for the excitement requirements, the authors suggest that culturally significant sounds or sonic art could be introduced to give a unique character and to increase excitement or attractive quality. The authors state that Kano model is successfully used in automobile industry for sound quality assessments. However, this model might seem confusing for spatial designers and planners .The authors also emphasize the need for application of the framework through real-life examples. Please see the reference for further details of the framework.

Solène (2011) applied cognitive mapping technique in her study on urban soundscapes. Participants were asked to draw graphical representations of sonic ambiences of three urban squares. They were also asked to describe their preference for ideal sonic environments in squares. Boundary was found to be an essential element in perceived ideal sonic environments since most participants described closed or semi-open squares. On the contrary, open squares were associated with negative sonic ambiance. Depending on the results, the author inferred that there is a strong bond between urban typology and perceived sonic ambiance. She concluded that sonic mind maps were appropriate for studying psychoacoustics of an urban environment.

Despite the short history of soundscape research on outdoor environments, there seems to be a variety of research techniques and methodologies which can be adapted to the researcher's objectives. Still, many issues such as interrelationships between factors that affect soundscape preferences and the effects of spatial design on acoustic comfort need to be further investigated. Spatial designers and planners can contribute to the soundscape research by developing new models and methodologies in order to display and emphasize their role on creating livable and high quality environments.

4. Conclusion

Our landscapes are natural and cultural heritage of our societies. With the rapid urbanization and development processes, change has become an inevitable outcome for our landscapes in global scale. Unfortunately, landscape change often occurs in negative ways. Loss of diversity and identity should be the main concern for future design and planning research for landscape architects. However, the role of perception and its effects on spatial behavior and attitudes must be realized first.

If landscape architecture aims to create livable and effective environments for people in the community, people's perception and interpretation of environments must be investigated. One can argue that landscape assessments should be made by experts because of their knowledge and experiences and general public can't judge environmental quality. On the other hand, environmental quality issues are still on the agenda. To some extent, local, national and even international authorities can be blamed for ignoring knowledge and suggestions of environmental designers and planners in sustainable development. Certainly, professionals have a lot to offer in terms of knowledge and skills. However, knowing and understanding the basic relationships between people and their environments is a necessity.

This chapter has summarized basic information and approaches on landscape perception both in visual and auditory context, aiming to provide an insight on perceptual and cognitive dimensions of environmental research. However, there are more to landscape perception research. Reference list can provide readers with valuable resources to read.

Although there is a vast amount of research on landscape preferences, there are still theoretical and methodological issues that have not been clarified yet. The outcomes of environmental psychology research can guide planners and designers in creating and managing our landscapes. Therefore, it is crucial to establish a multidisciplinary cooperation. The relevance and importance of landscape perception research has been neglected in Turkey so far. Most undergraduate curricula in landscape architecture programs do not cover perceptual dimension of the landscape. I hope this chapter draws an attention to the significance of the subject. Finally, I'd also like to emphasize the need for strengthening the role of landscape design and planning on landscape perception research.

5. References

Antrop, M. (2005). From holistic landscape synthesis to transdisciplinary landscape management. In: *Landscape Research to Landscape Planning: Aspects of Integration, Education and Application,* B. Tress, G. Tress, G. Fry, P. Opdam, pp.27-50, Wageningen UR Frontis Series No. 12., Springer, Heidelberg.

Baluch, F. Itti, L. (2011). Mechanisms of top-down attention. *Trends in Neurosciences,* Vol.34, No.4, pp.210-224, ISSN:0166-2236.

Barak-Erez, D. & Shapira, R. (1999). The Delusion of Symmetric Rights. *Oxford Journal of Legal Studies,* Vol.19, No.2, pp.297-312, Online ISSN 1464-3820.

Bell, P.A., Greene, T.C., Fisher, J.D. & Baum, A. (2001). *Environmental Psychology.* 5th Edition, Harcourt College Publishers, ISBN:0155080644, USA.

Berleant, A. (1992). *The aesthetics of environment.* Temple University Press, ISBN: 0-87722-993-7, USA.

Bishop, I.D. & Rohrmann, B. (2003). Subjective responses to simulated and real environments: a comparison. *Landscape and Urban Planning,* Vol.65, pp.261-277, ISSN: 0169-2046.

Burassa, S.C. (1988). Toward a theory of landscape aesthetics. *Landscape and Urban Planning,* Vol.15, pp.241-252, ISSN: 0169-2046.

Cain, R., Jennings, P. & Poxon, J. (2011). The development and application of the emotional dimensions of landscape. *Applied Acoustics,* In press, doi: 10.1016/j.apacoust.2011.11.006

Carles, J.L., Barrio, I.L., & de Lucio, J.V. (1999) Sound influence on landscape values. *Landscape and Urban Planning,* Vol.43, pp.191-200, ISSN: 0169-2046.

Carson, E. (2002). Locke's account of certain and instructive knowledge. *British Journal for the History of Philosophy,* Vol.10, No.3, pp.359-378, Online ISSN: 1469-3526.

Chang, H. (2009). *Mapping the Web of Landscape Aesthetics: A critical Study of Theoretical Perspectives in Light of Environmental Sustainability.* PhD Dissertation, North Carolina State University, Available from: http://www.lib.ncsu.edu/resolver/1840.16/3362

Clay, G.R. & Daniel, T.C. (2000). Scenic landscape assessment: the effects of land management jurisdiction on public perception of scenic beauty. *Landscape and Urban Planning,* Vol.49, pp.1-13, ISSN: 0169-2046.

Daniel, P. (2008). *Kant on the Beautiful: The Interest in Disinterestedness.* Date of access: 21/01/2012, Available from: http://arts.monash.edu.au/ecps/colloquy/journal/issue016/daniels.pdf

Daniel, T. C., (2001). Whither scenic beauty? Visual landscape quality assessment in the 21st century. *Landscape and Urban Planning*, Vol. 54, pp. 267-281, ISSN: 0169-2046.

Daniel, T.C. & Meitner, M.M. (2001). Representational validity of landscape visualizations: The effects of graphical realism on perceived scenic beauty of forest vistas. *Journal of Environmental Psychology*, Vol. 21, No.1, pp.61-72, ISSN: 0272-4944.

Downs, R. N. & Stea, D. (1973). Cognitive Maps and Spatial Behavior: Process and Products. In: *Image and Environment: Cognitive Mapping and Spatial Behavior*, Roger M. Downs & David Stea, pp. 8-26, Aldine, ISBN: 0202100588, Chicago.

Dunn, M.C. (1976). Landscape with photographs: testing the preference approach to landscape evaluation. *Journal of Environmental Management*, Vol. 4, pp. 15-26.

Echelberger, H.E. (1979). The semantic differential in landscape research, *Proceedings of Our National Landscape: a conference on applied techniques for analysis and management of the visual resource*, April 23-35, Nevada, pp.524-531.

Council of Europe, (2000). *European Landscape Convention*, Date of access: 15/12/2011 Available from: http://conventions.coe.int/Treaty/en/Treaties/Html/176.htm

Falk, J.H. & Balling, J.D. (2010). Evolutionary influence on human landscape preference. *Environment & Behavior*, Vol.42, No.4, pp. 479-493, Online ISSN: 1552-390X.

Fenton, D.M. & Reser J.P. (1988). The assessment of landscape quality:an integrative approach. In: *Environmental Aesthetics: theory, research & applications*, J.L.Nasar, pp. 108-119, Cambridge University Press, ISBN: 0521341248, USA.

Fisher, B.S. & Nasar, J.L. (1992). Fear of crime in relation to three exterior site features: prospect, refuge and escape. *Environment and Behavior*, Vol.24, No.1, pp.35-65, ISSN: 0013-9165.

Forster, P.M. (2010). *A brief introduction to environmental psychology*. Date of access: 22/01/2012, Available from:
http://www.scribd.com/doc/45853873/Introduction-to-Environmental-Psychology

Galanter, P. (2010). Complexity, Neuroaesthetics, and Computational Aesthetic Evaluation, *Proceedings of 13th Generative Art Conference GA2010*, pp.400-409, Italy.

Gibson, J.J. (1986). The theory of affordances. In: *The ecological approach to visual perception*, J.J.Gibson, pp.127-143, Lawrence Erlbaum Associates Inc, ISBN:0898599598, USA.

Göregenli, M. (2010). *Çevre psikolojisi: insan-mekan ilişkileri*. İstanbul Bilgi Üniversitesi Yayınları, ISBN:978-605-399-171-7, Istanbul.

Graham, L. (2008). Gestalt theory in interactive media design. *Journal of Humanities and Social Sciences*, Vol.2, No.1, pp.1-12, ISSN:1934-7227.

Gross, S.W. (2002). The neglected programme of aesthetics. *British Journal of Aesthetics*, Vol.42, No.4, pp.403-414, Online ISSN 1468-2842.

Hedfors, P. (2008). *Site soundscapes: landscape architecture in the light of sound*. VDM Verlag Dr.Müller Aktiengesellschaft & Co.KG, ISBN: 9783639094138, Germany.

Jennings, P. & Cain, R. (2012). A framework for improving urban soundscapes. *Applied Acoustics*, In press, doi:10.1016/j.apacoust.2011.12.003

Kang, J. (2007).*Urban Sound Environment*. Talor & Francis, ISBN:0-415-35857-4, Great Britain.

Kaplan, R. & Kaplan, S. (1978). *Humanscape: environments for people*. Duxbury Press, ISBN- 10: 0878721630, USA.

Kaplan, R., Kaplan, S. & Ryan, R.L. (1998). *With People in Mind*. Island Press, ISBN: 1-55963-594-0, USA.

Kaplan, R. & Kaplan, S. 1989. *The experience of nature: a psychological perspective*. Cambridge University press, ISBN:0521349392, USA.

Kaplan, S. (1988a). Perception and landscaper: conceptions and misconceptions. In: *Environmental Aesthetics: theory, research & applications*, J.L.Nasar, pp. 45-55, Cambridge University Press, ISBN: 0521341248, USA.

Kaplan, S. (1988b). Where cognition and affect meet: a theoretical analysis of preference. In: *Environmental Aesthetics: theory, research & applications*, J.L.Nasar, pp. 56-63, Cambridge University Press, ISBN: 0521341248, USA.

Kaplan, S. (1978). On knowing the environment. In *Humanscape: environments for people*, pp.54-58, Duxbury Press, ISBN- 10: 0878721630.

Kaplan, S. (1975). An informal model for the prediction of preference. In: *Landscape Asessment: values, perception, and resources*, E.H. Zube, R.O. Brush & Fabos, J.G., pp.92-101, Dowden, Hutchinson & Ross Inc., ISBN: 0470984236, Pennsylvania.

Knox, P.L. & Marston, S.A. (2003). *Places and Regions in Global Context: Human Geography*. 2nd Edition. Pearson Education Inc, ISBN: 0130168319, New Jersey.

Köhler, W. (1938). Physical Gestalten. In: *A sourcebook of Gestalt psychology*, W.D. Ellis, pp.17-54, Reprinted by Routledge in 2001, ISBN:0415209579, Great Britain.

Kroh, D.P. & Gimblett, R.H. (1992). Comparing live experience with pictures in articulating landscape preference. *Landscape Research*, Vol. 17, pp.58-69.

Kruse, F.E. (2007). Vital rhythm and temporal form in Langer and Dewey. *Journal of Speculative Philosophy*, Vol. 21, No. 1, pp.16-26.

Lee, S.A. (1973). Environmental perception, preferences and the designer, *Proccedings of the Lund Conference Architectural Psychology*, 26-29 June 1973.

Lekagul, A. (2002). *Toward preservation of the traditional marketplace: A preference study of traditional and modern shopping environments in Bangkok, Thailand*. PhD Thesis submitted to Virginia Polytechnic Institute and State University, Virginia.

Lothian, A. (1999). Landscape and the philosophy of aesthetics: is landscape quality inherent in the landscape or in the eye of the beholder. *Landscape and Urban Planning*, Vol.44, No.4, pp.177-198, ISSN: 0169-2046.

Lynch, K. (1960). *The Image of The City*, The MIT Press, ISBN: 0-262-62001-4, USA.

Lyons, E. (1983). Demographic correlates of landscape preference. *Environment & Behavior*, Vol.15, No.4, pp.487-511, Online ISSN: 1552-390X.

Martindale, C. (1996). How can we measure a society's creativity? In: *Dimensions of Creativity*, M.A.Boden, pp.159-198, MIT Press, ISBN:0262522195, USA.

Maulan, S., Sharifi, M.K.M. & Miller, P.A. (2006). Landscape preference and human well-being. *ALAM CIPTA, International Journal on Sustainable Tropical Design Research & Practice*, Vol.1, No.1, pp.25-32, ISSN: 1823-7231.

McVay, S. (1993). A siamese connexion with a plurality of other mortals. In: *The Biophilia Hypothesis*, Stephen R. Kellert & Edward O. Wilson, pp. 2-19, Island Press, ISBN: 1-55963-147-3, Washington DC.

Meinig, D.W. (1979). The beholding eye: ten versions of the same scene. In: *The Interpretation of Ordinary Landscapes: Geographical Essays*, D.W.Meinig & J. Brinckerhoff, pp.33-48, Oxford University Press, New York.

Mok, J.H., Landphair, H.C. & Naderi, J.R. (2006). Landscape improvement impacts on roadside safety in Texas. *Landscape and Urban Planning*, Vol.78, pp. 263-274, ISSN: 0169-2046.

Nasar, J.L. (1988a). Visual preferences in urban street scenes: a cross-cultural comparison between Japan and The United States. In: *Environmental Aesthetics: theory, research & applications*, J.L.Nasar, pp.260-274, Cambridge University Press, ISBN: 0521341248, USA.

Nasar, J.L. (1988b). Perception and evaluation of residential street scenes. In: *Environmental Aesthetics: theory, research & applications*, J.L.Nasar, pp. 275-289, Cambridge University Press, ISBN: 0521341248, USA.

Ode, Á. & Miller, D. (2011). Analysing the relationship between indicators of landscape complexity and preference. *Environment and Planning B: Planning and Design*, Vol.38, pp.24-40.

Pacione, M. (2005). *Urban geography: a global perspective*. Routledge, 2nd Edition, ISBN: 0-415-34305-4, USA.

Palmer, J.F. (2003). Research agenda for landscape perception. In: *Trends in Landscape Modelling*, E.Buchmann & S.Ervin, pp. 163-172, Hebert Wichmann Verlag, Heidelberg.

Palmer, J.F. & Hoffman, R.E. (2001). Rating reliability and representational validity in scenic landscape assessments. *Landscape and Urban Planning*, Vol.54, pp. 149-161, ISSN: 0169-2046.

Porteous, J.D. (1996). *Environmental aesthetics: ideas, politics and planning*. Routledge, ISBN: 0-203-43732-2, London.

Schafer, R.M. (1977). *The tuning of the world*. Knopf, ISBN:9780892814558, New York.

Scott, M.J. & Canter, D.V. (1997). Picture or place? A multiple sorting study of landscape. *Journal of Environmental Psychology*, Vol. 17, pp.263–281, ISSN: 0272-4944.

Sevenant M. & Antrop M. (2011). Landscape Representation Validity:A Comparison between On-site Observations and Photographs with Different Angles of View. *Landscape Research*, Vol.36, No.3, pp.363-385, ISSN: 0142-6397.

Shuttleworth, S. (1980). The use of photographs as an environmental presentation medium in landscape studies. *Journal of Environmental Management*, Vol. 11, pp.61-76.

Solène, M. (2011). Assessment of urban soundscapes. *Organised Sound*, Vol.16, No.3, pp.245-255, EISSN: 1469-8153.

Stewart, T.R., Middleton, P., Downtown, M. & Ely, D. (1984). Judgments of photographs vs. field observations in studies of perception and judgment of the visual environment. *Journal of Environmental Psychology*, Vol.4,No.4, pp.283-302, ISSN: 0272-4944.

Swanwick, C. (2002). *Landscape Character Assessment: A guidance for England and Scotland*. On behalf of The Countryside Agency and Scottish Natural Heritage, UK, Available from: http://www.naturalengland.org.uk/Images/lcaguidance_tcm6-7460.pdf

Szeremeta, B. & Zanin, P.H.T. (2009). Analysis and evaluation of soundscapes in public parks through interviews and measurement of noise. *Science of the Total Environment*, Vol.407, pp.6143-6149.

Taylor, J.G, Zube, E.H. & Sell, J.L. (1987). Landscape assessment and perception research methods. In: *Methods in Environmental and Behavioral Research*, R. B. Bechtel, & R. W. Marans, pp. 361-393, Nostrand Reinhold, ISBN: 0442211570 ,New York.

Ulrich, R.S. (1993). Biophilia, biophobia and natural landscapes. In: *The Biophilia Hypothesis*, Stephen R. Kellert & Edward O. Wilson, pp. 73-137, Island Press, ISBN: 1-55963-147-3, Washington DC.

Ulrich, R.S., Simons, R.F., Losito, B.D., Fiorito, E., Miles, M.A. Miles, M.A. & Zelson, M. (1991). Stress recovery during exposure to natural and urban environments. *Journal of Environmental Psychology*, Vol.11, pp.211-230, ISSN: 0272-4944.

Ungar, S. (1999). Environmental perception, cognition and appraisal. *Environmental Psychology 4 Lecture Notes*, Glasgow Caledonian University, Scotland.

Vandenabeele, B. (2012). Burke and Kant on the Social Nature of Aesthetic Experience. *The Science of Sensibility: Reading Burke's Philosophical Enquiry*, K.Vermeir & M.F. Deckard, pp.177-192 , Springer, e-ISBN: 978-94-007-2102-9.

Van den Berg, A.E., Kooole, S.L.. & van den Wulp, N.Y. (2003). Environmental preference and restoration: how are they related? *Journal of Environmental Psychology*, Vol. 23, pp. 135-146, ISSN: 0272-4944.

Wertheimer, M. (1938). Laws of organization in perceptual forms. In: *A sourcebook of Gestalt psychology*, W.D. Ellis, pp.71-88, Reprinted by Routledge in 2001, ISBN:0415209579, Great Britain.

Yang, B. & Brown, T. J. (1992). A cross-cultural comparison of preferences for landscape styles and landscape elements. *Environment & Behavior*, Vol.24, pp.471-507, Online ISSN: 1552-390X.

Yang, B. & Kaplan, R. (1990). The perception of landscape style: a cross-cultural comparison. *Landscape and Urban Planning*, Vol.19, pp. 251-262, ISSN: 0169-2046.

Yang, W. & Kang, J. (2005). Acoustic comfort evaluation in urban open public spaces. *Applied Acoustics*, Vol.66, pp.211-229, ISSN: 0003-682X.

Yu, K. (1995). Cultural variations in landscape preference: comparisons among Chinese subgroups and Western design experts. *Landscape and Urban Planning*, Vol.32, No.2, pp. 107-126, ISSN: 0169-2046.

Yu, L. & Kang, J. (20101). Factors influencing the sound preference in urban open spaces. *Applied Acoustics*, Vol.71, pp.622-633, ISSN: 0003-682X.

Zhang, M. & Kang, J. (2007). Towards the evaluation, description and creation of soundscapes in urban open spaces. *Environment and Planning B: Planning and Design*, Vol. 34, pp. 68-86, ISSN: 0265-8135.

13

Urban Landscape Design

Murat Z. Memlük

Ankara University
Turkey

1. Introduction

Peyzaj, the Turkish word for landscape, originates from French word "paysage" which means scenery. Nowadays, the word encompasses a wider and deeper meaning. While in the medieval period, "landscape" was used as a synonym for "region" and "territory" in most of the Germanic languages, beginning from the 15th century landscape became a pictorial genre (Tress & Tress, 2001). The use of landscape as a term in science is relatively new. Today, landscape refers to not only a phenomenon described and analyzed by scientific methods, but also a subjective experience which has perspective, aesthetical, artistic and existential meaning (Antrop, 2005a). It is dynamic and constantly changing. Antrop (2005b) identified four driving forces of landscape change; (i) accessibility, (ii) urbanization, (iii) globalization, and finally (iv) calamities. This chapter is about urban landscapes; therefore urbanization will be the beginning point of this study.

Urbanization has become a worldwide phenomenon after the second half of the last century. Today more than half of the world population lives in urban environments[1]. Urbanization is a complex and multidimensional concept with its spatial, ecological, economic, social and cultural aspects. While urbanization is widely accepted as a foundation of modernizing (Clarke Annez & Buckley, 2009), it has also caused environmental and socioeconomic challenges (Adams & Sierra, 2009). Consequently planning and design of urban areas are faced with challenges to create both ecologically and economically sustainable cities.

Natural landscapes have been dramatically transformed by the urbanization process throughout the world. Consumption of resources is highest in urban environments, which causes negative impacts on physical environment. Traffic, air, water and soil pollution, improper land use and greenhouse gas emissions are some of the major issues due to urbanization. The effects of urbanization process are not only limited to ecological damage, but changing sociocultural and economic structure also affects the quality of physical environment by influencing human behaviors and lifestyles. Indeed, there is a mutual relationship and interaction between physical environment and quality of life. Therefore planning and design of physical environments requires a holistic and comprehensive perspective.

---

[1] According to The World Bank data, 51% percent oft he world's population lives in urban areas as of 2010. (http://data.worldbank.org/indicator/SP.URB.TOTL.IN.ZS/countries/1W?display=graph)

Urbanization in Turkey gained momentum starting from 1950's as a result of intense immigration from rural to urban areas. This has caused vertical growth in cityscapes and increase of squatter settlements around planned areas which has lead to loss of character and identity. Currently there are 81 provinces in Turkey and the urban population is the 76.3% of the total population as of 2010 (TurkStat, 2011). Industrialization has always been the key driving force o the urban development in Turkey. While industrialization and urbanization have pushed the limits of environmental capacity, open and green spaces, or in other words the urban landscape has diminished within the city. Unfortunately, open and green spaces are hardly among the priorities of urban development in the planning system of Turkey. On the contrary, like Jacques-Menegaz (2006) points out; *"open and green spaces are not amenities but necessities in urban life"*.

This chapter aims to present the role and importance of urban landscape as a crucial part of urban environments, providing design basics and examples. Urban landscape elements, whether public or private property, are parts of the city's form and texture. Therefore design of urban landscape is inevitably a part of urban design. Hence, firstly, urban design concept and its bond with landscape architecture will be discussed. Then, benefits of urban landscape elements will be examined and basic urban landscape design principles will be presented.

2. Urban design

Urban design as a profession might be considered relatively new, but historically it has played a major role in forming cities (Arida, 2002). The concept of urban design has emerged as a bridge between planning and design in response to need for management of modernizing cities in the late 1950's (Krieger, 2009). However, there does not exist a commonly agreed definition of urban design yet, mainly due to its interdisciplinary character. While some try to create a precise and universal definition, some argue that it is unnecessary. Appleyard (1982) states that there shouldn't or can't be a single definition of urban design and points out if the existence of different kinds of urban design is recognized, then it is possible to get a better understanding of the nature of it (Rowley, 1994). There are many viewpoints on what urban design actually is. Traditionally, urban design has been regarded either as a subset of planning or as extension of architecture. On the other hand, one cannot abstract open and green spaces out of an urban environment. Hence, the role of landscape architecture in urban design needs to be understood and accepted as a key part in creating sustainable urban environments. This section does not aim to define urban design, but I believe a better understanding of scope and content of urban design will provide a conceptual basis for urban landscape design.

Urban design can roughly be defined as the art of creating and giving form to urban environments. Urban design involves many stakeholders whose interests and priorities may conflict and the physical product of urban design should serve the community's needs and expectations with its social, cultural and economic outcomes. This makes urban design a highly complex phenomenon; as a result the definition can or should not be limited to physical design.

Although examples of urban design could be traced back to Ottoman period, it, as a concept, was first introduced in Turkey in 1970s, starting from the adoption of the concept by university degree programs. However, the concept has not gone further than being an academic subject and a legal background does not exist, yet. According to Baş (2003) existing urban planning approach in Turkey is basically static and urban design is reduced to land readjustment. Furthermore, never ending controversies, mainly between urban planners and architects, over "who are the real urban designers" makes the cooperation difficult. On the contrary, compromise between related fields (i.e. landscape architecture, urban planning and architecture) is necessary in order to develop conceptual background for urban design in Turkey. This might also be the basic problem in defining urban design; arguing possession of the discipline rather than focusing on the extent. It is impossible to involve urban design in national development policies without lack of consensus.

In 2010, a national action plan titled "Integrated Urban Development Strategy and Action Plan 2010-2023, Turkey"(KENTGES in Turkish abbreviation) was created by the Ministry of Public Works and Settlement[2]. The action plan is a result of participation process in order to respond to the need for increased life quality and stronger socio-economical structure in urban environments. It is a national document which includes the strategies, policies and actions for sustainable urban development. In this document, the concept of urban design has been primarily linked to the second central axis of the action plan "increasing spatial and life quality of settlements". The actions related to urban design are summarized as follows:

- To establish interdisciplinary graduate degree programs in universities,
- To prepare guidebooks based on research and development projects' outcomes,
- To develop methodologies for quality urban design,
- To promote revitalization of urban centers and to establish maintenance, management, finance and participation approaches,
- To develop design solutions for disadvantaged groups.

The integration of urban design in such an action plan seems like promising. However, there is still much work to be done and it is too soon to tell whether the action plan will achieve its goals.

2.1 Urban design and urban life quality

Under the course description of "Urban Design: City-Building and Place-Making"at University California, Berkeley, urban design is explained as follows:

„The discipline of urban design is concerned with notions of the "good city." It is concerned with how urban environments work for people and support human needs, how physical designs may facilitate or hinder human behavior, how cities look, and what cities mean. It is concerned foremost with environmental quality, measured in many ways but particularly in terms of access, connectivity, comfort, legibility, and sense of place."

This statement supports the idea that urban design is strongly linked to life quality. Like urban design, there is no universally accepted definition of quality of life (QoL). The term was first used in USA in the post-war period and later was adopted by many fields such as education, health and, economic and industrial growth (Carr et al., 1996). Despite the technological development and increased income levels, it has been recognized that quality

---

[2] In 2011, The Ministry of Public Works and Settlement was closed down by a decree-law and KENTGES studies are now being carried on by General Directorate of Spatial Planning, The Ministry of Environment and Urbanization.

of life cannot be measured through material wealth (Pacione, 2003). According to United Nation's Environment Glossary (1997) the term is defined as the *"notion of human welfare (well-being) measured by social indicators rather than by "quantitative" measures of income and production"*. Meanwhile World Health Organization's (1996) definition of quality of life is more comprehensive and as follows:

"an individual's perception of their position in life in the context of the culture and value systems in which they live and in relation to their goals, expectations, standards and concerns".

QoL analyses can be a useful tool for identifying design goals and developing strategies in urban design. On the other hand, measuring QoL and development of quality indexes are challenging processes since they encompass many dimensions (e.g. economic, social, health, subjective, environmental etc.). Livability and sustainability are two basic concepts related to urban life quality. Livability of an urban environment is determined by physical environment as well as social environment conditions, hence urban life quality is a result of two kinds of input; physical/objective and psychological/subjective (Yıldız Turgut, 2007). For this reason, livability and consequently urban quality indicators might vary from one city to another. Parfect & Power (1997) suggest a situation assessment where weaknesses, deficiencies, inherent strengths and advantages are identified before identifying urban quality in urban design. Such an assessment could help to determine the priorities and deficiencies in urban development strategies and policies in both national and local context. Determination of priorities is also important in terms of finance, since cost of urban development is generally very high. As an example; United States Environmental Protection Agency (EPA), the U.S. Department of Housing and Urban Development (HUD) and the U. S. Department of Transportation (DOT) have formed a partnership to coordinate decision and policy making efforts in housing, transportation and energy efficiency. In 2009 the partnership identified six livability principles to guide the federal investments. Developing such principles regarding both natural and cultural values could be useful and guiding in urban design. The principles are as follows:

- **Provide more transportation choices:** Develop safe, reliable, and economical transportation choices to decrease household transportation costs, reduce our nation's dependence on foreign oil, improve air quality, reduce greenhouse gas emissions, and promote public health.
- **Promote equitable, affordable housing:** Expand location- and energy-efficient housing choices for people of all ages, incomes, races, and ethnicities to increase mobility and lower the combined cost of housing and transportation.
- **Enhance economic competitiveness:** Improve economic competitiveness through reliable and timely access to employment centers, educational opportunities, services and other basic needs by workers, as well as expanded business access to markets.
- **Support existing communities:** Target federal funding toward existing communities— through strategies like transit oriented, mixed-use development, and land recycling—to increase community revitalization and the efficiency of public works investments and safeguard rural landscapes.
- **Coordinate and leverage federal policies and investment:** Align federal policies and funding to remove barriers to collaboration, leverage funding, and increase the accountability and effectiveness of all levels of government to plan for future growth, including making smart energy choices such as locally generated renewable energy

- **Value communities and neighborhoods:** Enhance the unique characteristics of all communities by investing in healthy, safe, and walkable neighborhoods — rural, urban, or suburban.

Livability principles are strongly related to the concept of sustainability. The environmental damage caused by urbanization is in the heart of sustainability debates. It has become a key issue in development after The United Nations Conference on Environment and Development (UNCED) in 1992, which expressed the concerns for the future of the environment in global scale. Agenda 21 is one of the resulting documents of this conference. It is an action plan and comprises four main sections; (i) social and economic dimensions, (ii) conservation and management of resources for development, (iii) strengthening the role of major groups, and (iv) means of implementation which include detailed policies for sustainable development. Agenda 21, as a tool, emphasizes the importance of local participation in decision making and implementation processes. Besides, local participation is also necessary to find out psychological/subjective indicators for urban quality.

It is generally implied that the dynamics and rhythm of urban life decreases the life quality of citizens in many aspects. Contemporary urban lifestyles are fast paced and exhausting for many. Despite the fact that cities are the hearts of economic growth and cultural diversity, citizens might experience difficulties in enjoying the amenities of a city in context of time and space. During our busy daily schedule, mainly between work and home, we hardly find time to ourselves. When we have some spare time, we seek to enjoy what the city offers to us. On the other hand cities are densely built and populated environments which causes the feeling of "lost in space". People want to access and circulate through their living environments easily. Furthermore they need places where they can escape the stressful rhythm of urban life. This is where landscape architecture takes the leading role; design of open and green spaces to provide livable and accessible outdoor environments, as well as to support urban ecology. Further benefits of urban landscape, that affects the urban life quality, will be explained in the next section.

2.2 The role of landscape architecture in urban design

Landscape architecture is the art and science of creating and conserving outdoor environments with respect to cultural values and ecological sustainability. It uses both non-living and living materials for design and planning, therefore the outcome is always dynamic and changing. Until recently, urban design was associated mainly with architecture and urban planning, and the role of landscape architecture was neglected. Landscape architects have been criticized for their urban design practices with low density, little formal sensibility, and too much open space which at the end look like suburban environments (Krieger, 2009). Today, on the contrary, urban landscape is considered crucial to creating sustainable urban environments. Although the word "landscape" is often used to describe natural and/or rural environments, there is certainly more to it. A landscape is shaped by both natural and cultural dynamics which also influence human life styles. Therefore an urban landscape is not only about green spaces within an urban environment. It is comprised of various land uses such as streets and squares, playgrounds, railway and canal corridors, cemeteries, bicycle and pedestrian paths, and waterfronts. Even structures in a city influence the urban landscape character.

Today urban environments have also a vast pressure on rural environments due to decentralization which causes loss of boundaries between urban and rural environments. Therefore it has a negative impact on natural resources which provide goods and energy for the urban dwellers. That is why sustainability has become the most important goal in urban design and planning more than ever. Urban design approaches should not neglect the natural processes which shape and influence the quality of life in urban environments. Since ecological principles are fundamental to landscape design and planning, urban landscape design plays an influential role in creating sustainable urban environments in context of resource management.

3. Urban landscape

Urban landscape is basically formed of open and green spaces within an urban environment[3]. However, it is not totally independent from the surrounding buildings and structures. Altogether, they form the character and identity of a city, and sense of place. It contributes to the cityscape by means of aesthetics and function. It also supports urban ecology. It is dynamic and constantly evolving. According to von Borcke (2003) *it is not an add-on but rather forms the basis for creating places.* Urban landscape elements function as separator and/or connector agents between different land uses. They can form a buffer zone between conflicting uses (e.g. between industrial and housing areas) while they can facilitate movement of citizens throughout the city (e.g. greenways). They have the flexibility to serve for multiple uses and for different group of users in the community (Anonymous, 2009).

Urban landscape also contributes to the cityscape in terms of visual quality. Within dense built environments, it creates a sense of openness and more attractive places to live. Urban landscape helps to balance human-scale in city centers where vertical effect of buildings and structures dominates. It softens the "hardness" of buildings and structures. Well designed and managed urban landscape can improve citizens' quality of life in many other ways as well. The benefits of urban landscape are explained below.

Ecological and environmental benefits

Contemporary urban ecology assumes that urban areas are ecosystems since they have interacting biological and physical complexes (Cadenasso & Pickett, 2008). McHarg played a major role in emergence of ecological landscape design approaches in urban development. His work "Design with Nature"(1969) displays how nature and city might coexist together. However, ecology has been neglected in urban planning systems of most developing countries which mostly focus on the relationship between physical and socioeconomic aspects of an urban development. Urban green spaces are fundamental in sustaining the urban ecology. Some of the environmental and ecological benefits of the urban landscape are listed below:

- Urban green spaces provide flora and fauna with a habitat to live and therefore support biodiversity conservation.
- They also act as ecological corridors between urban, periurban and rural areas. They support movement of living organisms between these areas.

---

[3] From now on "urban landscape" and "open and green spaces" will be used as synonyms in the chapter.

- Vegetation cover in urban landscape helps to improve micro-climate of urban areas where climate is warmer than their surroundings due to dense built environment and human activities. Vegetation cover raises humidity levels, reduces the stress of the heat island and mitigates the less desirable effects of urban climate (Landsberg, 1981). Daytime temperature in large parks was found to be 2-3°C lower than the surrounding streets (GreenSpace, 2010).
- Vegetation helps to decrease carbon emission levels in cities. Through photosynthesis process in plants CO_2 in the air is converted to O_2. Therefore, urban vegetation cover helps to reduce excess CO_2 in the urban atmosphere. Although the degree of trees' drawing carbon emissions from the air is affected by their size, canopy cover, age and health, large trees can lower carbon emission in the atmosphere by 2-3% (GreenSpace, 2010).
- Vegetation cover also filters out other particles and dust in the air.
- Green spaces absorb and reduce the noise generated by human activities, especially trees act like noise barriers.
- Vegetation cover and soil in urban landscape controls water regime and reduces run-off, hence helps to prevent water floods by absorbing excess water.
- Trees can also act like wind breaker.

Social benefits

Humans are the dominating elements of an urban environment. Social interaction, as a basic need for humans, is essential in developing sense of community, belonging and security. Social interaction in cities is possibly the highest in public open and green spaces. Urban open and green spaces offer citizens various activity choices including recreational and sports activities which promote social cohesion. In 1992, researchers from Pennsylvania State University have conducted a nationwide study to investigate American public's perceptions of the benefits of local recreation and park services. They concluded that local parks and recreation services are linked to sense of community.

Furthermore people from different demographical backgrounds share public urban landscape in their everyday life. While today democracy is regarded as the only legitimate form of government throughout the world, urban open and green spaces possess the notion of democracy in their nature. These places are designed and serve to everyone in the community. On the other hand this raises the issues of accessibility, equity and participatory planning which will be discussed in the next section.

All people need leisure time for relaxation and self fulfillment. Especially people living in urban environments seem to be more stressed and need more leisure time for their physical and mental health. There is a strong relationship between lifestyles, physical environment and leisure (Oğuz & Çakcı, 2010). Most people engage in leisure activities so that they can socialize. Urban open and green spaces can be designed to serve the community's leisure needs. It has also an economical aspect; for example playing ball games in a park or picnicking with friends is much cheaper than having a membership to a sports club or going to a restaurant. Unfortunately, nowadays people seem to spend more time indoors rather than outdoors which also leads to physical health problems due to inadequate physical activity. Designers should consider the ways to attract people to open and green spaces so that every group in the community can enjoy social benefits of urban landscape as well.

Environmental education can be regarded amongst the social benefits of urban landscapes. Green spaces can be thought of as outdoor laboratories to observe and get to know about nature. Environmental education is necessary for developing awareness and responsible behaviors towards the natural environment, especially for children. Although environmental education is integrated within the most school curricula, without physical interaction with natural environments it is not possible to develop environmentally responsible behaviors and attitudes (Çakcı & Oğuz, 2010). Green spaces in urban landscape may play an important role in environmental education. For instance, botanical parks are where people can learn about different plants, their living conditions and observe the physiological changes during time. Even urban street trees present the seasonal physiological changes in plants. Green spaces are also found to have impact on reducing violent behaviors in urban environments. For example Sullivan and Kuo (1996) investigated the effects of plants on social behavior and concluded that urban forests can help to reduce domestic violence levels in cities (Jackson, 2003).

Health benefits

The degradation of natural environments inevitably affects human health in a negative way. According to World Health Organization (2012); *environmental hazards are responsible for as much as a quarter of the total burden of disease world-wide, and more than one-third of the burden among children.* The relationship between environment, particularly urban environments and human health is rather complex. There are too many environmental factors that influence human health. People are more likely to be exposed to pollution and infectious diseases in cities compared to natural environments. Besides, human behavior trends in urban environments facilitate microbial traffic (McMichael, 2000) and globalization expands the spread of epidemic diseases, mainly through global transportation of humans and goods. Moreover the changes in urban lifestyles lead to some serious health problems which decrease life quality. For instance, passive lifestyles of urbanized communities (i.e. low physical activity) are strongly linked to obesity which has become a major health problem throughout the world. Clinical studies have shown that there are many other serious health problems associated with obesity, such as diabetes (Mokdad et al., 2003), coronary heart disease (Flint et al., 2010), Alzheimer's disease (Profenno et al., 2010), reduced fertility (Brannian, 2011), depression (Luppino et al.2010), osteoporosis (Paula & Rosen, 2010) and cancer (Freedland et al., 2009). Likewise, stress, which is an inevitable outcome of fast paced urban lifestyles, is found to have an impact upon the immune, circulatory, and nervous systems (Esch et al., 2003). Moreover, urban citizens have less contact opportunities with nature which is also linked to health and well being.

Parks and gardens are where urban citizens can contact with nature in their daily life. Health benefits of being in contact with natural environments have been known for centuries. In his writings the Roman senator Pliny the Younger described the mental and physical therapeutic effects of exercising and spending time in his villa gardens (Bowe, 2004; Ward Thompson, 2005). In Europe, during the medieval ages, the cloister gardens of the monasteries were used as healing gardens where patients were treated, exercised and relaxed (Ulrich, 2002; Ward Thompson, 2005). However an urban environment contains too many stimuli which attract directed attention. According to Kaplan's Attention Restoration Theory (ART) (1995), intense directed attention causes mental fatigue and natural environments, where involuntary attention is attracted, help to recover from psychological fatigue and allow directed attention mechanisms to rest (Berman et al., 2008). Even viewing natural environments is suggested to have positive impacts on health. Kaplan (2001) points out that nature views from windows influences well being and residential satisfaction (Oguz et al., 2010). Similarly, Ulrich (2002) suggested that viewing nature or garden landscapes can reduce stress and improve effects of clinical treatments in hospitals. Grahn and Stigsdotter's (2003) research also supports the idea that people, who spend more time in outdoor environments, are less affected by stress.

In the last century urban parks were referred as being "lungs of the city", which emphasizes their physical health benefits for urban citizens. As mentioned previously, urban vegetation cover provides a cleaner environment. Besides urban open and green spaces offer citizens' environments to exercise. The positive effects of physical activity on human health are well known. The regular physical activity is associated with reduced rates of coronary heart disease, diabetes, hypertension, osteoporosis, colon cancer and depression (Powell & Prat, 1996). Wolch et al. (2011) found that body mass index (BMI) in children with better access to urban parks is less likely to show a significant increase. In their study, Maas et al. (2006) investigated the relationship between the amount of green space in living environments and perceived general health. They found that the amount of green space has a positive effect on health although the causes remain unknown. They also suggested that green spaces should be given more importance in spatial planning policies. However, according to Lee and Maheswaran (2010) there is weak evidence between green space and both physical and mental health, instead factors such as quality and accessibility influence the use of green spaces for physical activity purposes. On the other hand Takano et al. (2002) concluded that longevity of urban senior citizens is positively influenced by living in areas with green spaces in walking distances. Sugiyama and Ward Thompson (2007) also suggested that outdoor environments have an important role in older people's well being.

There are many research on understanding and explaining the effects of urban landscape on human health. Although the links between health and green spaces remain missing, the literature review given above clearly display that urban landscape has positive impacts on improving both mental and physical health in several ways. It is landscape architect's role to create outdoor environments which maximize the benefits of urban landscapes for citizens to relax, exercise and restore.

Economic benefits

Although the economic valuation of urban landscape is difficult, open and green spaces have economic benefits in several ways:

- Their aesthetic contribution to cityscape influences property values. In general, urban landscape elements increase the nearby property value and enhance marketability of real estate (Anonymous, 2010). Accessibility, quality and visibility are basic factors that determine economic value of urban landscapes in this context.
- Urban landscapes provide employment opportunities during their design, construction and maintenance. The construction and maintenance of urban landscapes also supports other sectors such as playground manufacturers and nurseries.
- The health benefits of urban landscapes which were summarized above can reduce the costs of national health expenses.
- Public urban landscapes provide environments for walking, sports and other recreational activities for no cost at all, especially for lower income groups.

- Green spaces can help energy saving. Right selection and planting of plants can provide cooler environments in summer and warmer environments in winter thus reduce air conditioning expenses.
- Urban landscapes can enhance tourism in cities by attracting people. Park Güell in Barcelona, Spain is a perfect example of how a park can become a global tourism destination.

4. Landscape design in urban environment

In this section landscape design is discussed in urban context. From a wider perspective, urban landscape is a part of urban matrix. Therefore design of urban landscapes should be considered as an integral part of urban design. Urban landscape design is clearly not urban design, but a crucial part of it. Hence, factors influencing urban design also influence the form and functioning of urban landscapes. It is advised that points stressed out in the previous heading "Urban design and urban life quality" should also be kept in mind while reading this section.

Design is a creative process influenced by designer's experiences, values, beliefs and vision. Hence it is mostly subjective, so is landscape design. Landscape design is the art of creating and designing aesthetic and functional outdoor environments. Because every landscape is unique, it is hard to define a universal guideline for design process. Nevertheless designing sustainable and liveable environments requires understanding of some basic principles which will guide the designer. Here, I will explain some of the principles which I believe are essential for urban landscape design.

Adaptability and sustainability

Urban landscape is as dynamic as urban life. It constantly changes. The design product is never finished due to both ever changing structure of urban realm and the living materials used in urban landscape design. Modern urban environments grow and expand so fast that efficient use of land becomes a necessity. Thus, any design should be capable of adaptation to changes through the time and space while maintaining identity. Adaptability and flexibility degree of a design product determines its lifetime. Therefore designing urban landscape requires a far-sighted approach. In landscape design adaptability can be achieved through selecting appropriate design elements (e.g. plants, water elements and construction material) that fit for site conditions (e.g. climate, soil and water resources) and creating multiuse or flexible outdoor facilities for activities for different groups in the community. On the other hand, if everything is designed to be flexible in order to achieve adaptability, then the design will fall apart without any sense of meaning or character. Hence some parts should be left permanent to provide a backbone for the design. Creating large open spaces is the easiest way to create flexibility in landscape design.

Adaptability is one of the key elements in achieving sustainability as it is about longevity. The concept of sustainability has been briefly explained in previous sections. Besides its strong relationship with life quality, ecological sustainability is fundamental to survival of all living organisms on the earth. Urban ecology, a relatively very new area of ecology, seeks to understand and explain ecological mechanisms within an urban environment. Embracing urban ecology in urban design and management is necessary to create sustainable environments. McHarg's outstanding work "Design with Nature" (1969) has triggered ecological perspective in landscape planning and design. His overlay method of site analysis (suitability analysis) aims to define potentials and constraints of an environment for land use planning. Although he has been criticized for neglecting cities and social dimensions, he promoted integration of the natural processes into planning and design. The suitability analysis of McHarg still constitutes a basis for contemporary landscape planning and design. Assessing the relationships between each component of an environment enables designers and planners to recognize the true potential of a site for various land uses. Neglecting natural values in design causes high costs of construction and maintenance (Memlük, 2009).

Ecological sustainability is a tough yet crucial challenge in landscape design. It is often hard for the designer to integrate his artistic desires with the ecological facts. Yet, ecological mechanisms can help and guide the landscape architect through the design, because landscape design mostly depends on natural resources. In their study, Cadenasso and Pickett (2008) presented and discussed five urban ecology principles in context of urban landscape design. Table 1 shows the summary of their work.

Principle	Principle basics	Design and practice implications
Cities are ecosystems	Cities are ecosystems by virtue of having interacting biological and physical complexes. Urban ecosystems include four components: organisms, a physical setting and conditions, social structures, and the built environment.	Design affects all four components of human ecosystems.
Cities are heterogeneous	Heterogeneity in urban landscapes can be caused by both biophysical and social structures and processes. In turn, biophysical and social processes respond to urban spatial heterogeneity.	Design should enhance heterogeneity, and its ecological functions.
Cities are dynamic	Change in the structures and flows within cities, and between cities and other ecosystems lend a dynamic element to urban form and morphology (Decker et al. 2000; Kaika 2005; Shane 2005)	Design must accommodate internal and external changes projects can experience.
Human and natural processes interact in cities	All landscape designs and management schemes should be judged for their ability to contribute to both social and ecological goods and services, and to reduce both social and ecological risks and vulnerabilities (Steiner 2002; Grove et al. 2007).	Design should recognize and plan for feedbacks between social and natural processes.
Ecological processes remain important in cities	Concepts and approaches basic to ecological research can be applied to urban areas in an effort to understand how the city itself functions as an ecosystem (Alberti et al. 2003).	Remnant ecological processes yielding ecological services should be maintained or restored.

Table 1. Ecological principles and design applications (adapted from Cadenasso and Pickett, 2008).

Some strategies for ecologically sensitive urban landscape design are:

- Support and preserve biotic diversity and create habitat corridors.
- Minimize energy use and promote use of renewable energy resources. Reduce energy costs by using solar and wind energy systems.
- Protect and improve quality of water resources.
- Reduce water and fertilizer use by selecting native and drought tolerant plant species.
- Reduce water runoff by decreasing the amount of hard surfaces and proper drainage design.
- Conserve aquifer recharge zones.
- Provide collection and storage of rainwater in order to use it for maintenance of green spaces.
- Support pedestrian and bicycle circulation within the city. People should move easily and freely within a city. Cities should not be designed for vehicle traffic. Connected urban open and green space systems could create a environmentally and people friendly transportation routes.
- Choose plants suitable for local climate and site conditions. Selecting right plant species will increase the survival chance of plants in harsh urban environment, success of design and decrease maintenance costs.

Coherence and legibility

Coherence of a landscape refers to the organization of landscape components; it is the degree of consistency between the components. According to Salingaros (2000) geometrical assembly of elements determines the coherence of an urban environment and connectivity at all scales leads to coherence. He also stresses that coherence is essential for vital urban living environments. Coherence and legibility are strongly related to understanding of a place and feeling of safety. Thus, level of an environment's coherence affects its legibility by its users. More coherent and legible an environment is, more people make sense of it, and more they feel safe.

Legibility in urban design was introduced by Kevin Lynch in his work "Image of the City" (1960) where he analyzed post-war North American cities' built environments. Lynch defines legibility as "...the ease with which [a city's] parts may be recognized and can be organized into a coherent pattern". According to Lynch, legibility is a key basis for sense of place. A sense of place evokes the feeling of "belonging" which makes a place psychologically comfortable. For a designer, it is important to turn empty "space" into a "place "with a meaning to experience. Although place identity is different from sense of place, character and identity of a place are two main aspects of sense of place. Therefore, giving character and identity to a designed place is essential for creating meaningful places for people. Antrop (2005b) states that coherence of particular properties defines identity and changes in coherence causes loss of identity or transforms the identity to a new one. According to Relph (1976) physical setting, activities and the meanings are three basic elements of place identity (in Turner &Turner, 2006). The degree of coherence of the first two components influence the meaning interpreted by the user, and as a result sense of place. Sense of place is unique to both individual and place since it results from the interaction of both. Turner and Turner (2006) identified the components of sense of place as:

- The physical characteristics of the environment,
- The affect and meanings including memories and associations, as well as connotations and denotations;
- The activities afforded by the place,
- The social interactions associated with the place.

Loss of identity and character in modern urban environments is one of the main challenges for designers today. Furthermore it causes loss of community attachment and community identity. Changes in land-use, globalization, decentralization, environmental pollution and changing socioeconomic structure are amongst the reasons for the loss of identity and character. Hence, solutions need to be developed first at planning scale. On the other hand, urban design can help with renewing and creating places with coherence, character and meaning.

As summarized above, the concepts of coherence and legibility are strongly linked to identity, character and sense of place. Therefore coherence and legibility should be adopted as important design principles in urban landscape design in order to create meaningful places which people enjoy to experience. Some key points in enhancing coherence and legibility in urban landscape design are given below;

- Coherence of landscape elements might be visual, functional or ecological. Hence, both natural and cultural landscape elements should be assessed in terms of coherence.
- Both history and cultural values of a place should be considered in attempt to achieve coherence and strengthening identity.
- Local architectural styles and materials should be taken into account in landscape design in order to provide integrity and coherence and preserve local identity.
- Visual quality of a landscape is important for reading and understanding the place. The entrances and exits of a place should be visually clear.
- Placing sculptures or other ornamental features can enhance the visual quality of the environment.
- Human scale is essential in legibility.
- Landscape construction materials should be selected to support or enhance visual coherence overall and between different landscape elements.
- Designed environments should support perceived safety. Open spaces offer people a sense of security since they have a wider perspective of a place to see what is going on around them, however people also needs to find a "niche" for themselves to have some privacy or to hide if they feel insecure.
- Spatial definition also helps people to feel more secure. Knowing boundaries increases confidence. Hence, edges should be clear and visible.
- For ecological coherence, green networks throughout and around the city should be created and connected to each other.
- Time is another variable of coherence. Forms, textures and colors of an urban landscape vary through time. Therefore changes through time and alternatives for the future should be taken into account in design process.
- Coherence between user needs and expectations, and proposed activities should be evaluated. Participation of local people in design process may help to create more coherent design alternatives.

- Diversity is essential for both healthy functional landscapes and creating more appealing environments for people. Diversity of landscape elements without coherence may cause chaos.
- Connectivity between different landscape elements enhances coherence.

Equity and accessibility

Urban environments are where people from various ethical, social, cultural, economic backgrounds, ages and gender live together. Segregation of urban communities leads to both spatially and socially divided urban environments. As a result, equity becomes an important issue in achieving sustainable community development. Thus, designing public spaces for everyone is crucial in today's communities for developing community identity and preventing social fragmentation. Urban landscape design helps to create accessible environments for everyone in the community in terms of public open space. Public open spaces are places where people from different backgrounds get together and experience the urban environment equally in democratic communities. Thus, the activities and opportunities offered in these places should serve the needs and expectations of different groups in the community. Open and green spaces should be distributed evenly throughout the city. Special design techniques are needed to be taken into account for accessibility of disadvantaged groups, such as disabled and elderly. Not only the activities but also the landscape material should be selected appropriately to ensure easy accessibility and safety. Children should also be amongst priority in landscape design. Encouraging children to spend their free time in urban open and green spaces is a completely different research area itself. It is an interdisciplinary subject involving many dimensions from child development to landscape design. On the other hand, there is a clear need for promoting children living in urban environments to get involved in outdoor activities for healthier lifestyles as well as developing sense of community. Therefore, urban landscape design should give importance to creating safe, enjoyable and creative environments for children, and of course for their parents where they feel safe to let their children out.

Urban open and green spaces offer a unique opportunity for integration of different social groups and individuals. Hence, designers should seek ways to support social cohesion through space. Community involvement in planning and design schemes is necessary in order to have an idea of existing problems, needs, and expectations of different groups within the community.

Community involvement

Community involvement in urban planning and environmental impact assessment has gained widespread recognition in the last few decades. In urban design, community participation takes place mostly during decision making process in urban renewal projects. Involvement of community in design process increases their confidence in the project and their responsibility. Through information and knowledge exchange both local community and designers can be inspired. As a result, local community is more likely to support the project and to embrace the designed environment. This leads to satisfaction and long lasting use of the designed place. Therefore, community involvement should always be considered by the designer as an essential part of the design process.

We design outdoor environments for public benefit. Hence, understanding public opinions on their living environments could be helpful in creating livable urban environments. On the other hand, designers should realize that it is impossible to please everyone and respond to everyone's needs and expectations. Therefore, community involvement should only be accepted as a tool to understand the possible effects of design proposal on its future users.

Cost

Implementation and maintenance costs should be taken into account during design process. Cost efficient design strategies should be adopted in order to maintain durability and longevity of the designed environment. Urban open and green spaces can be designed and managed to minimize expenses. For example, stored rainwater in ponds and lakes could be used in irrigation, or use of fertilizers can be minimized by decreasing amount of lawn areas. Solar energy panels could be used to store energy during daytime which could be turned into electricity for lighting at night time. Multi-functional spaces also reduce cost by letting various activities take place in one place. Selection of durable paving material and site furniture will also decrease the maintenance costs in long term.

5. A case study - Cumhuriyet Square, Afyonkarahisar (Turkey)

In this section a square design project, which was submitted to a national urban design competition in 2011, will be briefly explained. The proposed project has been designed by a multi disciplinary team involving landscape architects, architects and urban planners. The competition was opened by the Municipality of the province Afyonkarahisar which is located in central Anatolia, Turkey. The aim of the competition was to promote design solutions for revitalization of the square, namely Cumhuriyet Square, which has lost its identity and function due to surrounding densely built environment and vehicle traffic.

The square is located in the city center and surrounded with many historical structures which constitute a vital part of the city's identity. The city Afyonkarahisar has been home to many civilizations from the Hittite Empire to Phrygians, from Byzantines to Ottomans. Because it has a rich historical background, the built structures which have historical significance have been preserved in the design proposal. One of the biggest problems within the site is lack of open spaces. This also has a negative impact on pedestrian circulation within this historical site. Therefore, open spaces have been given importance in design process in order to promote pedestrian circulation and create a place for citizens to gather (Figure 1).

The existing urbanization pattern has caused loss of connectivity between focal features of the city. The design proposal has aimed to link the square to other key locations of the city, such as the Afyonkarahisar Castle, Ulucamii (a historical mosque), the train station, and the university campus (Figure 2a). Architectural restoration has been suggested between these landmarks to strengthen the historical identity. The roads surrounding the square are the main traffic routes in the city. This prevents integration of the square into the urban form, thus an underground passageway for vehicle traffic has been created in the proposed design. Furthermore, all the historical streets have been pedestrianized (Figure 2b) and designed for easy accessibility.

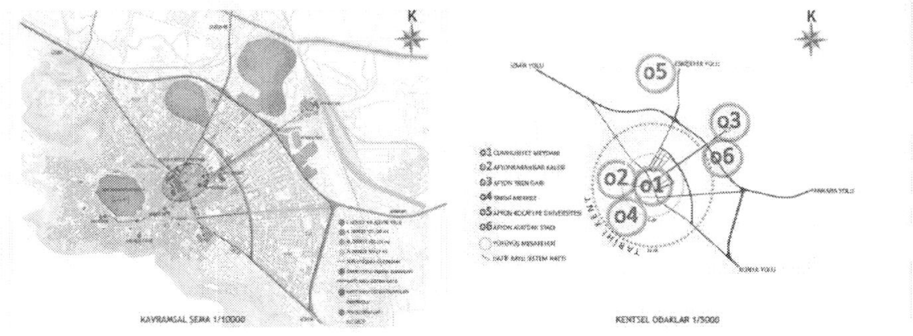

Fig. 1. The conceptual framework study.

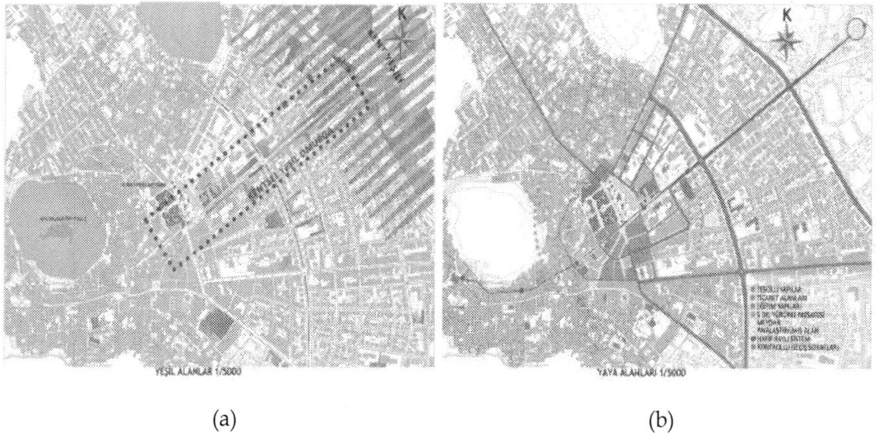

(a)   (b)

Fig. 2. (a) Green network (b) Pedestrian areas

Due to its historical values, Afyonkarahisar is also a tourism destination. To provide accessibility to historical city center, parking areas and other serving facilities have been developed. Moreover, a cultural center has been designed by the square (Figure 3) and the city center is transformed into a historical and cultural asset for the city.

the sake of human future. While debates over new urbanism vs. landscape urbanism remain in the agenda, I believe we will witness willingly cooperation of planning and design disciplines to create sustainable and quality urban environments in the next few decades.

7. References

Adams, J.W & Sierra, K. (2009). Foreword, In *Eco² Cities: Ecological Cities as Economic Cities (Conference Edition)*, Hiroaki Suzuki, Arish Dastur, Sebastian Moffatt & Nanae Yabuki, pp. xiii- xiv, The World Bank, Washington DC.

Anonymous. (2009). Urban spaces- enhancing the attractiveness and quality of urban environment, Date of access: 12/12/2011, Available from:
http://www.urbanspaces.eu/files/Working%20Paper%203.1.2_FINAL.pdf

Anonymous. (2010). The economic benefits of open space, recreation facilities and walkable community design, *Active Living Research*, Date of access: 13/12/2011, Available from:
http://www.activelivingresearch.org/files/Synthesis_Shoup-Ewing_March2010.pdf

Antrop, M. (2005a). Handling landscape change. *Proceedings of ECLAS 2005 Conference: Landscape change*, ISBN: 975-482-688-9, Ankara, September 2005.

Antrop, M. (2005b). Why landscapes of the past are important for future. *Landscape and Urban Planning*, Vol.70, No.1-2, pp. 21-34, ISSN: 0169-2046.

Arida, A. (2002). *Quantum City*, Architectural Press, ISBN: 0-7506-5012-5.

Baş, Y. (2003). *Designing urban space with the tools of the development legislation*, Master of Science Thesis, Middle East Technical University, Ankara.

Berman, M.G., Jonides, J. & Kaplan, S. (2008). The cognitive benefits of interacting with nature. *Psychological Science*, Vol.19, No.12, pp. 1207-1212, Online ISSN: 1467-9280.

Bowe, P. (2004). *Gardens of the Roman World*. Jç Paul Getty Museum, Los Angeles, ISBN-10: 0892367407.

Brannian, J.D. (2011). Obesity and fertility. S D Med, Vol. 64, No.7, pp. 251-254, ISSN: 0038-3317.

Cadenasso, M.L. & Pickett, S.T.A. (2008). Urban principle for ecological landscape design and management: Scientific fundamentals. *Cities and the Environment*, Vol.1, No.2, Article 4, The Berkeley Electronic Press, Available from:
http://digitalcommons.lmu.edu/cgi/viewcontent.cgi?article=1024&context=cate

Carr, A.J., Thompson, P.W. & Kirwan, J.R. (1996). Quality of life measures. *British Journal of Rheumatology*, Vol. 35, No.3, pp. 275-281, ISSN: 0263-7103.

Clarke Annez, P. & Buckley, R.M. (2009). Urbanization and growth: setting the context, In *Urbanization and Growth*, Michael Spence, Patricia Clarke Annez, Robert M. Buckley, pp. 1-45, The World Bank, ISBN: 978-0-8213-7573-0.

Çakcı, I. & Oğuz, D. (2010). Is environmental knowledge enough to motivate the action? *African Journal of Agricultural Research*,Vol.5, No. 9, pp.856-860, ISSN: 1991-637X.

EPA. (2009). HUD-DOT-EPA Partnership for Sustainable Communities. Date of Access: 15/12/2011, Availbale from:
http://www.epa.gov/smartgrowth/partnership/index.html

Esch, T., Fricchione, G.L. & Stefano, G.B. (2003). The therapeutic use of the relaxation response in stress-related diseases. *Med Sci Monit*, Vol. 9, No.2, Date of access: 15/12/2011, Available from:
http://www.MedSciMonit.com/pub/vol_9/no_2/3454.pdf

Flint, A.J., Rexrode, k.M., Hu, F.B., Glynn, R.J., Caspard, H., Manson, J.E., Willett, W.C. & Rimm, E.B. (2010). Body mass index, waist circumference, and risk of coronary heart disease: A prospective study among men and women. *Obesity Research & Clinical Practice*, Vol. 4, No.3, pp. 171-181, ISSN: 1871-403X.

Freedland, S.J., Banez, L.L., Sun, L.L., Fitzsimons, N.J. & Moul, J.W. (2009). Obese men have higher-grade and larger tumors: an analysis of the duke prostate center database. *Prostate Cancer and Prostatic Diseases*, Vol. 12, No.3, pp. 259-263, EISSN: 1476-5608.

Grahn, P. & Stigsdotter, u.A. (2003). Landscape planning and stress. *Urban Forestry and Urban Greening*, Vol.2, No.1, pp.1-18, ISSN: 1618-8667.

GreenSpace. (2010). *Blue Sky Green Space: Understanding the Importance of Retaining Good Quality Parks and Green Spaces, and the Contribution they make to Improving People's Lives* (Draft). Acess date: 12/12/2011, Available from: http://www.green-space.org.uk/downloads/GreenLINK/BlueSkyGreenSpaceFullReportDraft.pdf

Jacques-Menegaz, M. (2006). Amenity or necessity: parks and open space in San Fransisco, In *Urban Action 2006*, Michelle Jacques-Menegaz & Nina Haletky, pp. 35-41, http://bss.sfsu.edu/urbanaction/ua2006/ua2006.pdf

Jackson, L.E. (2003). The relationship of urban design to human health and condition. *Landscape and Urban Planning*, Vol.64, No.4, pp.191-200, ISSN: 0169-2046.

Krieger, A. (2009). Where and How Does Urban Design Happen? In *Urban Design*, Alex Krieger & William S.Saunders, pp. 113-130, University of Minnesota Press, ISBN: 978-0-8166-5639-4.

Lee, A.C.K. & Maheswaran, L. (2010). The health benefits of urban green spaces: a review of the evidence. *Journal of Public Health*, Vol.33, No.2, pp. 212-222, Online ISSN: 1741-3850.

Luppino, F.S., de Wit, L.M., Bouvy, P.F., Stijnen, T., Cuijpers, P., Penninx, B.W.J.H. & Zitamn, F.G. (2010). Overweight, obesity and depression: A systematic review and meta-analysis of longitudinal studies. *Arch Gen Psychiatry*, Vol.67, No.3, pp. 220-229, Online ISSN: 1538-3636.

Lynch, K. (1960). *The Image of The City*, The MIT Press, ISBN: 0-262-62001-4, USA.

Maas, J., Verheij, R.A., Groenewegen, P.P., de Vries, S. & Spreeuwenberg, P. (2006). Green space, urbanity and health: how strong is the relation?. *J Epidemiol Community Health*, Vol.60, No.7, pp.587-592, Online ISSN 1470-2738.

McHarg, I.L. (1969). *Design with Nature*. Doubleday, Garden City, NewYork.

McMichael, A.J. (2000). The urban environment and health in a world of increasing globalization: issues for developing countries. *Bulletin of the World Health Organization*, 78 (9), pp.1157-1126, ISSN: 0042-9686.

Memlük, M.Z. (2009). Design with nature in urban areaas: The case of Ankara- Bademlidere. PhD Thesis submitted to Ankara University Graduate School of Natural and Applied Science, Ankara.

Mokdad, A.H., Ford, E.S., Bowman, B.A., Dietz, W.H., Bales, V.S. & Marks, J.S. (2003). Prevalence of Obesity, Diabetes, and Obesity-Related Health Risk Factors, 2001. *JAMA*, Vol.289, No.1, pp. 76-79, Online ISSN: 1538-3598.

Oguz, D., Cakci, I., Sevimli, G. & Ozgur, S. (2010). Outdoor environmental preferences in nursing homes: Case study of Ankara, Turkey. *Scientific Research and Essays*, Vol. 5 No. 24, pp. 3987-3993, ISSN: 1992-2248.

Oğuz, D. & Çakcı, I. (2010). Changes in leisure and recreational preferences: A case study of Ankara. *Scientific Research and Essays*, Vol.5, No.8, pp. 721-729, ISSN: 1992-2248.

Pacione, M. (2003). Urban environmental quality and well being- a social geographical perspective. *Landscape and Urban Planning*, Vol. 65, pp. 19-30, ISSN: 0169-2046.

Parfect, M. & Power, G. (1997). *Planning for Urban Quality: Urban design in towns and cities.* Routledge, London, ISBN: 0-415-15967-9.

Paula, F.J.A. & Rosen, C.J. (2010). Obesity, diabetes mellitus and last but not least, osteoporosis. *Arq Bras Endocrinol Metab.*, Vol. 54, No.2, pp. 150-157, ISSN: 0004-2730.

Powell, K.E. & Pratt, M. (1996). Physical activity and health, *BMJ*, 313, pp. 126-127.

Profenno, L.A., Porsteinsson, A.P. & Faraone, S.V. (2010). Meta-Analysis of Alzheimer's Disease Risk with Obesity, Diabetes, and Related Disorders. *Biological Psychiatry*, Vol. 67, No.6, pp. 505-512, ISSN: 0006-3223.

Rowley, A. (1994). Definitions of urban design: The nature and concerns of urban design. *Planning Practice and Research*, Vol.9, No.4, pp. 179-197, Online ISSN: 1360-0583.

Salingaros, N.A. (2000). Complexity and urban coherence. *Journal of Urban Design*, Vol.5, No.3, pp. 291-316, Online ISSN: 1469-9664.

Sugiyama, T. & Ward Thompson, C. (2007). Older people's health, outdoor activity and supportiveness of neighbourhood environments. *Landscape and Urban Planning*,Vol.83, No.2-3 , pp. 168-175, ISSN: 0169-2046.

Takano, T., Nakamura, K. & Watanabe, M. (2002). Urban residential environments and senior citizens' longevity in megacity areas. The importance of walkable green spaces. *J. Epidemiol. Comm. Health*, Vol. 56, No.12, pp. 913-918, .

Tress, B. & Tress, G. (2001). Capitalising on multiplicity: a transdisciplinary systems approach to landscape research. *Landscape and Urban Planning*, Vol. 57, Issues:3-4, pp. 143-157, ISSN: 0169-2046.

Turgut Yıldız, H. (2007). Kentsel yaşam kalitesi: Kuram, politika ve uygulamalar. In *Mimarlık Dergisi*, Date of access: 15/12/2011, Available from: http://www.mimarlarodasi.org.tr/mimarlikdergisi/index.cfm?sayfa=mimarlik& DergiSayi=53&RecID=1325

TurkStat (Turkish Statistical Institute). (2011). Date of access: 10/12/2011, Available from: http://www.turkstat.gov.tr/PreTablo.do?tb_id=39&ust_id=11

Turner, P. & Turner, S. (2006) . Place, Sense of Place, and Presence. *Presence*, Vol.5, No.2, pp.204-217, ISSN: 1054-7460.

Ulrich, R.S. (2002). Health benefits of gardens in hospitals, *Proceedings of Plants for People International Exhibition Floriade*, Floriade, 2002.

United Nations, Department for Economic and Social Information and Policy Analysis. (1997). *Glossary of Environment Statistics*, United Nations, Series F, No.67.

University of California, Berkeley. (2011). Urban Design: City-Building and Place-Making, In *Course Threads*, Date of access: 15/12/2011, Available from: http://coursethreads.berkeley.edu/courses/urban-design-city-building-and-place-making

Von Borcke, C. (2003). Landscape and nature in the city, In: *Sustainable Urban Design: An environmental approach,*Randall Thomas & Max Fordham LLP, pp. 33-45, Spon Press, ISBN: 978-0-415-28123-2, USA.

Ward Thompson, C. (2005). Common threads in a changing world: landscape and health. *Proceedings of ECLAS 2005 Conference: Landscape change,* ISBN: 975-482-688-9, Ankara, September 2005.

Wolch, J., Jerrett, M., reynolds, K., McConnell, R., Chang, R., Dahmann, N., Brady, K., Gilliland, F., Su, J.G. & Berhane, K. (2011). Childhood obesity and proximity to urban parks and recreational resources: A longitudinal cohort study. *Health & Place,* Vol 17 (1), pp. 207-214, ISSN: 1353-8292.

World Health Organization. (1996). *WHOQOL Bref,* Available from:
http://www.who.int/mental_health/media/en/76.pdf

14

Irrigation

Mehmet Sener
Namık Kemal University
Turkey

1. Introduction

The water is basic source for all living in the world and it covers two third of the earth, it is the lifeblood of plants and its permanent of temporary shortage can cause serious damage to plants yield and quality. However, despite the importance of water for the continuation of life, 97.5 % of existing water is in ocean and high seas water that cannot be used. The remaining just 2.5 % is fresh water and can be used by living things; however, the vast majority, like 90% of this fresh water is in the poles and underground water and it doesn't have usable form (William, 2007). Available water resources in the world not showing continuity in time and space, the increasing effect of global warming and the three main water users (agriculture, industry and domestic sectors) creating a growing demand and pressure for water supply increases the importance of water.

Irrigation has an important place in terms of water usage, its environmental impact and reuse of urban waste water. Irrigation; firstly begin with people becoming sedentary and agricultural practices, it has been made to improve the quality and quantity of crop production. Especially since the 1950s, people living in the cities started to be the owner of a garden, studies conducted in the urban landscape water is needed in these areas with different types of grass and ornamental crops, in order to ensure the sustainability of irrigation, which resulted in entering of irrigation into urban life. Today, irrigation emerges as an indispensable element in Landscape architecture.

2. Irrigation water quality and reuse of waste water for irrigation

2.1 Irrigation water quality

Irrigation water quality is important for two reasons for irrigation engineers. The first , depending on the quality, irrigation method and irrigation system components are determined, the needed additional equipment lead to an increase in total cost of the system. The second, water quality affects adversely plant growth and soil fertility after the system operates. Irrigation water characteristics are examined in three groups; physical, chemical and biological (Wilko & Short, 2007).

2.1.1 Physical properties

In the physical properties, suspended matter is one of the most important features. Suspended matters in natural waters are usually composed of plaques erosion, parts of organic matter and planktons (Ayyıldız, 1990). Suspended matters are one of the most important things because they could cause obstructions in emitters in the sprinkler and drip irrigation. This kind of water should be filtered appropriately and then should be given to the system.

2.1.2 Chemical properties

The most important chemical properties in terms of irrigation is pH, the amount of solid matter dissolved in water (electrical conductivity), sodium adsorption ratio (SAR), residual sodium carbonate (RSC), Na % and amount of boron. pH is a measure that shows the acidity or alkalinity of water and it is required to be between 6.5 to 8.0 in irrigation water. The amount of dissolved solids is the degree of salinity in irrigation water. The degree of salinity in irrigation water is expressed as electrical conductivity (Ayyıldız, 1990). Salt in the irrigation water is important in terms of physical and chemical properties changes of the water and soil, making toxic and physiological effects on plant. Most water of acceptable quality for turfgrass irrigation contains from 140 to 560 μmhos/cm soluble salts (Cockerham & Leinauer, 2011). It is necessary to be careful when irrigating with highly salinity waters, especially higher than 2250 μmhos/cm.

Sodium is important in terms of blocking soil colloids and prevents the formation of a available air water balance in the soil. Ratio of sodium ion on total cations (Na %) and its ratio to magnesium and calcium ions (SAR), the two indicators are used in evaluation of irrigation waters. In order to create a condition of suitable soil in the root zone of the plants, the percentage of Sodium should not be more than 50-60%.

Sodium adsorption ratio is another important factor. As the increase of the sodium amount in the irrigation water breaks its physical properties, it is also important in terms of making alkaline the soil (Smith, 1996). In the analysis of irrigation water, due to the higher total salt concentration the higher effect of SAR will be, while estimating the effects of the SAR, the total concentration in the irrigation water should also be considered.

Boron is another important criterion, which is usually not available as element in nature, it is usually found as sodium borate or calcium borate. More than 0.5 mg/l in irrigation water may cause toxic effect on plants.

2.1.3 Biological properties

Because of bacteria threaten the human health, care should be paid in the use of such polluted water in areas where people live. As specified in the regulation of water quality, the fecal coliform values up to 200/100 ml of water can be used for irrigation purposes (Harmancioglu et. al., 2001).

2.2 Reuse of waste water in irrigation

Water is polluted physically, chemically and bacteriological by various wastes especially in industrial and agriculture as well as household use. Water contaminated in such ways is called waste water. In the regions where clean water resources is limited or water charges are high, irrigation water requirements are tried to be solved with the use of waste water contaminated with industrial and specially with urbane use (Novotny et. al., 2010). However it should be remembered that the waste water can be used for irrigation purposes if it is treated to the degree that it meets the quality regulations and communiqués criteria. In case of irrigation with waste waters, it is essential to identify an irrigation method and program that will no cause pollution of clean water as a result of deep seepage or runoff. Especially before using bacteriological contaminated waste water for landscape irrigation, the size of the area, population density, pollution degree of water must be analyzed very well (Surampalli & Taygi, 2004). Drip irrigation method should be preferred to sprinkler irrigation in order to minimize the damages to the people around since water is applied to the plants through atmosphere in sprinkler irrigation and to prevent black logs on the plant during irrigation of polluted waters. If sprinkler irrigation method is required to be applied, irrigation will possibly minimize the health and environmental effects in windless conditions.

3. Quantitative characteristics of irrigation water

3.1 Evapotranspiration

Plants, like all living things require water to sustain their lives. In order to fulfill the basic functions, enough water is needed in the soil containing roots. Otherwise, plant growth decline and deaths occur in later stages. Evapotranspiration (ETc) is consisting of two basic elements. These are: transpiration through stoma of plant leaves and evaporation of open surface of the soil around the plant. Many ways including direct and indirect ways are used in determination of ET. However, reference evapotranspiration (ET_0) are used as pre-projecting factor reference. A mathematical relationship is used for this purpose as given bellow. (Doorenboos and Pruit, 1977):

$$ETc=ETo*kc \tag{1}$$

Where,
ET_c: Evapotranspiration of the crop, mm
ET_0: Reference evapotranspiration, mm
k_c: crop coefficient

ETc is calculated based on plant that is taken as reference in ET_0. This plant is usually alfalfa or meadow pasture. There are many improved equations to calculate the reference Evapotranspiration. These include Penman FAO modification, Penman Montheith, Hargraves, Blaney Criddle and Class A evaporation tank etc. (Tekinel and Kanber, 1988).

The second factor affecting the Evapotranspiration is crop coefficient. Crop coefficient is defined as a ratio of ETc to ETo. Crop coefficient varies depending on factors such as plant type, age, growth period, soil moisture etc (Allen et. al., 1978). Grass plant is usually grown in the fields of landscape. Although this plant has many types, its k_c value has been reported to vary between 0.7 and 1.05.

3.2 Net irrigation requirement

Surface water resources are not distributed in a homogeneous way in many regions of the world, irrigation is applied in these areas in order to ensure vegetation sustainability due to irregular and inadequate rainfall. Irrigation water requirement is calculated with the help of the following equation.

$$IR: ETc-Peff \tag{2}$$

Where,
IR= Net Irrigation requirement, mm
ET_c= Evapotranspiration, mm
P_{eff}= Effective rainfall, mm

The irrigation requirement occurred as a result of evapotranspiration can be meet by precipitation. Precipitation usually occurs in the form of rain in plant growing season. Therefore, in determining of irrigation water requirement, the rainfall used by plants must be taken into account. However, a certain amount of rainfall is used by plants through surface runoff and deep infiltration. The rainfall that is stored and used in the root zone of the plants is called effective rainfall.

3.2.1 Net irrigation requirement for each irrigation

In the irrigation scheme, in the event that evapotranspiration is not met with natural rainfall, the deficit should be met with irrigation. In a operation unit, the quantity of irrigation water is determined with the help of the following formula.

$$dn=(FC-WP)/100)*Ry*Yt*D \tag{3}$$

Where,
dn: Net irrigation requirement for each irrigation, mm
FC= Field Capacity, %
WP= Wilting point, %
Ry= Allowable soil water deficit % (0.3-0.4 can be taken for turf)
Yt= bulk density, g/cm^3
D= the root depth, mm

As shown in the equation 3, there are two basic features affecting quantity of irrigation water. These are soil and plant features. In the irrigation area, due to the fact that the plants in each operation unit have different evapotranspiration, the quantity of the bounden irrigation water is also changing. At the same time, depending on soil characteristics, the water-holding capacity representing the rest quantity of water between field capacity and wilting point, also changes while water-holding capacity is low in a sandy soil, it is high in a clayed soil. In contrast to the sandy soils, due to the high water-holding capacity of clayed soils, more irrigation water will be applicated. Therefore, in the preparatory stages of recreation areas, when creating operation units, ensuring the collection in the same area of the same kind of plants and the fields with the soil structure will provide great convenience to users in the operation of irrigation system. This planning is the only way to ensure the uniform water distribution in the system without causing the overuse of excess water. In case of ignoring this situation, in the operation units that exits different plants or they are nested, some plants will be overwatered, and also some plants will be watered insufficiently. In addition, depending on soil characteristics, while the pondings may be in some areas, there will be losses of deep-seepage in some areas.

3.2.2 Gross irrigation requirement for each irrigation

Even in optimal projecting condition, there is no storage of all of the water in the root zone of plants. During the delivery and application of irrigation water, the water losses are experienced. It is considered that there is no delivery losses in water pipes transmitted with closed pipes. The total quantity of the required irrigation water for each irrigation is calculated by the following formula.

$$dg=dn/Ea \qquad (4)$$

Where,
d_g: gross irrigation requirement, mm,
d_n: net irrigation requirement, mm,
E_a: irrigation application efficiency %.
E_a can be taken as 0.8 when the planning for sprinkler irrigation

3.2.3 Irrigation interval

It is an expression how many days elapsed between consecutive two irrigation. It can be determined with the help of the following formula.

$$Ti: dn/Etc \qquad (5)$$

Where,
Ti: irrigation interval, day
d_n: Net irrigation requirement for each irrigation, mm
ET_c: evapotranspiration mm /day

Due to heavy textured soils have greater water-holding capacity, the irrigation interval increases depending on the net irrigation water, and it decreases in sandy soils. In suitable soil conditions, as reducing by half the irrigation period of a week or ten day period of irrigation, increasing irrigation efficiency can be achieved.

3.2.4 System capacity

The system capacity turned from water supply or irrigation duration is determined by the following formula.

$$Q=A*dt/3.6*T \qquad (6)$$

Where,
Q:System capacity, L/s
A: Irrigated Area, da
T: Irrigation duration, h

4. Irrigation methods

The way of irrigation water supplied to the crop root zone is called as irrigation method. In determining the irrigation method for use, soil and topographic features of the land, plant type, irrigation water supply, labor and technical skills conditions, facilities and operating costs and climatic data of region are factors that are needed to be considered. Basically, there are two groups as surface irrigation methods and pressurized irrigation methods. Surface irrigation methods are evaluated amongst the traditional irrigation methods and transmission of water is carried out with the help of energy provided through geographic height difference completely between the source and target parcels. The basin, furrow, border and uncontrolled flooding irrigation methods are also involved in the surface irrigation method.

The purpose of Irrigation planners is to re-increase the optimum level the decreasing water levels in the root zone with a minimum of irrigation water by using an efficient irrigation system. The pressurized irrigation methods are preferred than surface irrigation methods because of its need more less water and staff per unit area. In pressurized irrigation methods, water is transmitted via the pipe system from the source to the relevant parcels under a certain pressure. A pressurized irrigation system that will be installed in general, is composed of pump unit, control unit, the main pipe line, sub main pipe line, manifold, lateral and water emitters (Phocaides, 2007) (Figure 1).

4.1 Sprinkler irrigation

This method is the systems which water is supplied from the source such as stream, lake, dam, or drinking water system in the form of droplets sprayed through atmospheric air to the plant with the help of sprinklers (Ingels, 2003). These types of systems which are high degree-uniform-water distribution can be supplied with proper planning in all kinds of soil conditions and slope and rough terrains. Sprinkler irrigation is the most suitable and common method for landscape irrigation. It is suitable for irrigation by using any type of water source on the condition of using appropriate equipment. Sprinkler irrigation elements: water supply, pumping unit, control unit, the main canal, sub main canal, manifold, lateral and sprinkler heads. Sprinkler irrigation method is the most preferred method in landscape irrigations. Advantage and disadvantage of sprinkler irrigation method are given below.

The advantages of sprinkler irrigation method

- The larger areas can be irrigated with smaller quantity of water than the method of surface irrigation.
- In problematic areas from topographic aspect, the desired performance can be achieved through appropriate design and arrangement.

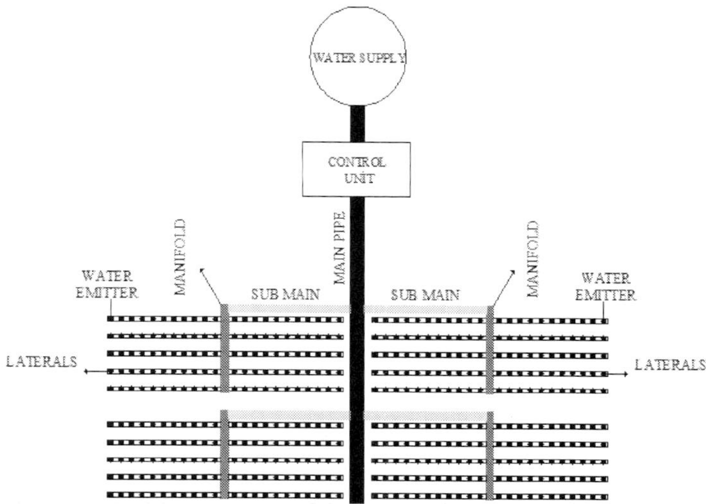

Fig. 1. Main components of pressurized irrigation systems

-    High uniform distribution of water can be provided,
-    Irrigation efficiency is high,
-    Fertilizing can be applied during irrigation,
-    Frost protection can be achieved.
-    Irrigation labor is less than in surface irrigation.
-    Its initial investment cost is lower than the drip irrigation.
-    It prevents the salt deposit in the root zone of plants by the effect of leaching

The disadvantages of sprinkler irrigation method

-    The realization problems of uniform distribution of water are experienced especially in
     windy conditions.
-    Blasts may occur in the leaves in applications during sunny conditions
-    It leads the occurrence of an illness (crown gall) or spread of disease in the wet foliage
-    The evaporation loss is occurred at noon.

Components of sprinkler irrigation system are given below.

4.1.1 Water supply

In sprinkler irrigation systems, rivers, lakes, well water, dam can use as a source of water. Urban drinking water schemes and wells are generally used in Landscape irrigation.

4.1.2 Pump units

In sprinkler irrigation system, water should be sprayed from the headings under a certain pressure. This pressure is called as a head-operating pressure. In addition in the system, depending on equipment used between water supply and irrigation area, in case of an extra pressure requirement due to the difference of levels with terrestrial and permanent head loss, this requirement is met by pumping station that will be installed on the system. The pump units can be electric or diesel engines. Centrifugal pumps are suited if the height between the pump and the water supply is less than 8 m if not deep well pumps are suited.

4.1.3 Head control

The head control unit that is located between the water supply and the delivery line, is composed of hydro cyclone, sand and gravel filter, fertilizer unit, mesh filter and pressure regulator from the pump to main pipe line. In addition, the components such as check valve, a shut-off valve, pressure gauge, water meter and the equipments such as the nipple, ell and tee combining these components are involved.

4.1.3.1 Fertilization unit

In areas where sprinkler irrigation systems are installed, the fertilization process can be carried out with the aid of a pump unit such as venture tubes or injection pumps, the fertilizer tank that will be placed in head control unit. Due to different fertilization requirements of different plants, this process is used more easily and effectively in areas where there is especially the only type of plants such as grass fields.

4.1.3.2 Filter

Depending on the source of water used in sprinkler irrigation system, irrigation water contains substances that can clog heads such as silt and organic matter. The filtration is performed to prevent clogging problems in short-and long-term in the system. Depending on the nature of water resources, it can be used to avoid clogging the heads of sand separator, hydro cyclone and disc filter. Due to mostly use of drinking water for landscape irrigation, filtering process is carried out with the disc or mesh filters. In sprinkler and drip irrigation systems, 80-120 mesh filters are usually preferred.

4.1.3.3 Valves

The sprinklers are the equipments which the water quantity supplied to the processing and accordingly are used to adjust the pressure in irrigation systems. In general, the valves divided into manual and automatic valves. Manual valves can be listed as ball valves, gate valves, butterfly valves. In landscape irrigation, irrigation control is carried out with the help of automatic valves. The control valves used in the different points of landscape control the direction to various processing of the water in the system according to different requirements (Figure 2). The check valves that are installed sprinkler heads, they prevent back flow of water when the system is turned off.

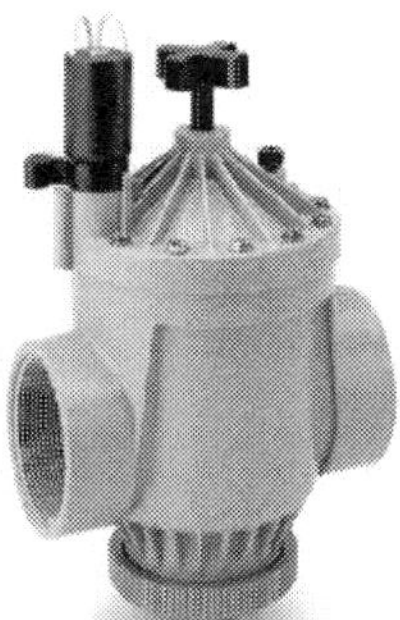

Fig. 2. Control Valve and Application Sample

4.1.3.4 Pressure gauge

Operating pressure of sprinkler head within system requires to be obtained steady. Since changes such as increase or decrease to be occurred in operating pressure affect flow value to be ejected by heads, they prevent corresponding water diffusion from being provided. Due to over pressure to be applied in irrigation units, it causes occurrence of breakdown. In addition, probability of occlusion in fertilization tank can be detected through pressure gauge placed in front of and at the end of fertilization tank. Thus, pressure value is constantly kept under control through manometers to be placed in system control unit.

4.1.3.5 Pressure regulators

Pressure regulators are equipments that provide irrigation in constant volume, preventing pressure fluctuation within system. Inappropriate choice of pressure regulator for system causes to be defected by loading excessively over system units. Also, irregular, inadequate or high pressure cause breakdown of corresponding water diffusion and unnecessary additional irrigation. Pressure regulator generally used in control unit may be used before manifolds in main systems if required.

4.1.3.6 Control units

Operating units generated in pressurized irrigation systems have possibility of being operated on time and duration which is desired by the help of control units. As well as control units are used in garden-type small irrigation area having only one operation, the more developed ones can be done by means of a program which can operate several operations at the same time if requested. Electrical or wireless models of control units are used in areas in which electrise is problematic (Orta, 2009). If required, program used can regenerate irrigation program again by help of rain, wind and soil sensors placed in control units. Shut-down of system can be obtained by itself on the instant of rain through programming rain shutoff on control units. Also, it can prevent irrigation in case of soil`s having enough moisture by programming soil moisture sensors (Cardenas-Lailhacar et. al., 2008). Moisture sensors to be installed within working area provide more economic use of water, generating irrigation program.

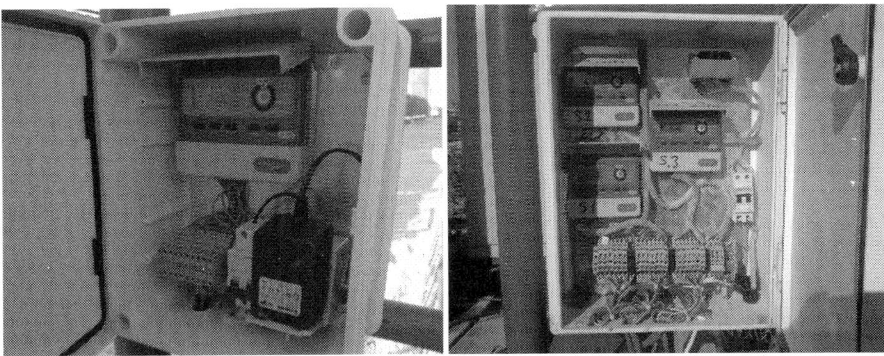

Fig. 3. Control unit panel having 8 Station used in landscape irrigation Systems

4.1.4 Delivery line

Consist of pipes which deliver irrigation water from pump units to water emitters. A typical delivery line consists of main and lateral pipes and equipment which are used to tie those pipes. But submain and manifold are also included in great irrigation area.

4.1.4.1 Main line

They are pipes which transmit water from control units of water to submain canals if any, or to manifolds. Main canals carrying highest flow in system are manufactured as PVC and PE. Generally, 63-160 mm diameters are used depending on the size of system. They are the pipes in which frictional loss is at the most.

4.1.4.2 Submain line

Submain pipes are the same type of pipes with main pipes. The highest cost in irrigation systems is cost concerning pipes and mainly main pipes. In choosing of irrigation pipes, pipes which have minimum capacity to meet requirements of flow in system should be chosen. By purpose of water distribution to be provided in different regions in major irrigation systems, dimensions of pipes are changed from a point which is broken into pieces from main pipes and by this way, disposals can be provided, using pipes which carry as need. These pipes transmit water taken from main pipes to manifolds.

4.1.4.3 Manifold line

Manifolds placed under soil are smaller-scale pipes as compared to sub main pipes as other pipes. Manifolds are placed vertical to parcel edges and maintain laterals. Manifolds mostly consist of HDPE pipes.

4.1.4.4 Lateral line

They are the smallest scale pipes within system. Lateral pipes are responsible for transmitting water from manifold pipes to sprinkler pipes. Laterals are placed under soil in landscape irrigation. All pipes in system are tied to each other by help of connectors.

4.1.5 Sprinkler heads

The most important parts of sprinkler irrigation system are certainly sprinkler heads. Sprinkler heads are placed at certain intervals over laterals. As well as sprinkler heads is placed on a riser over lateral, they are generally embedded underground in landscape irrigation and known as pop up. Pop up heads do irrigation, moving over soil surface when water is given in system. There is two type of sprinkler heads divided as spray and rotor. Spray heads irrigating without turning are used for confined space (Figure 4). Wetting dimensions of them are 2-4.5 m and working pressures are 1-3 bar (Orta, 2009). Spray heads throw water in higher amounts in more narrow space as compared to rotor. Rotors are used in wider areas with lower precipitation rate (Figure 5.). Rotor type heads are generally used in landscape irrigation. Unit price of spray heads are lower as compared to price of rotor heads but facility costs of sprays are higher as compared to the ones of rotors. Factors such as soil infiltration rate, type of plant, wind conditions, limitations of pressure, demands of owner of system should be taken into account in choice of heads. Also, lowest flow of system, low operating pressure and widest spacing of sprinkler head are considered while making a choice between heads which can meet necessary requirement of water in system. Technical properties related to pop up head are presented in schedule (Table 1).

4.1.6 Water distribution in sprinkler irrigation method

Water throws towards to air at a certain angle in sprinkler irrigation. As a result of this, head as being in center, causes a circular area to be wetted. This area is called as wetting area of sprinkler head. Big water grains drop around head during irrigation and size of water grains dropping over soil get smaller as head goes far. Each head constitute a water distribution depending on dimension of orifices over which each head is equipped and operating pressure (Melby. 1995). In case of head pressure`s decreasing under optimum pressure or increasing it causes deformation of water distribution in Figure 6. Changes of water distribution under different pressures are given Figure 6.

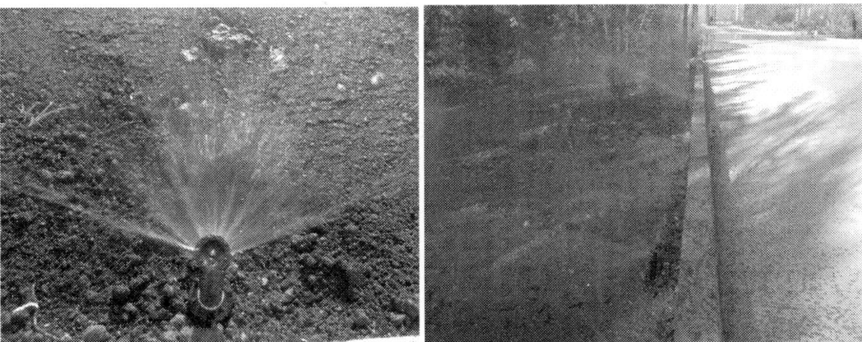

Fig. 4. Samples of using spray

Fig. 5. Samples of using Rotor

It is not possible to provide uniform distribution in irrigating area by singular heads in sprinkler irrigation system. Uniform distribution is obtained by using more than one heads. Factors which affect water distribution are the sprinkler nozzle, operating pressure, flow rate, speed & uniformity of rotation, spacing of the sprinklers, pattern of the sprinkler grid and wind. What is required to be done after choice of appropriate head by purpose of providing uniform distribution is defining sprinkler pattern and sprinkler space. Heads are commonly lay out in shapes of triangle, square and rectangular. Triangular shape is used most commonly in landscape irrigation. In choice of triangular shape better water distribution is obtained in areas having equilateral triangle as compared to square shape. Water disposal is obtained in decreasing waste water due to its water distribution pattern`s being better as compared to square shape. Head in less numbers are used since sprinkler heads are placed to more distant area in triangular shape. Triangular shape presents better performance in irregular areas as compared to square shape (Melby. 1995). Another factor is distance between heads. For a good water distribution head intervals on lateral shouldn`t be more than 50% of dimension of wetted area. Also lateral intervals affect water distribution.

Lateral intervals in line of main pipes shouldn`t be over 65% of wetting area. Technical charts presented by producing company are available for each head. In these charts head flows and wetting area in different orifice dimensions and operating pressure are stated. These charts are used in choice of appropriate heads in irrigation area (Yıldırım. 2008).

Nozzle Model	Pressure (Psi)	Radius (m)	Flow (l/min)	Precip. Υ (cm/h)	Precip. Δ (cm/h)
360°	1.4	2.1	6.51	6.58	9.59
	2.1	2.1	8.06	8.13	9.40
	2.8	2.7	9.39	9.47	10.95
	3.4	2.7	10.52	10.62	12.27
270°	1.4	2.1	5.15	6.93	8.0
	2.1	2.7	6.25	8.41	9.70
	2.8	2.7	7.15	9.63	11.13
	3.4	2.7	8.06	10.85	12.52
180°	1.4	2.7	3.29	6.65	7.67
	2.1	2.7	4.05	8.18	9.45
	2.8	2.7	4.66	9.40	10.85
	3.4	2.7	5.22	10.54	12.17
90°	1.4	2.7	2.0	8.10	9.35
	2.1	3	2.42	9.78	11.30
	2.8	3	2.73	11.00	12.70
	3.4	3	2.95	11.91	13.77

Table 1. Technical schedule related to pop up head

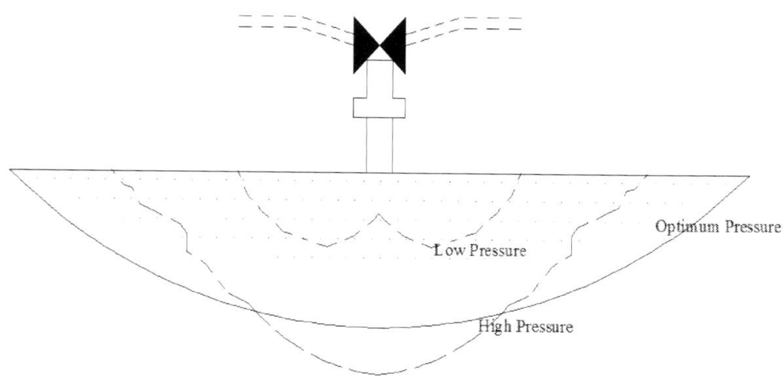

Fig. 6. Water distribution histogram obtained under different operating pressures (Gungor et. al. 2010).

Since sprinkler heads make irrigation by throw water towards to air at a certain angle they are highly affected from windy air conditions. By aiming at obtaining the best distribution in windy conditions by irrigation system a number of arrangements in choosing of sprinkler are required (Barrett et. al. 2003). These arrangements may help to perform a better distribution since water which is sprinkled from less wetting scale heads drop closer area and their interaction with contact of wind is less in areas which are exposed to wind excessively. Having the smallest throw angle should be used. Optimal throw angle is 32° for sprinkler irrigation under normal circumstances. However it is required to reduce this angle in windy conditions. Adjusting throw angle as 22-27º in areas where wind speed is higher than 2.6 m/sec shall give better results. However heads are required to be placed in a closer area for corresponding water distribution since fire distance is decreased in this case. Also laterals should be placed vertical to wind direction.

Water distribution uniformity is a parameter that questions situation of hydration of irrigated area with corresponding amount of water. This parameter is one of the main factors which is useful for decreasing system costs thanks to consume of less electric, water, fertilizer and pesticide in connection with its irrigation performance. Different parameters are used by purpose of determining uniformity of water distribution. Parameters which are commonly used are sprinkler uniformity Christiansen's coefficient of uniformity (CU), Distribution Uniformity (DU) and Scheduling coefficient (SC).

4.1.6.1 Distribution uniformity (DU)

Distribution uniformity is one of the methods which is preferred in determining uniformity in irrigated area. In this method, amount of precipitation is measured by means of caps placed in irrigated area. Then uniformity value is determined by estimating average catch amount belong to quarter which has least water with average catch amount of irrigated area (ASCE.1978). For example, 5 caps Where has the lowest catch amount represent average catch in the low quartile in a test where there are 20 measurement caps (Walker and Skogerboe 1987). Distribution uniformity (DU) should be greater than 75% and if greater than 85% is excellent and acceptable for any sprinkler irrigation. The value of Distribution Uniformity coefficient is calculated using the following expression.

$$DU=(MQl/M)*100 \tag{7}$$

Where.
DU: Distribution uniformity, %
M_{Ql}: Average collected volume of lower quarter of catch cans, l
M: Average collected volume of all catch cans, l
Distribution uniformity should be at least 70% in rotor heads and 50% in spray heads. As being different from DU, CU and SC parameters can be determined.

4.1.7 Christiansen coefficient of uniformity

Christiansen Coefficient of Uniformity is used commonly in irrigation sector. Christiansen (1941) developed a formula which depends on average value of irrigation water amount and standard deviation value which are measured through caps placed in irrigated area. CU is calculated by means of formula below.

$$CU=100((1-(\Delta \bar{q})/\bar{q}) \tag{8}$$

Where.

CU: Christiansen coefficient of Uniformity, in %
$\Delta \bar{q}$: th average absolute deviation from the mean, m³/h
$\bar{q}$: mean application rate, m³/h

While evaluating CU parameters values which are over or under average are considered as similar. This coefficient is developed for agricultural areas and is not highly useful for turf area. Total variation is applied to define applications which are not uniform. However visual quality should be spread over all area in turf area (IA. 2003). CU values being higher than 84% are suggested as acceptable (Anonymous. 2009).

4.1.8 Scheduling coefficient

Scheduling coefficient helps us to define how much critical dry area shall be left in irrigated area and irrigation duration being necessary for its application to eliminate this area (Zoldoske. 2003). In calculation of SC. it is useful to be benefitted from computer program. Due to limitations in DU approach SC gains value especially in turf and golf industry (Wilson and Zoldoske, 1997). The driest area is usually user defined as 1. 2. or 5 percent of the coverage area. SC generally varies between 1.1 and 1.4. An efficient irrigation system should aim to achieve a scheduling coefficient less than 1.3.

$$SC = \text{Average catch overall / Average catch in the critical dry area} \tag{9}$$

4.2 Supplemental irrigation methods in landscape irrigation

Drip and micro sprinkler irrigation are used in areas here sprinkler irrigation is not appropriate in landscape irrigation. Drip irrigation system is installed in two different types such as under soil and over soil. Water is transmitted to root region through a pipe network and applied here by a means of drippers. Vaporization, runoff and deep seepage into deeper are prevented due to water`s implication by low flow rate in drip irrigation system. Therefore it is irrigation method of which application performance is the highest (Schwankl & Prichard. 1999). Also, drip irrigation minimizes sickness and insect damages because it makes irrigation without wetting leaves. It has less operating pressure. It applies water at fewer amounts as compared to sprinkler irrigation. Problem of weed occurs less because certain part of area is wetted. Also, drip irrigation performs at high efficiency without runoff on hilly terrain. This method is used in soils where area to be irrigated is narrow and saltiness ratio is problematic and resource for water is limited and soil is highly inclined in landscape irrigation (Dines & Brown. 2001).

Micro Sprinkler can be carried out by placing them singularly under trees in situation that drip irrigation can`t provide adequate wetting area. Throw distance of these heads is highly low. However, flow rates changes between 30-300 l/h. accordingly wetting ratio which is requested is provided in a more economic way. Micro sprinkler is developed by the purpose of linking good sides of sprinkler and drip irrigation system.

5. Planning stages of sprinkler irrigation systems

Steps of sprinkler irrigation project are carried out a number of calculations starting from choice of head and to flow of pump.

5.1 Sprinkler flow

Sprinkler heads changes according to properties of cross-sectional area of orifice, operating pressure and processing property of orifice. Because of head losses, sprinkler head loses pressure as goes from beginning of lateral towards to end of lateral. Therefore decrease in flows occurs (Anonymous, 2010). Difference of maximum 10% in flow and 20% in pressure across lateral line should be allowed to provide a suitable corresponding water distribution. Sprinkler flow is calculated by help of formula below

$$q = 3600CA\sqrt{(2gh)} \tag{10}$$

Where.Q:Sprinkler flow m^3/h
C:effective coefficient (0.80-0.95)
A: nozzle cross-section area m^2
g:gravitational acceleration m/s^2
h:operation pressure of sprinkler head. m.

5.2 Precipitation rate

It is defined as water amount given per unit time in irrigation area (Connelan. 2002). It is generally expressed as mm/h. Main factors which affect precipitation rate are sprinkler flow, distance between sprinkler and distance between laterals. The average precipitation rate is calculated with the following equation.

$$Pr = 1000 * q/S * L \tag{11}$$

Where.
P_r:The average precipitation rate. mm/h
1000: a constant which converts meters to mm.
q:the total flow applied to the area by the sprinklers. m^3/h
S:the spacing between the sprinkler along lateral. m
L: The spacing between rows of sprinkler. m

The flow rate of sprinkler heads automatically changes in case of their making irrigation in different angles. For example, when sprinkler angle decrease from 360° to 180° degree, flow rate increases doubled. Therefore in the system where heads having different angle values are used, the average precipitation rate is calculated by means of the following formula.

$$Pr = 360000 * q/\phi S * L \tag{12}$$

Where.
Pr: Average precipitation rate of sprinkler. mm/h
360000:a constant related sprinkler's angel.
q: flow rate of sprinkler. m^3/h
ϕ: Working angel of sprinkler. ^{0}

Precipitation rate requires considering soil infiltration rate in order to prevent runoff and deep seepage to be determined.

5.3 Irrigation duration

Irrigation duration refers to required time for each irrigation. This duration is a function of net water application depth, precipitation rate of sprinkler and irrigation efficiency. Irrigation duration is calculated by the following formula

$$Ta=dn/(Pr*Ea) \tag{13}$$

Where.
T_a: Irrigation duration. h.
d_n: net water application depth. mm
P_r: Precipitation rate. mm/h
E_a: Application efficiency. % (80% can be taken for sprinkler)

5.4 Lateral flow

Lateral flow changes according to the number of sprinkler which is planned to be placed on the lateral and their flow rate. Lateral flow which they need to carry increases similarly as number of heads and flow rate are increased. The lateral flow inlet is determined by equation 14.

$$Ql=qs*ns \tag{14}$$

Where.
Ql: lateral flow. l/s
qs: sprinkler flow. l/s
ns: the number of sprinkler on lateral

Sprinkler heads having wetting area in shape of circular, semi-circle and quarter circle on the same lateral are often used depending on geometrical shapes of irrigation area. In case of using heads having different flow rate on a lateral, heads which have the same flow rate are grouped and number of heads are multiplied and the total flow rate of lateral is defined.

5.5 Operation unit

Unit which is constituted from heads making irrigation in irrigation area is called as operating unit. Maximum operating unit is calculated by means of following formula.

$$Nmax=(Tg/Ta)*SA \tag{15}$$

Minimum operating unit is also calculated by means of following formula.

$$Nmin=\Sigma q/Q \tag{16}$$

Where.
q:sprinkler flow m³/h
Q:system flow m³/h

N_{max}: maximum number of station
Tg: achievable irrigation duration per day, h/day
Ta: irrigation duration, h
SA: irrigation interval, day.
Operation unit is considered a number between N_{min} and N_{max} in planning step.

5.6 Hydraulic calculation

Two different head loss are occurred in duration of water`s reaching from resource to plant in delivery of water in an irrigation system. One of them is friction head loss. Friction head loss occurs due to friction within pipes and it is related to length, dimension and coefficient of roughness of pipes. Friction losses are calculated by means of formula of Darcy-Weisbach below.

$$hf=\lambda(L/D)(v2/2g) \tag{17}$$

Where.

h_f:friction head loss m
λ:friction coefficient
L: length of line (m)
D:Inner diameter of pipe work (m)
v: velocity of fluid (m/s)
g: acceleration due to gravity (m/s^2)

Second local head losses occurring in system is called as geodetic and changes depending on type of equipment used. Local head losses are calculated by means of formula below.

$$hl: kf*V2/2g \tag{18}$$

Where.
h_l: local head losses, m
k_f: friction coefficient of irrigation equipment
V: velocity of fluid, m/s
g: acceleration due to gravity, m/s^2

5.7 Matching water flow and pressure with pipe size

Flow in pipes is defined as functions of dimension of pipes and velocity of flow. While determining dimension of pipe in irrigation system investment costs can be minimized through choosing the possible smallest dimension of pipe. However keep in mind that reducing dimension of pipe shall increase the velocity as seen in following formula. Increase of velocity means increase of loss of frictions occurring in pipes.

$$Q=A*V \tag{19}$$

Where.
Q:system flow (m^3/h)
A: line cross-section area: m2
V: velocity of flow. m/s

Possible minimum dimension of head should be chosen according to limitations of friction loss which are allowed. While choosing dimension of main pipe it is required not to exceed 15% of pressure of pumper. Pressure difference between starting and ending point in lateral shouldn`t be over 20%. Also total head losses in system shouldn`t exceed half of orifice pressure.

5.8 Total dynamic head of the system and power requirement

Pressure which is necessary for irrigation system is generally carried out through a pump installed at the beginning of system excluding the situations in which adequate elevation is not available between resource and irrigation area. Pump should be chosen at power which provides optimum pressure in the last head in the line which is called as critical line and has head losses at excessive rate in this system. Total dynamic height of pump in a system is calculated by means of following formula;

$$Ht=hs+hh+he+hsuc \qquad (20)$$

Where.
H_t: total dynamic head of pump, m
h_s: sprinkler operation pressure, m.
h_h: total head losses, m.
h_e: elevation difference between highest point in irrigated area and pump, m
h_{suc}: suction line height if there is elevation difference between pump and water supply suction line height should be considered.

Also engine power of pump is required to be determined. Pumping Energy which is appropriate for system is determined by means of following formula.

$$Np=Ht*Q/75* \mu1\mu2 \qquad (21)$$

Where.
N_p:Pump power, HP
H_t: Total head loss, m.
Q:system capacity, l/s
$\mu1$: pump efficieny, %
$\mu2$: driver efficieny, %

Pumps which provide necessary pumping power are included in system by help of catalogues of relevant firms.

6. Conclusion

Irrigation is one of the main factors on plant growth and quality. Irrigation is applied in areas where evapotranspiration is not met by rain especially in semi-arid or arid climate. Pressurized irrigation methods are used generally in landscape irrigation.

The most common method between pressured irrigation methods is sprinkler irrigation method. Sprinkler irrigation method is preferred by reasons such as providing high corresponding water distribution, its use easily in any kind of soil and area where plants are grown, its low labor costs and fertilization proceeding`s being easily applied . Determining system units completely according to requirements in project step provides system performance being in high level. Drip and micro sprinkler irrigation methods should be considered as an alternative method in where sprinkler irrigation system is inadequate or ineffective.

7. References

Allen. R.G. Pareira. L.S. Raes. D. Smith.M. 1998. Crop Evapotranspiration. Guidelines for Computing Crop Water Requirements. *FAO Irrigation and drainage paper 56*. Rome.

Anonymous (2009). *Sprinkler Irrigation Systems*. Minister of Agriculture and Rural Areas Agricultural Production Directorate of Operation of Adana and Education Center. Adana. Turkey.

Anonymous. (2010). *Sprinkler Irrigation Systems and Design*. Arılı Plastic Indistury. Istanbul. Turkey.

ASCE. (1978). Describing irrigation efficiency and uniformity. *J. lrrig. and Drain. Engr.* 104(1). 35-41.

Ayyildiz. M. (1990). *Irrigation Water Quality and Salinity Problems*. University of Ankara Agriculture Faculty. Ankara. Turkey.

Barrett. J. Vinchesi. B. Dobson. R. & Roche. P. (2003). *Golf Course Irrigation: Environmental Design and Management Practices*. John Wiley & Sons. NY. USA.

Cardenas-Lailhacar. B. Dukes. M. D. & Miller. G. D. (2008). Sensor Based automation of irrigation on bermudagrass. during wet weather conditions. *J. Irrig. and Drain. Engr.* 134 (2008). pp 120- 138.

Christiansen. J.E. 1941. The uniformity of application of water by sprinkler system. *Agric. Eng.* 22: 89-92.

Cockerham. S. T. & Leinauer. B. (2011). *Turfgrass Water Conservation. (2 nd Edition.)*. University of California Agriculture & Natural Resource. CA.USA.

Connellan. G. (2002). *Efficient Irrigation: A Reference Manual for Turf and Landscape*. Irrigation Best Management Training Program. Melbourne. Australia.

Dines. N. & Brown. K. (2001). *Landscape Architect's Portable Handbook. McGraw-Hill*. New York. NY. USA.

Doorenboos J. & Pruit. W. O. (1977). Crop Water Requirements. *FAO Irrigation and Drainage Paper No 24*. Rome.

Gungor. Y. Erozel. A. Z. & Yildirim. O. (2010). *Irrigation (4 th Edition.)*. University of Ankara Agriculture Faculty. Ankara. Turkey.

Harmancioglu N. Alparslan N. & Boeele E. (2001). *Irrigation. Health and the Environment: A Review of Literature from Turkey (Working paper)*. International Water Management Institute. Colombo Sri Lanka

Ingels. J. E. (2003). *Landscaping Principles and Practices(6 th Edition.)*. Thomson Delmar Learning. NY.USA.

Irrigation Association.(2003.) *Certified golf irrigation auditor*. Falls Church. VA: Irrigation Association.

Melby. P. (1995). *Simplified Irrigation Design: Professional Designer and Installer Version. (2 nd Edition.)*. John Wiley & Sons. NY.USA.

Novotny. V. Ahern. J. & Brown. P. (2010). *Water Centric Sustainable Communities Planning. Retrofitting and Building The Next Environment*. John Wiley & Sons. USA

Orta. A. H. (2009). Irrigation in Landscape Area. University of Namik Kemal Agriculture Faculty. Tekirdag. Turkey.

Phocaides. A. (2007). *Handbook on pressurized irrigation techniques.* Food and Agriculture Organization (FAO). Rome.

Schwankl. L. & Prichard. T. (1999). *Drip Irrigation in the Home Landscape.* University of California Agriculture & Natural Resource. CA. USA

Smith. S. W. (1996). *Landscape Irrigation: Design and Management.* John Wiley & Sons. NY. USA.

Surampalli. R. & Taygi. K. D. (2004). *Advanced in Waste and Waste Management. American Society of Civil Engineers.* Reston. Virginia. USA.

Tekinel. O. & Kanber. R. (1988). Results of the trickle irrigation experiments carried out by the University of Çukurova. *Center for Advanced Mediterranean Agronomic Studies.* Bari. Italy. 22 pp.

Walker. W. R. & Skogerboe. G. V. (1987). *Surface irrigation theory and practice.* Prentice-Hall. Englewood Cliffs. N.J

Wilko. J. B. & Short. R. (2007). *Water Quality.* In: Valentine M. editor. Water Resources Planning Manual of Water Supply Practices(2th Edition.). Denver. CO: AWWA. pp:171-187.

William. O. (2007). *Introduction.* In: Valentine M. editor. Water Resources Planning Manual of Water Supply Practices(2th Edition.). Denver. CO: AWWA. pp:1-12.

Wilson. T. P. & Zoldoske. D. F. (1997). *Evaluating Sprinkler Uniformity.* CATI Publication. California. Washington D.C.

Yıldırım. O. (2008). *Desing of Irrigation System.* University of Ankara Agriculture Faculty. Ankara. Turkey.

Zoldoske. D. F. (2003). *Improving Golf Course Uniformity: A California Case Study.* Center for Irrigation Technology. Fresno. CA.

15

Private Plantation Techniques

Ömer Lütfü Çorbacı[1] and Murat Ertekin[2]
[1]Ankara University, Faculty of Agriculture
[2]Bartın University, Faculty of Forestry
Turkey

1. Introduction

The visual value of a town increases directly proportional to the density of her open and green spaces. Vertically and horizontally formed greenery is an indispensable part of urban design. However, with the advanced technology during the 20 th century, wide construction areas, highways, agricultural and industrial zones have developed in an unplanned manner, and natural resources were abused in an unsystematic way. Unfortunately, the number of natural elements in towns has decreased rapidly in recent years, and with the help of uncoordinated urbanization, the situation has turned for the worse for green areas. If we were to analyze this fact with figures, the example of Ankara, Turkey would prove to be more than enough. In physiological terms, according to oxygen exchange and leaf surface calculation, there is a theoretical need of 25–40 m^2 green area per person in an urban area. But in Ankara, this figure was 5.1 m^2 in 1950, 2.8 m^2 in 1965, and 1.8 m^2 in 1979. However, the urgency of the matter has been realized during recent years, and inner urban greenery works have been started.

Improving the environmental conditions of the indoor and outdoor places where humans live, and also to arrange them to become suitable for living, has become a foremost priority. Today, extreme urbanization has become ever fast growing, and inner urban tree planting techniques are changing and improving accordingly too. Nowadays, it is necessary to make use of all new developments in technology and find ways to meet the ever increasing demands of modern life.

The first time when large plants were uprooted and transferred to somewhere else was during the Munich Olympic games in Germany. Back then, a whole new Olympic village was created with immense greenery. At that time, this transplantation process was realized with much more labour force and time, also simpler techniques were used. However, the same could be done today with much time effort and time spent, through the use of modern techniques. A very important aspect of landscaping works is the time needed until it reaches an effective power, or in other words, the dimension of time. Trees and landscaping elements need on average 30-40 years to reach an effective power in terms of physics, visual, climactic etc. aspects. Therefore, it is very important to foresee the needs of the future, and do the landscaping planning accordingly. This is a difficult and compulsory responsibility to do. However when a planning is done, people of today believe that reaching a necessary green area needs to be done as rapidly as other advancements.

Through the transplantation works done with this purpose, the inner urban areas to be planted become green very rapidly, compared to the years spent on planting seeds.

Due to industrialization and domestic immigration during the last thirty five years, Turkey has entered a fast urbanization phase and as a result of this, modern people have lost their opportunity to live in a natural environment. Therefore, they try to fill this gap by planting within as much as possible. Tree planting in urban areas is a very new application in Turkey. As well as transplantation works with simple tools, a machine for tree planting and uprooting is also used for the last few years.

Giving a short explanation on the meaning of the term "transplantation" would be useful in preventing any confusion on the meaning. The term transplantation is used in some departments of science in such a degree that it has become a cliché. For example, the term "transplantation" in medical literature means the transplantation of any organ from a person to another person, where all physical, biological and technical conditions are suitable. When an organ transplantation is to be carried out, high importance is paid for the organ of the donor to be transplanted to match certain criteria of the receiver, such as biological structure, physical conditions etc, which is important for the receiver to maintain his life healthily. And the term "transplantation" used in landscaping architecture means the re-planting of a plant from one place to another. However, it would not be right to use the term "transplantation" for all types of plants. Just like in medical terminology, the term "transplantation" in landscaping architecture does not mean the transfers of plants at an early age, but at their more mature periods. Again similar to other branches of science, transplantation process here; is a process which is realized in line with certain steps and in consideration of some basic principles, and in line with the necessary technical conditions.

Just like in all other landscaping applications, tree transplantation works also require a controlled monitoring during all phases and meticulous and well arranged implementation principles. In generally, trees and shrubs are transplanted when purchased or planted. These plants often grown in the field, and harvested in the form of bare-root, balled and burlapped (ball of soil and roots wrapped in burlap), or containerized. In nursery, trees and shrubs are often grown using cultural practices, such as root pruning, to prepare them for harvesting and transporting to the sales area. Nursery plants may have 75% of their root system intact after they are dug, nevertheless wild plants may only have 25% or less of their root system intact. When woody plants in the landscape are transplanting, they are exposed to stress because of any of the special procedures used in nurseries before the transplanting day. The increased stress on plants can make the difference between an attractive or healthy. Nursery stock grown in containers is often much more tolerant to transplanting than field-grown or wild grown plants (Anonymous, 2012 a).

2. Historical development of the transplantation of larger plants

2.1 Transplantation of larger plants in the world

It is a known fact that the Egyptians during the ancient times have carried trees with boats from distances as far as 1500 miles. They did it in order to cool down the dry climate of the Nile River Basin, and to create some shade. Plants in Egypt are being arranged in a formal way. Fruits, vegetables and medical plants are alongside other decorative plants within the gardens. The most commonly used plants are Phoenix, Palm Tree, Lotus and Papyrus.

Ancient Greeks in particular, have worked on issues regarding tree transplantation and tree protection. In relation to this, Theophrastis has carried out a research in 300 BC, on necessary methods to wholly protect the root system during plant transplantation (Nadel, 1977).

From the 15th Century, with the start of the Renaissance wave, meaning "Re Birth", the dark view of the medieval times were broken in the west, great changes were made to beliefs, and fast advancements were made in science and arts. With these changes, living spaces also went through improvements, and trees were once again considered to be used in living spaces. During the 17th Century, tree was considered as a sign of royalty in France, and people belonging to higher classes had planted large amounts of trees at their living spaces. That way, the transplantation techniques of larger trees have developed, and machines to lift and carry these trees have been developed. During this time, tree transplantation has become very important in England. There are also rumours that, thanks to the new methods and machines developed, hundreds of years old oaks have been transferred. Therefore, as early as the beginning of 19th Century, France and England have made great advancements in transferring of trees (Mayer, 1982). On the other hand, many written sources appeared regarding trees during the 17th Century. British author William Lawson has written in 1618 "A New Orchard and Garden" which was mainly about maintenance, repair and aesthetical values of trees. This book is important because it was the first to mention about the appropriate planting intervals. And in the book, "Sylvia", written by John Evelyn in 1664, information has been given on growth features and maintenance principles of trees. Frenchman Le Notre has implemented the tree planting details given at this book in the famous Versailles Palace. During those days, having a large number of trees inside the palace gardens was considered to be a sign of civilization (Nadel, 1977).

During the 17th and 18th Centuries, a connection was also started to be made in England, between settlement areas and the nature. Great squares or open spaces have started to appear during the 17th Century, and they were surrounded by large buildings. Another century later, these squares became the dominant element of London settlement and led to the trees being used extensively within urban areas. When squares were being built, tree transplantation was widely used. Trees started to be considered alongside with urban planning only after the 18th Century. Tree transplantation works at that time were generally used for planting trees alongside the roads within the city. With this purpose in mind, engineer Baron George Houssman was assigned by Napoleon III in 1853, and he started re planning all over the city of Paris (Nadel, 1977). For the tree planting works at that time, 82.000 trees of different types and with a height of 10-12m. were transplanted, which was a real success (Altan and Önsoy, 1982). Fredeick Law Olmstead, who was the father of Landscaping Architecture and the designer of New York Central Park, which was held in 1858, had given works about urban forestation. In these works, he talked about forestation programs, particularly at the road sides in New York and San Francisco (Nadel, 1977).

In modern cities of the 20th Century, there also have been changes and improvements of the tree transplantation principles and methods. During the first half of the century, USA in particular has shown some improvements. There have been academic works in Russia, regarding plant transplantation.

Landscape architects that attended American Fair in Moscow on 1959 have made some researches in order to carry out their transplantation works. In one of these researches, they succeeded in planting a 25.40cm (10 inches) lime tree in the middle of winter when there was frost until 1,22m (4feet) depth. As the soil was frozen, soil fescue was cut with air powered saw, no fastening or molding was needed. Tree pits were also formed by chainsaw. After preparing the pits, metal covers were placed on them and a fire was lighted in it for a few days in order to ensure the heat to stay inside the pit when the soil around it was heated, root fescue was put into the pit and the process was completed. Another interesting event was that birch was to be transplanted in the middle of July. A regular maintenance and irrigation guaranteed the continuousness of the life of tree (Zion, 1968).

The landscaping design made for Munchen Olympiads in 1972 which covered the entire village. 3 years before the Olympiads, 12-15 birches that were 30-40 years old were transplanted successfully. As a result, when 1972 Olympiad games started, it looked as if landscaping in the area had started 30-40 years ago (Ürgenç, 1998).

An island system will be built 5-7 km distanced from shores of Dubai, the capital city of the United Arab Emirates. There will be 1060 small houses on the island, 5 thousand people will reside in the houses and 12 palm trees will be planted. The complex is palm shaped which has 17 branches in the middle part; it will increase the length of Dubai shore beaches to 120 km (Anonim 2012 b). Transplantation of trees is much easier today thanks to the techniques and machines that are developed with modern technology. Bigger areas can be planted in shorter times successfully.

2.2 Transplantation of larger plants in Turkey

After the Industrial Revolution in Europe, while changes in economical and social structures affected physical appearance of cities, Ottoman society were different from the societies western regions in terms of development dynamics and city types. As Turkish society wasn't directly in mechanization process, the need for public domain and green places couldn't be realized for a time until the proclamation of republic when city plans started to be done in a more organized way (Şahin, 1989).

In order to discuss plant transplantation works in Turkey, we should first talk about the understanding of open land and green land and importance given to green lands. Turkish cities were built on the basis of three elements; streets, gardens and houses. Public buildings formed most of the physiognomy of cities in Ottoman Empire, while green lands were used as parts of house gardens. In fact there were no organizations serving for the protection of public green lands in local public institutions.

In Ottoman Empire era, there were some recreation spots such as public gardens, and coppice forests that were used by the society in big cities such as İstanbul, İzmir, Edirne and Manisa (in Turkey). Besides this, royal houses and houses of high class people were organized for special use (Caner, 1976). According to the literature about the era, there were not many special plant transplantation works during the era. But in plantation of some palace gardens, parks, and roadsides roots of trees were removed from the soil and transplanted.

In old Turkish cities, similarly there were transplanted trees in squares and lined up on roadsides. But properness and professionalism in these plantation processes is controversial.

After proclamation of republic, new buildings in cities were built and systematic greening processes started in cities. As bushes and small trees can be more easily transplanted than bigger trees, they were preferred for greening applications. Tree transplantation was especially used in central refuges. But still there weren't many works that were done for this purpose. Tree transplantation works were very few when compared to the other works and methods preferred for greening processes.

Efforts for transplanting big sized plants were successfully carried out when technology wasn't developed by taking some precautions in the eras. For instance in İstanbul Sedef Island, very old plants that were put into boxes was successfully transplanted. Similarly, in Bahçeköy garden and plantation fields, maintenance processes were carried out, lime trees and horse chestnuts were successfully transplanted. Barbaros Boulevard, Maçka, Tophane, Kabataş, Şemsi Pasha parks, Beyazıt Square and Saraçhane were greened by İstanbul Municipality. Many species and types of plants such as pines, magnolia, sycamore, horse chestnut, cedar and oak were planted with simple method and positive results were reached (Ürgenç, 1998).

Palms in Kalamış Marine in Kadıköy, İstanbul were removed from the soil without any damage with the decision of Tree Transplantation Commission and transplanted into the places of a dried Palm placed in Sarmaşık Park in Kozyatağı and an area in front of Kalamış Youth Center (Anonymous 2012 b). In our country, palms have been used in transplantation processes in many facilities built especially in Mediterranean and Aegean Regions.

Artvin Çoruh University Faculty of Forestry and Foundation for Combating Desertification and Erosion General Directorate has carried out a project called "Protection of Endemic and Non-endemic Rare Plants that will submerge Çoruh Valley Deriner Dam Water Mirror". The aim of the project was to save the species that will submerge and extinct because of the dam project. In scope of this project, with the contributions of Artvin Regional Directorate of Forestry crews, 400 pieces of 18 rare plant species were removed, potted and transferred to Artvin Çoruh University Faculty of Forestry Greenhouse (Anonymous 2012c).

Transplantation processes have been used more in landscaping processes that have been carried out in recent years. This process is accelerated especially with the increase in the number of tree drawing and planting machines. Today, landscape design works and plant transplantation are made and the areas that are planned are filled more professionally with green plants.

3. Transplantation of large plants and plantation techniques

Transplantation of large plants, especially trees, has been carried out with different methods until today. These plants have been moved as bare roots, in balloons completely leaving the roots out, within tied sacks, by wrapping the plant on a wire cage, wrapping it with a tie beam or with ratchet devices. The attention and care that we show during transplanting plants ensures the healthy continuity of its life. Pruning roots during 3 years before the transplantation, digging a wide root circle, careful wrapping and binding, carrying the plant with big and detailed devices, giving attention to the preparation and maintenance of plantation area leads the plant's adaptation to its new place and live healthy.

It is assessed that whether or not to be a successful transplant before transplanting a tree or shrub. Stresses in transplanting of trees and shrubs may cause plants to die or to become unattractive. Plants are already in advanced stages of decline, particularly likely to succumb to transplantation stress. Generally, if a young nursery-grown plant than older growth plant will provide more long-term benefits in the new planting area so younger plants better than older plants. Also shrubs have better transplant tolerance than trees, deciduous plants better than evergreens and shallow rooted species better than deep rooted species. When deciding whether or not to transplant a plant, consider the species transplant tolerance, transplanting season, new planting site conditions, the equipment and follow-up care (Anonymous, 2012 a).

3.1 Principals to consider during application

Applied methods have both advantages and disadvantages. The success and failure of the transplantation depends on: species of the chosen plant, present conditions, and cultivation aspects of the natural place of the plant besides the aspects of the place it will be transferred. Besides the care and attention during in all these processes, the transplantation process itself is a crucial factor in success (Zion, 1968).

3.1.1 Choosing plant

Almost all kinds of plants can be transplanted. But every plant species have a different sensitivity level. Transplantation of plant species changes according to the aspects of plants during the time period necessary for plants' adaptation to the environment conditions. Transplantation of bushes is much easier than the tall trees. We can divide and analyze the criterion that should be taken into consideration while choosing plants during transplantation.

3.1.1.1 Species and age

Studies in the field showed that some plant species can be transplanted more successfully than others. Plants with roots closer to the stem, the one that are more fibrous can be generally transplanted more successfully than less fibrous and deep rooted plants. Besides, success in transplantation generally decreases from small bushes to tall trees.

The most easily transplanted plant species are: *Acer* sp. (Maple), *Alnus* sp. (Mountain Alder), *Castanea* sp. (Chesnut), *Celtis* sp. (Hackberry), *Fraxinus* sp. (Ash Tree), *Malus* sp. (Apple Tree), *Ulmus* sp. (Elm), *Paulownia* sp., *Platanus* sp. (Sycamore), *Populus* sp. (Poplar), *Robinia* sp. (Locust), *Salix* sp. (Willow), *Tilia* sp. (Lime Tree); and plants known as summer-growing plants which are: *Phoenix canariensis (Palm)*, *Washingtonia filifera* (Desert Palm), *Washingtonia robusta* (Mexican Palm), *Chamaerops excelsa* (China Palm) *and Olea sp.* (Olive Tree). Besides these, some other easily transplanted plants are: *Gleditsia* sp. (Honey Locust), *Abies* sp. (Fir), *Juniperus* sp. (Juniper), *Picea* sp. (Spruce), *Pinus* sp. (Pine), *Betula* sp. (Birch), *Cornus* sp. (Cornelian Cherry), *Eleagnus* sp. (Elaeagnus), *Ginkgo biloba (China Gingko Biloba)*, *Quercus palustris* (Swamp oak) *and Pyrus sp.* (Pear) (Turhan, 1994).

Juglans sp. (Walnut), *Quercus* sp. (Oak), *Carya* sp. (American Walnut) *and Fagus* sp. (Beech Tree) are the plants that are known to be difficult to transplant. While there are different opinions on the transplantation of *Aesculus* sp. (Horse Chestnut) species, there has been some successful transplantation of medium-sized *Aesculus* sp. (Horse Chestnut) species (Ürgenç, 1998).

The species whose transplantation can be easily done are: *Malus* sp. (Apple Tree), *Fraxinus* sp. (Ash Tree), *Ulmus* sp. (Elm), *Tilia* sp. (Lime Tree), *Platanus* sp. (Sycamore), *Populus* sp. (Poplar), *Salix* sp. (Willow) *and Celtis* sp. (Hackberry). Mild-climate plants are not included in this study. Some of the plants which are the most difficultly transferred are *Juglans* sp. (Walnut), and some *Pinus* sp. (Pine) species. Another important point that should be paid attention is that plants that have soft roots generally are not strong enough to be carried by frozen root skein too (Himelick, 1981).

As a general rule, no matter how big their sizes are, bushes can be much easily and successfully transplanted than trees; and deciduous trees can much easily be transplanted than evergreen trees and coniferous trees. But the success of transplantation is also related with the health of the plant (Turhan, 1994).

In order to successfully transplant the tall plants, necessary information about their root systems, root distribution depths, distribution styles, roots' activity times should be known. Plants' root systems are divided into 3 groups as taproot, heart root, shallow root (Figure 1).

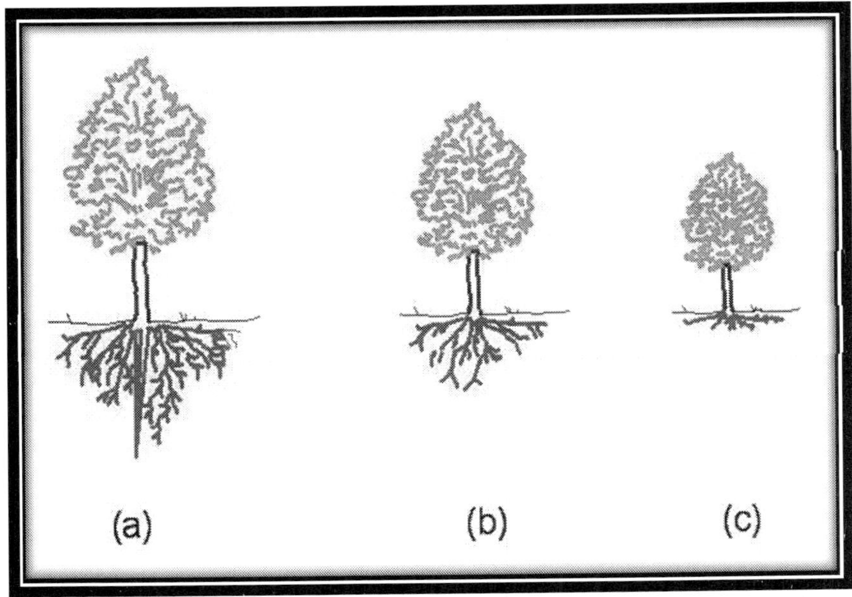

Fig. 1. Different root systems of plant species (a: taproot, b: heart root, c: shallow root).

Taproot System: *Juglans* sp. (Walnut), *Quercus* sp. (Oak), *Pinus* sp. (Mountain Pine), *Castanea* sp. (Chesnut tree) *and Cedrus* sp. (Cedar). Heart root system: *Fagus* sp. (Beech Tree), *Acer* sp

(Maple), *Tilia* sp. (Lime Tree), *Magnolia* sp. (Magnolia), *Liriedendron* sp. (Tulip Tree), *Robinia* sp. (Locust), *Quercus coccifera* (Red Oak), *Pinus strobus* (Vermouth Pine). Shallow Root System: *Betula* sp. (Birch), *Abies* sp. (Fir), *Picea.* (Spruce), *Acer saccarinum* (Sugar Maple) *and Salix* sp. (Willow). But besides the differences between the species in the same system (for instance *Abies* sp. (Fir), species belong to shallow root system while their roots aren't as shallow as *Picea.* (Spruce) species) there are some differences in the same species. For example although natural *Quercus* sp. (Oak) species have deep and taproot system, *Quercus rubra* (Red Oak) have the heart root and *Quercus palustris* (Swamp Oak) have the shallow root system. Surely the environment they grow has a big impact on this situation. *Quercus palustris* (Swamp Oak) grows in humid climate and has shallow root in order to ease the oxygen intake. All these factors should be taken into consideration while determining the plantation field. Especially roots' growth periods should be known in order to know if the plantation time is appropriate or not. So, these growth periods should be evaluated in terms of the region's aspects and years (Ürgenç, 1998).

Almost all plants can be transplanted, but some requires more time and attention. In addition, it should be kept in mind that young plants' transplantations are more successfully made when compared to older ones.

3.1.1.2 Plant characteristics

Generally, small sized plant species can be transplanted much easier than bigger sized plant species. Besides this, plants that are not very old and whose height are 1-2 m. can generally be transplanted successfully as their root systems don't grow very much. The taller the plant is, the more difficult it becomes to carry; and it has less chance to adapt to its new place. As plants that grow in nursery are taken care more than the other ones, their roots are more fibrous and together. They have more attractive upper parts when compared to the ones that aren't grown like them. Big plants are transplanted when the soil changes, during road construction and extension and when they are too big for the place to live in. Although it is very difficult to remove the plants that are squashed because of their structures, transplanting them to better places is important for their health and life (Harris, 1983).

3.1.2 Characteristics of site condition and transplantation field

Root systems of plants in fertile and well aired soil are thicker and fibrous when compared to sandy, barren or slimy and clayey soil whose underneath is watery. Roots of the plants that grow in sandy, slimy and clayey soil have a few twigs on sideways or as have a small root close to the stem. While in sandy soils, roots have the tendency to be close to the deep, they are closer to surface and broader in clayey and slimy soil. Trees growing in soil which don't have any stone or other obstructive substances can be more easily transferred. It is very difficult to transfer trees from wet and slant soil to empty fields vertically (Harris, 1983). There may be some difficulties in removing plants from slant soils to smooth fields. In these cases, one part of the soil is higher than the other part which obstructs the adaptation of plant to smooth soil. The field of transplantation shouldn't be too slimy or dry in order to use the transplantation devices properly and make a successful transplantation. Sidewalks, cables, wires, pipes and natural gas piping systems cause difficulties in removing and planting the plants. In such cases, helicopters are used for plantation and removing processes.

Most of the planting spaces in cities are rather harmful for newly transplanted big plants. Paved roads, structuring cause increase in air temperature and radiation density. Buildings can cause wind tunnels and airflow corridors. These circumstances can make it difficult for newly transplanted plants to get enough water. Regular irrigation and proper weeding can be useful for these kinds of plants (Harris, 1983).

3.1.3 Soil characteristics

Different root characters occur according to soil structure. Root enlarges, spreads and grows with its small roots and enlarges and deepens in well aired and sandy soil. Shallow and distributed root bodies are formed in silty, clayey or drained sub soil. Roots of some plants' same species have different characteristics in different soil structure. It is difficult to transplant plants in an areas furnished with solid construction material or densely vegetated with plants (Kim, 1988).

Appropriateness of soil aspects from which the plant is removed is as important as the plant species. Some soils can be as effective as the plant species in growing a root system which is compact and rich in terms of hairy root. As plants that grow in sandy soil forms deep and complicated root systems, they are more risky in transplantation when compared to the plants removed form clayey soil. But as there is not enough oxygen in clayey soil, capillary roots that are very important in new root formation don't grow enough. Because of this, ventilation of soil with different methods increases the level of success. Deep soil without any rocks, logs etc. are better in removing big plants (Ürgenç, 1998).

3.1.4 Transplantation time

Some definite periods of a year are much better for transplantation of plants. But this situation doesn't mean that plants can only be transplanted in these periods. Successful plantation can be made with a more careful digging, planting and after care processes if the transplantation isn't made in these definite periods.

It is very important to determine the weather conditions that will affect the placement stage of plant's plantation and transfer period. This important factor increases the level of success in transplantation of special plant species. In addition to this, in order to make a good development during transplantation, preparations should be completed before planting, landscaping programs and lists should be made and reviewed. Spring season is preferred more in regions that have cold climate. A plant that is transplanted in early spring regains some of its sections that it lost before the weather becomes warmer and it renews itself although partially. A plant that is transplanted in autumn has to be very strong and endure the winter season before completely recover from the shock of transplantation. August is generally preferred for planting evergreen plants in cold climate regions. Transplantation of *Betula sp.* (Birch) species is preferred to be made in early spring. As roots of *Magnolia sp.* (Magnolia Tree) are damaged during transplantation, they are exposed to fungus disease. This is why transplantation process should be carried out in spring when the plants are awake and their physiological activities are more alive; thus they are more resistant to these diseases. In such cases, the best thing to do is to take professional opinions into consideration (Harris, 1983).

Although antiperspirant sprays are used in plant transplantations that are made off-season, these transplantations shouldn't be made as much as possible especially when the plants have just started to stool. A proper digging process is one of the most important factors in the success of transplantation. Studies until now have shown that digging for the transplantation of a tree in leaf foliated should be made in two stages. Firstly, bottom roots should be dug and irrigated, then after waiting for 7-10 days; all roots should be dug and taken out.

Divaricated and in leaf foliated plants are transplanted mostly at the beginning of autumn and at the end of spring. If the winter is very mild in a region, the transplantation can be done towards winter; but soil should be prepared separately and should be prevented from becoming mud. Transplantation in winter has the advantages of cool and cold weather. But if it becomes too cold, plants may be affected and be damaged. During spring plantations, trees should absolutely be protected from cold weather and soil should be moist.

Plantations at the end of summer and in autumn have a big advantage which is related with the warmth of soil. Soil warmth lead the plant roots grow healthy and distribute. Some plantations during summer gave better results than the spring plantations in terms of longevity.

3.1.4.1 Transplantation time for non-evergreen plants

It is more appropriate for non-evergreen plants to be transplanted before the leaves start to fall and change color, before the soil is frozen in early winter or before the growth starts in spring (Kim, 1988).

3.1.4.2 Transplantation time for evergreen plants

Coniferous trees are generally transplanted during early autumn or late spring. The proper time for the plantation of Latifolius – Broad leaved evergreen plants is generally spring and autumn (Kim, 1988).

3.1.5 Effects of seasons on transplantation

When the ground is not frozen, some species may transplanting any time during the year but woody plants are generally moved in the spring but also they may moved in the fall after leaf drop and before the ground freezes. Fall planting should take place soon after leaf drop. Before the ground freezes in the fall, evergreens are especially prone to winter browning. Therefore, they should be moved late in the summer to early fall. Antitranspirants applying may help reduce the effects of winter desiccation in some species. Fall transplant success may be increased by transplanting hardy plants into sites with good soil moisture and wind protection. When shoot growth is peak, it's shown that the greatest transplant injury so woody plants are transplanted in late spring and early summer (Jakson et al., 1998).

Spring: Shoot growth in plants prevents them from being damaged from cold weather. This situation will promote root growth before Top growth starts. But as plantation during active growth period will cause various negations, if it is possible, plantations shouldn't be made during that time. Because when the plants' roots or branches are pruned, plants loose more water from these parts when compared to the other seasons. Because of these, transplantations shouldn't be made during the middle of spring when fast growth occurs and at the beginning of summer months (Kim, 1988).

Summer: In summer, plants actively absorb water that is passing through the plants' xylem. This is why too much sap loss will occur in the cut places of roots during plant transplantations in summer. It is determined that when plants grow in spring and complete their development, some of them accommodate better to the summer transplantation. We don't need to worry about the sufficient water amount in plants as active transpiration occurs more in hot air (Kim, 1988).

Autumn: Towards the end of summer season and during autumn, there are generally warm weather conditions that prevent root growth. As the days shorten and weather becomes warmer, plant transpiration decrease. Autumn season is the best season for most of the plants' transplantation. In this season, plants don't loose too much sap. *Citrus sp.* (Lemon), *Hibiscus sp.* (Hibiscus), *Bougainvillea sp.* are the plant species which can be damaged easily without placing their roots. It is better to transplant this kind of fragile plants in spring (Kim, 1988).

Winter: As the weather is warm and cold during winter, plant activity decrease which is a big advantage for transplantation. Plants can make use of the cold weather in winter. Transplantation can be done if the freezing level of soil is about 30 cm in big trees. But special attention should be given to pores in order not to freeze, and to roots root ball in order not to be broken. Transplantation at about 3 °C weather is proper as plants can be damaged in other weather conditions. No matter what the season is, plants should be protected from freezing and drying. Planting pits should be filled with water a few times before transplantation. Transplantation area should be mulched; after the area is filled with mulch, other irrigation process can be done although the ground is still always wet. Drainage system is crucial for increasing the success of transplantation process (Kim, 1988).

3.2 Preparation of plants for transplantation

Plants that are grown in nursery are rich in terms of capillary and hairy roots as they get all the necessary elements. These plants that have compact structures are transplanted very successfully. On the other hand, transplantation of plants that grow in rural areas and forests where maintenance process aren't made is very difficult. In this scope, root of a plant that will be transplanted should be nurtured 1-3 years before the transplantation and other maintenance processes should be completed. Transplantation will be successful if these conditions are carried out.

3.2.1 Preparation of large bushes and shrubs for transplantation

Bushes higher than 3-4 meters and shrubs that can reach 8-10 meters are in this group. While root structure of a plant that grows naturally in nature varies, there are root systems that are scattered to the sideways, elongated, and moved into deeps. If nutrient is abundant in the field where these plants grow, these kinds of plants don't need dense and capillary roots. If we try to transplant these kinds of plants without any process, only a part of the root will fit into the root soil and as a result of this, root/body will become unbalanced in the new planting site. This will increase the risk of plant's drying. This is why; root pruning should be done 1-2 years before transplantation. Root pruning should be done before the start of root activities when significant root growth occurs. Too much grown roots are cut with a sharp knife according to the size of the plant; for example, for small plants that have 4-6 thickness, 30-40 cm radius circle is drawn and roots around this circle is cut. In this way, new roots grow more strongly near the area of cut root. This increases plant's chance to adapt the new place (Figure 2).

Fig. 2. Deeply spading the plant that has very long and many side roots from x and y points and forming a new, strong and more compact plant root system.

3.2.2 Preparation of middle sized plants for transplantation

Trees that are 10 - 20 m tall are in this group. A denser root nurturing is necessary in transplanting these plants, or else the chance of plants to continue life decreases. This is why; preparations should start 1-2 years before transplanting the plant. Firstly tap part of the plant is pruned strongly but according to the rules. A pit about 30-40 cm is dug around the plant by taking plant's tap corolla. Especially the depth of pit is significant; all side roots of the plant should fit in to the pit. Roots that appear in the soil that is dug are cut with a sharp device, if the thickness of root is over 1 cm, puttying should definitely be done.

Organic substances, compost, qualified and slight soil mass is filled into the pit in order to ensure the easy growth of roots and accordingly ensure plants to form a strong root system. In this way, many new thin and capillary roots develop. New roots tie soil mass stronger and minimize the risks in transplantation (Figure 3).

While forming the pit, digging part by part can increase the level of success. Digging and filling the pit is extended over 2 years. The area around the plant is divided into 6 equal parts. 3 of these parts (A) are filled as can be seen in (Figure 4), while the other 3 parts (B) are dug and filled in the second year. The aim of this is to prevent the collapse of tree because of a possible wind effect.

Mulching during this process will be helpful for strengthening the root system placed between ditch and plant body. Soil should also be aerated and substances that will enhance development should be used. In this way, plant can be removed easier and better during transplantation. Digging process of a plant that will be removed should be done from a direction that will prevent root ball from splitting (Figure 5).

Fig. 3. Strengthening root system by digging a pit (a: Before pruning long side roots; b: Pruning and shortening long side roots in the dug pit; c: Filling the pit with materials that promote rooting and grown roots).

Fig. 4. Root nurturing in two phases and preparation of plant for transplantation in 3 years.

Fig. 5. Digging direction of the plant.

3.2.3 Preparation of big sized plants for transplantation

Plants that are taller than 20 m are in this group. The pit dug for developing root structure of these plants can be formed as 2 or 3 yearly periods. In 2 yearly period, during the first year, A,C,E parts are dug and filled while in the second year B, D, F parts are dug and filled. In 3 yearly period, in the first year A, D parts are dug and filled, in the second year B, E parts are dug and filled and in the last year C, F parts are dug and filled (Figure 6). When the last filling process is completed, necessary pruning is done and after 1 year, plant becomes ready for transplantation.

Fig. 6. Preparation of a plant for transplantation in four years with root nurturing in three phases.

3.3 Transplantation methods

While determining the transplantation methods of plants: aspects of the natural soil and transplantation area, distance between these areas, aspects of the settlements around these areas, the amount of time between removing and planting the plant, devices that will be used during the process and finance factors have significant roles. In order to make vegetal design of big areas and make successful plantations time, money, protectors and development methods are needed.

Removing the plant which will be transplanted should be made in overcast, rainy weather rather than in windy, sunny, too dry or too cold weather; nights are preferable for transplantation as microorganisms that revitalize root development are protected at night. Microorganisms are damaged and sometimes die because of direct exposure to sun and because of dryer winds. If plants are grown in nurseries, transplanted a few times, while being replicated their roots are pruned and capillary roots are increased, their plantations will be more successful. Although plants whose roots have never been maintained before, plants that are grown in a compact area, taller plants, sensitive and precious species can be transplanted in far fields, they should be protected very carefully and more precautions should be taken as their roots can fall apart (Ürgenç, 1998). Transplantation methods of plants are divided into three categories; bare roots, root balls, sacks, boxing and mechanical plant transplantation.

3.3.1 Transplantation of bare rooted plants

These are the plants whose body diameter is under 5 cm. In removing the small bare rooted plants, firstly a pit is dug which can taken all the root system in; the distance between the body and pit should be proper, or else roots can be damaged. Beginning from the edge of the root, the ground is dug until main root system appears. In order to ensure plants' adaptation to their new environment, soil parts between roots should be protected as much as possible. If the plant has a taproot system, this part should be laterally cut with a cutter and the plant should be released. The removed plant should be wrapped loosely with a piece of cloth in order to protect it from the wind and sun and create a humid environment, and then it must be transferred (Ürgenç, 1998).

Bare Rooted transplantation method is used more for large surfaced special trees. It is more successful during winter. Transplantations during mild climate winter are more successful. Plants should be waited in sandy soil and the ground they will plant shouldn't be too far. The important point is to ensure the development of plant roots. Necessary precautions should be taken in order to protect plants from fog and smoke (Harris, 1983).

After carrying necessary soil into the pit the plant will be planted and it the plantation is completed, it is covered up with sandy soil. In order to ensure the continuity of plants' lives, systematic irrigation and development conditions are crucial. Plants in Disneyland Amusement Center in the United States of America were planted with this method and have been very healthy through years (Harris, 1983).

3.3.2 Transplantation of plants with bare soil

Some contended plants can be transplanted without bare soil mass. Although depending on the plant species, generally many bushes and trees can be transplanted even in summer season. But the soil mass shouldn't lose its humidity in a short time. Before the plantation, a few holes should be made on soil mass in order to enable the water pass through the soil and prevent the humidity (Ürgenç, 1998).

3.3.3 Transplantation of plants by freezing soil mass

This method is used in regions with cold climate. In order to use this method, the soil should be frozen at least 30 cm and deeper. Firstly, a pit is dug and soil around the root is prepared for removal. Soil is often irrigated during frost period in order to freeze the soil. Freezing can be accelerated by using carbon dioxide. If it is -7 C° during daytime, wrapping during digging and plantation is unnecessary. In a *Pinus slyvestris* (Scots Pine) forest, a tree at the age of 40 can be successfully transplanted (Ürgenç, 1998).

3.3.4 Transplantation as root balls, sacks and boxes

Transplantation with soil is the best way to transplant the evergreen, needle-leaved, big trees and other plant species that drop leaves no matter how big they are. While transferring a plant from a green area to another, it is carried with a soil mass and wrapped in order to prevent it from falling apart. If the width and depth of the pit isn't enough, plants' chance to live is very little. Plant roots can become smaller or bigger with the effect of weather and development. In such cases, preparations should be done by taking possible difficulties into consideration. The most proper method is to dig a pit around the plant in order to leave it a little smaller than the size it will be at the end. In the first year, an area equal to the half size of the root of plant is dug and filled with organic substance and soil mixture in order to promote root rooting. In addition to these, the plant can be dug as big as tits root ball and prepared or root balls are promoted for growing together with soil (Harris, 1983). If root balls aren't pruned according to the pit, this can pose a risk in terms of getting wet. The second method is growing roots without damaging the environment very much. Studies have shown that the first method is better than the second method. On the other hand, it is known that pruning the roots before transplantation cause loss of time (Zion, 1968).

The plants that are thicker than 10 cm should be transplanted with soil. Although it is possible to plant small sized plant species that fall leaves during winter, in vegetation period without soil, it is more proper to plant them with soil. Otherwise their chance to live will decrease. These are the plants that should especially be taken care of; *Fagus sp.* (Beech), *Betula sp.* (Birch), *Cornus sp.*, *Ginkgo sp.* (China Ginko Biloba), *Liriedendron sp.* (Tulip), *Magnolia sp.* and *Quercus sp.* (Oak). The amount of digging starting from the roots is a very important point in order to ensure the health and continuity of plant life. In Kim's study in 1988, a formula is made for root ball's diameter and depth:

For big-sized plants:

R= Diameter of root ball and the height of root ball diameter (cm.)
DS= Diameter of the stem (cm.)
(8)=Constant.
R=(8 + 8) x DS
For bushes and small sized plants (generally smaller than 3 m. height)
R= (6 + 2) x DS
For instance: root ball diameter and removal depth of a plant that has 15 cm stem diameter will be as such:
R= (8 + 8) x DS
= (8 + 8) x 15 =240 cm.
For instance: root ball diameter and removal depth of a plant that has 4 cm stem diameter will be as such:
R= (6 + 2) x DS
= (6 + 2) x 4 =32 cm.

This formula can be used for pits. In order to add mulch, pits should be dug around the plant. This is the minimum diameter needed for cutting the root system from the bottom of stem. After calculating the diameter of the area that will be dug, the digging should be made clockwise. If the digging isn't made clockwise, root system can be mixed and it would be difficult to make a root ball. After cutting the widest root, root bark should be peeled by a knife. In order to cut the peeled root from the middle, sharp devices should be used. When root systems are peeled with sharp devices, cambium cells promote root system more (Kim, 1988) (Figure 7).

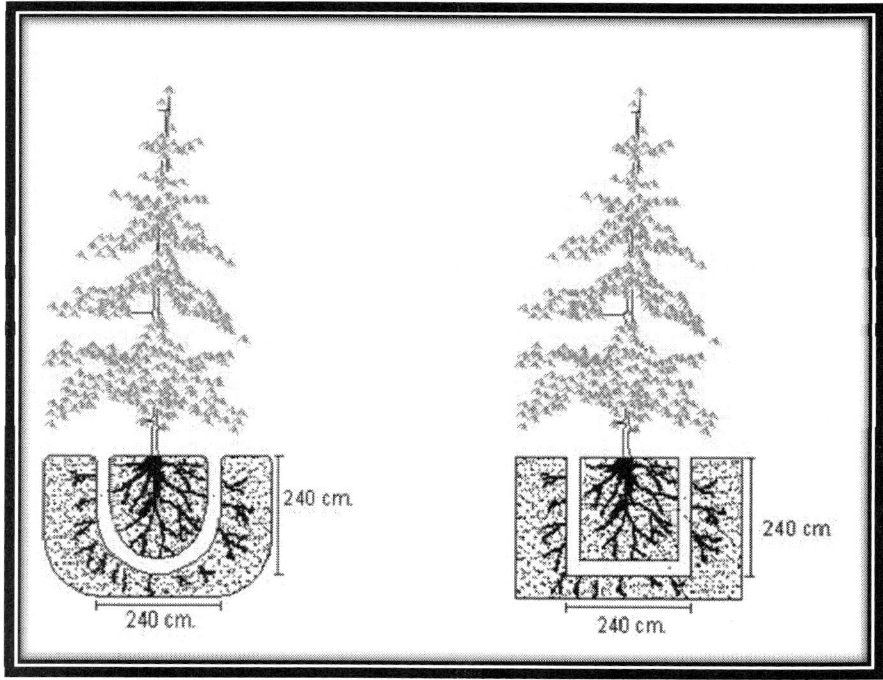

Fig. 7. Root ball diameter of a plant that have 15 cm body diameter and its removal depth.

As a general rule, the size of the soil mass around the root should be 8-12 times bigger than the diameter of the stem at the level of the chest height of the plant. While this measure can be more in smaller plants (for instance 12 times), it can be less in bigger plants (for instance 8 times). If this formula is taken into consideration, a plant with 25 cm diameter should be removed with a 100 cm soil mass. The depth of the soil mass shouldn't be less than 75 cm no matter how big is the size of the plant. In small sized plants, the mass should be %75 of the soil mass; in big sized plants, it is %40. For example, a 25 diameter plant is removed with 100 cm. radius or 200 cm. diameter soil mass, the soil mass depth should be at least 80 cm. (Ürgenç, 1998).

Digging process should start when the soil is moist. The digging start from 7.5-12.5 cm. outer and the pit is dug outwards. But some people take the roots out starting from the sides towards the inner part although their pit is wide enough; they sometimes exceed the level of soil which should be dug and come close to the stem. In this case, soil mass is smaller. As a general rule, in the soil-root mass; the soil amount should be as little as possible while root amount should be as much as possible. The sack that covers the roots should be moist. Roots that are out during planting, should be prepared in their natural positions and planting should be completed (Ürgenç, 1998). Moist material covering the root balls should be fastened with rope and protected carefully. If steel rope will be used in this process, root surface should be dried carefully in order to protect the rope from rust. Spray is used during transplantation in order to prevent leaves from falling. Concentration that will be used in the spray should be carefully chosen and applied. Short branches of the plant should be pruned and then the plant should be carefully tied (Kim, 1989). After wrapping and fastening of the plant that will be transplanted, special attention should be given in order not to touch the plant and clean the underside of the soil after the plant slants. If there is a steel cable or another fastening material around the underside of the ball, they should be cut in order to free the big soil roots. During transplantation of plants, chains or cables shouldn't be connected to root ball or basic stem. In order to transplant big sized plants, steel ropes can be attached to plants' root balls. But steel ropes shouldn't be used in short distanced transplantations (Zion, 1968).

Roots of the leaves can sometimes get smaller or bigger than the development sizes calculated according to the effect of time and air. In such cases, roots should be pruned 1-2 years before the transplantation in order to avoid ant possible difficulties (Harris, 1983).

In his research in 1988, Kim mentions transplantation of a 10 m. height coniferous plant's transplantation in Korea only with human power to 50 m. distance. He said that he needed 5 people for this plantation and summarized the method as such:

Firstly, the soil is dug and root ball is made, the root ball is then fastened. After that, the plant is bended to one side with 15-30 angles which cause emptying of the other bottom part of the root ball. The bottom part of the root ball is filled with the soil that is dug while preparing the root ball. The plant is moved to the other side and bended to one side again. The hollow that is created with removing the root ball without using heavy devices is filled with soil. This process continues until the bottom part is filled with soil. Finally, the plant is moved out of the pit and gathered in order to protect it from any damage in case it is bended to different directions. While 2-3 people push the plant, the other two holds the stem. After that, the plant is rolled to transplantation area. The stem is hold in a bended position and special attention is given in order not to lay it n the floor. Because the plant is heavy and it is difficult to move. This is the easiest way to transplant a plant without using heavy machines or devices. On the other hand, if the plant will be moved to a long distance, it can be wrapped with sack and tied (Kim, 1988) (Figure 8).

Transplantation is made with boxes if the soil is sandy or easy to scatter. Plants that are put into boxes in the shape of tetragonal prisms, with wider tops and narrow bottoms are transplanted. The soil mass of the plant whose around is opened with pits are put into strong cases or boxes. When crating is completed, bottom board is put under the root system and the process is completed. Crating can be started even at the process of preparation. The 5.5-15 cm. space between box and soil mass is filled with fertilizer, compost and highly nutrient soil and kept waiting for one year. In this way, a rich capillary root system is produced. In order to fertilize this process, crating is prepared in 3 years firstly by preparing the 2 faces in the first, the other 2 in the second year; a safer transplantation is made in this way (Turhan, 1994).

Plants' root development should be measured and determined in boxing method too. The width of the pit from which the plant will be removed is determined (*which is in the shape square cr rectangle rather than hemisphere shape in root balls with boxing method*) by taking the situation and development of the root area into consideration. If the width and depth of the pit is not proper, the plant might die. Firstly, the soil is dug and the mass that will be moved is prepared. This mass is surrounded by wooden material and boxing process is started. The plant is bended 15-30 angle with the method mentioned above; the space left is filled. The same process is applied to other side and can be moved without using crane or any other technical devices. When plants are bended to a side, needed pruning is done without harming roots, the bottom part of the prepared part and boxing is done (Ürgenç, 1998)

Fig. 8. Transplantation of plant without using heavy instruments and equipments.

Points that should be given importance during plant transplantation can be summarized as (Kim, 1988):

- Root balls of the plants should be measured and determined, and the number of healthy roots should be as much as possible.
- The ground should be dug when it is warm.
- Roots should be clean cut, wide root skin should be whittled with a knife and then saw should be used.
- Big scissors should be used for small roots.
- Irrigation should be well.
- Pruning reduces transpiration.
- Alginate hormone should be used for roots.
- Water should be sprayed for moist.
- Fertilization should be done with 0.225 kg for 0.08 m^3 area.

Apart from plantations made by big companies, company owners, municipalities, it is difficult for private garden owners to use heavy tools. They may not find the chance to use complicated machines. Using human power is the only choice in such situations.

3.3.5 Transplantation with mechanic methods

There are some models developed for digging, removing and transferring of plants. In a model called "Tree Nursery", carriage of plant can be done with an articulated lorry or truck. Plants are wrapped and hydro alkali booster four scoops are stuck on the ground; after shaping the root ball, the bottom of the plant is lifted with the scoops and moved out of the soil. If the weather is proper and the distance is not very long, the plant can be carried vertically or can be bended toward the front and carrying can be done. Pits in which plants will be put should be dug beforehand. Planting should be done by taking the slope and the shape of the ground into consideration (Harris, 1983).

The second method used in plant transplantation is called "Skin with Cohesive". This skin is made of a scoop and there are sharp parts on the tips. This sharp part is installed before starting to dig the plant and locked. Transplantations that are done with this method is applicable for big sized plants. Tree nursery technique is preferred for smaller sized plants (Harris, 1983).

The third method is making a hole by digging the plant roots with one scoop and removing the root balls. Scoops are clamped together and the root ball is lifted with crane. Each of the scoops are for digging soil. There are two kinds of curved scoops; 80 cm. and 200 cm. the weight of device is totally 115 kg. Compressor that works with tractor or mobile gas produces hydraulic power. Scoops are numbered according to their usage methods (Turhan, 1994).

Functions of tree lifting are (Harris, 1983):

- Digging planting pits.
- Removing the plant with its roots neatly and in a way that fits the planting pit.
- Carrying the removed plant to the planting place without harming and planting it properly.

Tree removing and planting machine is made of 2 strong lifter arms connected to truck frame, digging knife unit, hydraulic system etc. hydraulic pump is provided by the vehicle to which the machine will be connected. Rotary digging knife unit is connected to lifting arms and can be controlled while working, transferring in vertical position and horizontally. Digging knife unit is made of at least four knives which are arranged around the knife frame. The number of knives can vary according to the aspects of vehicle. The unit is openable and closable; knives can be pushed downwards and can be pulled back upwards (Harris, 1983).

3.4 Transplantation of palms

Palms are monocotyledon and they have either one body or many bodies that are not branched out. Any damages on the body leave a scar as cambium shell doesn't exist. Most of the bodies of palm species are in the shape of filaments that grow at the bottom of body. If a root is cut or broken, the plant generally dies. Although palms can be transplanted at any time of year, warm spring and summer months are generally preferred as root growth is fast in these months (Harris, 1988). Transplantation in our country should be made in summer because of climatic conditions; transplanted palms' roots die during the process and soil heat is needed in order to ensure the development of new roots. According to the research in Mediterranean Region, it was determined that transplantations should be done on April, May and June; and if there is not a chance to make transplantation in these months, then the period between September-November should be the second choice (Anonymous 2012 d).

Before carrying palm fronds (each one of pieces that constitute the crusty structure of tree body), leaving 6 or 8 per body pruning is a general rule as each body has an eye and it should be protected. Annuluses of leaves besides the eye are generally separated in order to prevent from the pressure in eye when fronds are tied. Less leaves are suggested for urgent transplantation of palms. Long sticks should be tied to body in order to support palms that have long thin bodies (*Phoenix redinata* Jacq. Fragm. and *Acoelorrhaphe wrightii* (Griseb. & H. Wendl.) (Himelick, 1981).

Plantation of palms is similar with the other trees. Planting pit is dug wider than the root ball. Crumbled stones that are removed from the pit are used for refilling. Palm should sit on the palm pit which is very important for palms as they have curved body. Palms should be put in 75-125 mm deeper than their original position in order to prevent development of new root. Many kinds of mixtures can be used in filling the planting pit such as sand and soil with vegetal mixture. These soils should be around the root and be irritated very well (Harris, 1988). Leaves on shoot should be tied around the leaf that grown the latest in a way to protect the leaf (Anonymous 2012 d).

In order to ensure immobility, palm should be tied with steel topes but nails or screws shouldn't be stuck into the body. Protector bars that are placed around the body protect them from stuck (Harris, 1988). If the bottom part of body moves, thin roots that newly develop split and the plant cannot grow. In order to prevent this, palm body thickness should be measured and metal circle should be prepared; then, this circle should be mounted onto the body and the tree should be set from 4 sides. Steel circle is made of 2 hemicycles that can be affixed on 1/3 bottom part of the tree. These hemicycles are connected to one another with two screws. Additionally, 4 ringers for setting steel rope on the circle (Anonymous 2012 d). Single body palms are piled on Lorries and treys and carried. Palms that have wide and many stems should be laid on truck haulage and be tied with one stem in order to increase stability and the width of the load (Harris, 1988).

Newly planted palms should be irrigated very well in the first season. They should be irrigated slowly, twice in a week for 8 hours during the first two months, after that period, they should be irrigated once in a week for 12 hours. This process can be carried out in well drained soil. Fronds of newly planted palms pale especially because of alkali soil, cold weather or due to the lack of manganese or iron. This is why; necessary mineral elements and fertilizers should be added into the soil (Harris, 1988).

4. Transplantation of trees in campus of Bartın University (case study)

Transplantation processes of large trees in Campus of Bartın University were analyzed in this study. The aim of this study was to ensure a modern appearance and qualification to Campus of Bartın University in terms of plantal arrangement. A commission was formed by Deanship of Faculty of Forestry. Necessary literature review and field analysis studies were done by the commission and some alternative projects were prepared. At the end of the meeting with deanship, the most proper project in terms of practicability and cost was chosen. The chosen project was detailed and prepared for application. Mechanic tree mover was rented from Karabük Municipality. The project was applied on May 2004 and transplantation of 10 plants that were of 6 species was completed.

Material: Faculty of Forestry has been put into service on 1993-94 academic years. Campus was built at Gaffar district, Ağdacı Village, Bartın approximately 5.5 km distanced from city center on 1.14 hectare land. 900 m village road on Bartın-Kozcağız highway was used for going to the campus.

Pinus pinea L. plantation in Boğaz district situated in Bartın Central Forestry Operation Directorate was the area on which working area existed; transplantation process, removal and plantation of trees were carried out on this area. Application area, plantation, was an inclined land on the right side of campus entrance, situated at the north of library building in Faculty of Forestry. The area was approximately 450 m².

Campus's map section in 1/500 scaled base map was used in order to draw the project and determine the definite boundaries of application area; digital camera was used in order to view the sections of transplantation process in the area.

Method: The method used in this study can be summarized as:

- Determination of the area on which transplantation will be done in the campus,
- Determination of area's map section on 1/500 scaled base map,
- Transferring present plan from paper map sheet to computer ,
- Determination of plants which will be transplanted,
- Preparing projects of the area on which transplantation will be done,
- Renting mechanic tree mover and preparation of necessary equipment,
- Making land survey. Within this scope:
    - Digging of the pit in which plants will be placed,
    - Completing removal and transfer processes of plants,

- Planting plants,
- Maintenance of plants that were transplanted.

A study group was formed by Deanship of Faculty of Forestry in order to carry out transplantations. At the end of the meetings held in the group and with Deanship authorities, processes and necessities were determined. Application process started after determining the priority of these necessities and processes.

4.1 Transplantation process in campus of Bartın University

It was determined that the plants that had been planted during ten years history of Faculty of Forestry would be transplanted as they couldn't create a beautiful effect; especially big sized plants should be transplanted in order to increase the scenery quality of the faculty. In this context, meetings were done with Karabük Municipality and tree mover was rented for two days in order to use it in transplantation process.

4.1.1 Determination of the transplantation area

We can classify the factors that play role in the process of choosing field:
Soil characters of species that will be brought to the field,
Enable the mechanic tree mover to enter the field and make transplantation easily,
Choosing the best places for plants in terms of aesthetical and functional aspects,

By taking these criterions into consideration, the area of the campus transplantation area was determined to be: At the north of the library building, southwest of former Forest Engineering and Forest Industrial Engineering building, northwest of Landscape Engineering and Vocational School of Higher Education and the area on the right side of campus entrance was chosen.

4.1.2 Choosing plants

Time is crucial in transplantation process. The plant that is removed should be transferred to the plantation area as soon as possible. This is why, the species that would be brought to campus were chosen in meetings held with Bartın Forestry Operation Directorate. Chosen areas were in the area that belonged to Bartın Central Forestry Operation Directorate, Boğaz district *Pinus pinea* L. plantation. Special importance was given in order to enable mechanic tree mover move freely; the chosen area was a level land as much as possible, there were no other plant species that were very close to the working area.

While choosing the species, it was very important to choose plants whose habitats were similar with the area that they will be planted; on the other hand, their colors, forms, texture and line of these plants were very important in terms of creating the desired effect in the field.

Removal and planting, namely transplantation of some species of trees are easier than the others. Tree species that have shallow roots with hairy, thin and compact root systems are more successfully transplanted than the species that have thin roots and taproots which go deep in the soil. But if the maintenance of the species which have thin root and taproot system is done carefully, they can be transplanted successfully and they can live long. By taking these criterions into consideration, at the end of researches and analysis in Boğaz district *Pinus pinea* plantation it was determined that 3 *Pinus pinea* L., 1 *Pinus nigra* Arnold., 3

Cedrus libani A. Rich., 1 *Prunus avium* L. and *Malus floribunda* Sieb. were chosen to be planted (Figure 9, 10 and 11).

The *Picea punges* Engelm. that was placed at the north of Bartın University public housing, at the south of information building in Faculty of Forestry was a decorative species, but it had lost its aesthetical aspect in the area. Because of this, in order to use the tree more aesthetically, it was determined that it should be transplanted. North side of plant was marked with oil color and it was transplanted according to these marks.

4.2 Characteristics of the species used in the transplantation

Pinus pinea L. (Stone pine or umbrella pine):

Stone pine grows in the Mediterranean countries and their picturesque shape of straight trunk and domed crown. Leaves are in pairs (Brain and Valerie Proudley, 1976). This tree grows in native to south-west Europe around the Mediterranean to Greece and Asia. The seeds are eaten raw, roasted like peanuts or added to stews a ragout, a traditional Italian dish. Remains of husks have been found in Roman camps in Britain, indicating a long history of their use. Height may reach 30 m (100 ft) but in the open it forms a much lower umbrella-shaped tree. Flowers open in June, males golden and clustered, females pale yellowish green, about 1-2 cm (1/2 in) long. Cones are large, about 2-5 cm (5 in) long and heavy. They remain closed for 3 years. Needles are dark green, thick, slightly twisted and pointed. They are in pairs and often rather sparse. Bark is reddish brown with deep dark cracks forming long plates (Roger, 1979).

Fig. 9. *Pinus pinea* L. plantation in.Boğaz district, Bartın, Turkey.

Cedrus libani A. Rich. (Lebanon cedar):

Lebanon cedar is native to Taurus (south Anatolia) and near east (Lebanon). The trees are planted as an ornamental in Europe and N.America. This is the familiar slow-growing tree of our parks and large gardens. Younger ages, it is pyramidal in shape gradually becoming flat-topped with age. The widespread branches of clear gren foliage sometimes suffer damage during heavy falls of snow and should be propped where possible (Brain and Valerie Proudley, 1976). It is height to 24-36 m (80-120 ft). Male flowers are abundant 1cm (1/2 in) and pale green through the summer expanding to 5 cm (2 in) to shed pollen in November. Females appear in November and develop in to large purplish green cones 9-15 cm (3.5-6in) which taper to the top. Foliage is made up of dark green needles up to 2cm (3/4in) long. Young twigs are almost hairless (Roger, 1979).

Lebanon cedars grow at elevations of 4,264-6,888 ft. They grow best in deep soil on slopes facing the sea. The trees require a lot of light and about 40 inches (1000 mm) of rain a year. They form open forests with a low undergrowth of grasses (Anonymous, 2012e)

Fig. 10. Transplantation of *Picea pungens* Engelm.

Pinus nigra Arnold. (Black pine):

Black pine grows in native to Austria, Italy, Yugoslavia, Greece and Anatolia. It is planted in Britain for shelter and for ornament. Height may reach over 30 m (100 ft) in the forest. Flowers open in late May, males is golden yellow, females is red, about 0.5 cm (1/4 in) long. Cones are 5-7,5 cm (2-3 in) long and the scales open to release winged seeds. They are arranged in pairs in dense clusters separated by bare lengths of twig. (Roger, 1979). The rough bark is brown to dark brown and the cones solitary or in clusters up to 8 cm (3 in.) long. It is a useful shelter belt tree for dry, chalky soil (Anthony, 1973).

Its habit is broad and vigorous and the long needles are of a delight ful, dark green colour, givig this pine a sound and luxuriant look all the year round. It tolerates wind and poor soil, and will grow most attractive in a light, sunny localitiy but needs plenty of space in order to unfold in all its glory (Eigil, 1973).

Malus floribunda Sieb. (Japanese or Shoey Crab Apple):

A Japanese tree, probably a hybrid rather than parks and streets, as it flowers profusely every year. It has height to about 6-9 m (20-30 ft). Flowers open in late April and early May, each about 2,5-3cm (1-1,25in) wide in clusters of 4-7. Fruits are about 2cm (3/4in) in diameter, ripening yellow in October (Roger, 1979).

They would look nice as specimens on the lawn or with other bushes in the front garden, where they give the entire road a festive look in the spring. The trees can be bought in the shape of ordinary bushes or standarts. In many instances a young bush specimen will grow into a handsome tree with many slightly cooked trunks of much better effect then one long, straight trunk. This should be as for ordinary apple trees, i.e. good, deep soil, rich in humus without stagnant water in winter. Pruning should consist of a suitable thinning-out of the branches at an interval of a few years, always done in such a manner as to maintain then natural shape of the top (Eigil, 1973).

Prunus avium L. (Sweet Cherry):

A very fast-growing, ornamental tree which will not produce berries owing to its double flowers (Eigil, 1973). Its fruits trend to be bitter but it is one of the parents of most European cultivated cherries. The wood is reddish-brown with a very straight grain, and used in cabinet making, and for anything requiring a straight bore such as pipes and musical instruments. To be found in hedges and woods, gardens and parks in most of Europe, also cultivated and naturalised in eastern N. America. It has height to 18 m (60 ft) or more. Flowers open in mid April, each about 2-5cm (1in) across in clusters on previous year's growth. Fruits are about 2 cm (3/4 in) across and may be light or blackish red, sweet or bitter. Leaves have stems red above and yellowish beneath with 2 or more glands or lumps near the base of the leaf blade and colour yellow and red in autumn. Bark is reddish Brown and clearly marked by lenticels in horizontal lines and broken by large cracks (Roger, 1979).

Picea punges Engelm. (Colorado spruce or Blue spruce):

Colorado spruce grows in native to the Rocky Mountains in western N. America, particularly at the south end of the range grown for ornaments and sometimes for timber in northern and central Europe. Height to about 30 m (100 ft) but may reach 45 m (150 ft) in favourable conditions. Flowers open in May, males about 2 cm (4/5 in) long, females twice that. Cones are a distinctive pale colour, with wavy toothed scales. They are about 7,5-10 cm (3-4 in) long. Needles are 4-sided with whitish-blue buds on each face. They are spine tipped when young, becoming blunter with age. Bark is purplish-grey, breaking into coarse plates. (Roger, 1979).

They are seldom as high in cultivation where the fine glaucous foliage forms are more common. Because of their brilliant colouring these named clones are some of the most desirable of conifers suppliers being hard put to meet the demand for established specimens (Brain and Valerie Proudley, 1976).

Fig. 11. *Malus floribunda* selected in order to transplantation.

4.3 Determination of planting area of transplanted species

After choosing the species to be transplanted, the project was prepared in order to place the chosen species on the field and the project was drawn on computer (Figure 12). According to this project, in order to determine the exact places of plants, piles on which plant species were written were penetrated one by one on the ground and application work was carried out.

4.3.1 Features of mechanic tree mover

Working style and features of tree removing and planting machine can be summarized as: The machine works with a hydraulic system; hydraulic pump of the system is placed in the vehicle on which the machine is mounted. The machine is composed of 4 strong lifter arms, digging unit, hydraulic system and a trailer system by which the tree will be carried and some other parts. Digging knife lifting arms are jointed and while it is hold vertically during working, it is hold horizontally during transfers. Digging knife unit is made of 4 knives put around the knife frame. It has an extensible and closable frame. Knives can be pushed downwards and pulled back upwards.

4.3.2 Making planting pits of the plants

Size of the planting pits in which plants will be transplanted depends on the mouth size of mechanic mover. Plant pits have to be opened again with the same device while mechanic tree mover removes and transfers plants (Figure 13). Size of the pit from which plant is removed and into which it will be planted is significant in terms of adaptation to new place, being affected form strong winds and regular root development.

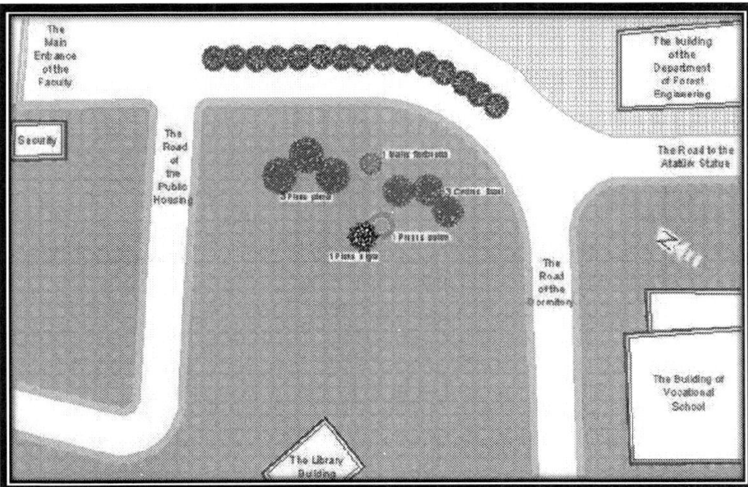

Fig. 12. The project of transplantation tree.

Fig. 13. *Picea punges* Engelm. planting pit.

In the transplantation made by mechanic tree mover, firstly planting pit is opened and irrigation must be done as a rule; because when tree mover transfers the plant, no time should be wasted and plants should be protected from damage caused by time. This is why each time mechanic tree mover comes to the campus, planting pits were opened and then the plants that would be transplanted were brought.

4.3.3 Removal and transfer of the plants

Time effect has a big importance in the success of removal and transfer of plants. As a general rule, overcast and flurry weather should be preferred. But as the tree mover rented from Karabük Municipality had to be returned maximum two days later, the process had to be carried out in sunny days.

Transplantation season is also important in order to continue a healthy life and adapt quickly to the new soil. Early spring is the most proper season for transplantation as vegetation time doesn't completely start in this season. More careful removal, transfer and maintenance are needed for plantations that are made in the seasons except transplantation seasons. Mechanic tree mover used in the project studies were required from Karabük Municipality on March, but it could be taken from the Municipality on May because of its workload. As the season wasn't very appropriate enough for transplantation, processes during transplantation were carried out very carefully.

Tree removing mechanism was put on the ground from the back of the vehicle and surrounded the tree that will be removed with the opened frame. Knives moving with the electronic system penetrated into the soil according to the points determined by the frame and they were united in a way to shape a dome. The tree and soil was picked up with tree removing device and was placed onto platform and booth vertically, and the tree was ready to be transferred (Figure 14). Distance between the campus and Boğaz district *Pinus pinea* L. plantation was approximately 20 km. Transfer process was very carefully made as the distance was long.

4.3.4 Plantation and maintenance process of the plants

Plantation: Plants were planted into the pits where piles were situated according to the project. Trees whose northern sides were marked with oil paint were brought to the planting field and they were placed in a way to face the same direction and stand straight. Pits were irrigated with plenty of water before plantation.

Refilling removal pit after plantation: Pits that were formed during plantation were filled with soil. In this way roots could be ventilated and they were protected from drying and dying. Soil that was used for filling had a pervious structure and was mixed with organic substance.

Protection of Stem: Stems were wrapped with sacks in order to protect newly grown trees from sunburn, frost cracks, damages of winds and cold.

Supporting of root fescue: Newly transplanted trees were supported in order to make them get used to their new places and protect them from environmental pressures. In the supporting works, trees were roped with rubber hoses by leaving 2-3 cm spaces from stems of trees according to the incline of the field. Rubber hoses were roped to trees from 3 sides tightly with 45° angle (Figure 15 and 16).

Fig. 14. *Cedrus libani* A. Rich planting processes.

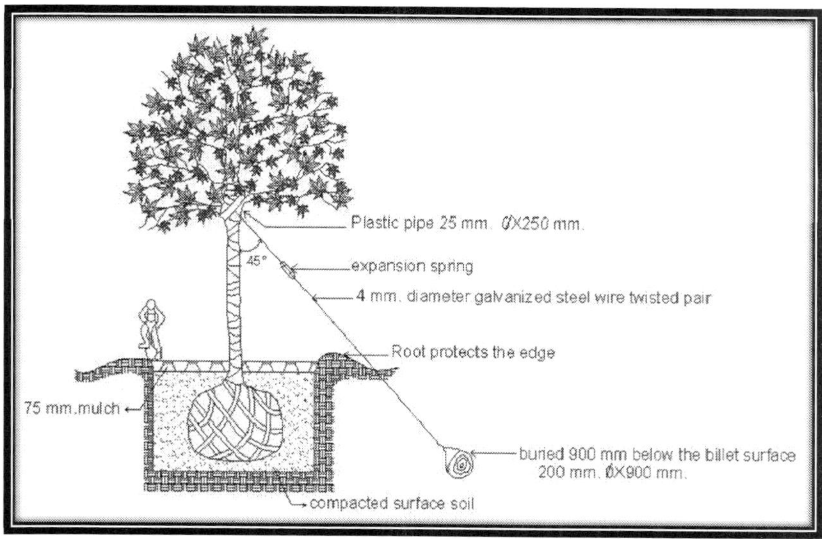

Fig. 15. Detail of supporting made with buried piles under the ground.

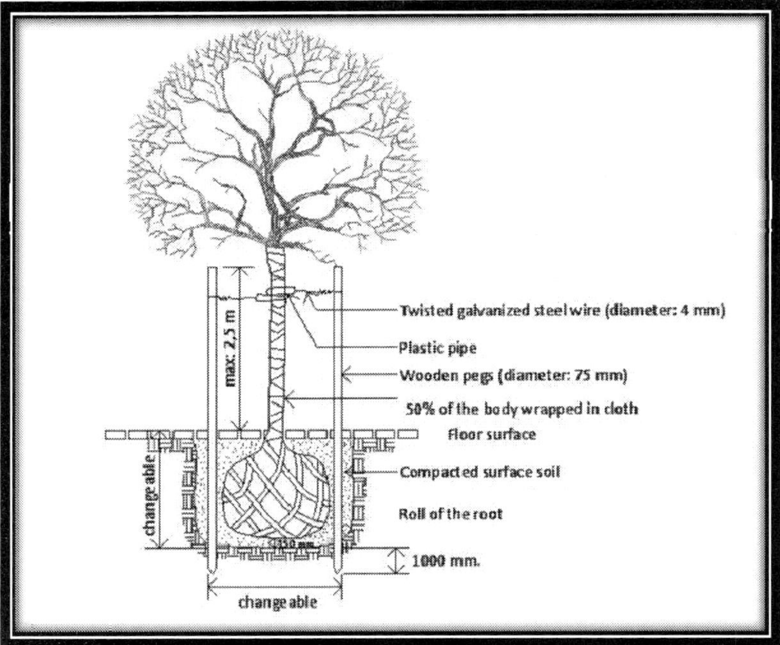

Fig. 16. Detail of supporting made with vertical piles.

Pruning: Branches of trees that were broken and dried were pruned. This had a big effect in preventing water loss. 1/3 – ¼ of unnecessary branches were pruned and growth direction was controlled.

Irrigation and Mulching: Mulching with 5-10 cm thick rotten leaves and well developed fertilizer applied on the bottom and all around the plants that were transplanted; mulching process decreased evaporation in the soil around roots thus protected the humid in soil. On the other hand, mulching balanced the heat of soil and made a positive effect on the life of the plant. Plants that were transplanted were irrigated periodically in order to develop the relation between root and soil and give necessary moisture to the soil.

5. Conclusion

Transplantation is the process of removing plants that are at a certain age and planting them to other places in order to create the desired effect. When places such as public gardens, youth centers, schools, Olympiad villages should be designed in a short time, especially trees that are at a certain height and form are preffered. Plantation of young bushes doesn't create the same effect created by the older ones. The importance of transplantation in terms of landscaping as to minimize the loss of time until small bushes grow and have enough strength in the field. Younger bushes create the desired effect in 20-30 years while planting big trees that have aesthetic, functional and climatic effects will create the desired appearance and will balance size and space in a short time.

Transfer time is very important in the process of removing and transferring plants. The most proper time for transplantation is the stable period between October and March. Mechanic tree mover rented for transplantation was requested from Karabük Municipality on March. But as the municipality had used the vehicle for its works, it was sent to the project on May. As the weather was very sunny and hot on the day of transplantation and the vehicle was rented only for two days, there was no other choice but to complete transplantation in two days. This caused plants to adapt their new places difficultly. As the proper period for transplantation, namely early spring had passed and it was very hot, a more intensive care was necessary. Not making enough maintenance caused drying of two *Cedrus libani* species. Dried cedars were removed from the field.

6. References

Anonymous, 2012 a. Transplantating Trees and Shrubs,
(http://www.ag.ndsu.edu/pubs/plantsci/trees/f1147w.htm),
(accessed 19 April 2012).
Anonymous, 2012 b. Kalamışın Palmiyeleri, (http://www.sabah.com.tr/Yazarlar/baydar
/2011/07/04/kalamisin-palmiyeleri-39255257939), (accessed 19 April 2012).
Anonymous, 2012 c. Bitki Nakli, (http://wap.ntvmsnbc.com/Haber/Goster/25326022)
(accessed 19 April 2012).
Anonymous, 2012 d. Palmiye, (http://www.palmiyemerkezi.com/palmiyelerhak.htm),
(accessed 19 April 2012).
Anonymous, 2012 e. Lebanon Cedar, (http://www.blueplanetbiomes.org/
lebanon_cedar.htm), (accessed 19 April 2012).
Altan,T. and Önsoy, C. 1982. Road Trees in Cities and Problems. Nature and Human
Magazine, Year: 16, Issue: 1, Ankara.
Anthony, H., 1973. Evergreen Garden Trees and Shrubs, ISBN 0 7 137 0621 X, London.
Brain and Valerie Proudley, 1976. Garden Conifers İn Clour, ISBN 0 7137 0 807 7
Caner, M. 1976. A. Recreational Approach and The Green Areas and Open Spaces In The
City of Ankara. Middle East Technical University. Master Thesis. Ankara.
Eigil, K., 1973. Garden Scrubs and Trees, ISBN 0 7 137 0649 X, London.
Harris, R.W. 1983. Arboriculture: Care of Trees, Shrubs and Vines in the Landscape, Prentice
– Hall, Inc. Englewood, Cliffs, N.J.
Himelick, E., B. 1981 Transplanting Manual for Trees and Scrubs. Intl. Soc. Arbiriculture,
Urban, III.
Jakson, M., Harsel, B., Fornes L., 1998. Transplanting Trees and Shrubs,
(http://www.ag.ndsu.edu/pubs/plantsci/trees/f1147w.htm) (accessed 19 April
2012).
Kim, H., 1988. Green World. Green Grower Publushing Company, Moreno Valley, USA.
Mayer, F.H., 1982. Baume in der Stadt. Verlag. Ulmer. 380.
Nadel, I.B.1977. Trees in the city. New York: Perggamon Press. USA.
Roger, H., 1979. Trees in Britain, Europe and North America, ISBN 0 330 2548 04
Şahin, Ş. 1989. Ankara Kenti Yol Ağaçlarının Sorunları ve Peyzaj Mimarlığı Açısından
Alınması Gerekli Önlemler. Ankara Üniversitesi Fen Bilimleri Enstitüsü, Yüksek
Lisans Tezi. Ankara.

Turhan, 1994. Peyzaj Uygulamalarinda Büyük Bitkilerin Transplantasyonu ile ilgili Sorunlar ve Çözümlerine İlişkin Ankara'da Yapılan Çalışmalar Üzerine Bir Araştırma. Ankara Üniversitesi Fen Bilimleri Enstitüsü, Yüksek Lisans Tezi. Ankara.

Ürgenç, S. 1998. Genel Plantasyon ve Ağaçlandırma Tekniği. İstanbul University, University Issue No: 3997, Faculty Issue No: 444, ISBN 975 – 404 – 443 – 0.Istanbul.

Zion, L.R., 1968. Trees for Architecture and the Landscape. Van Nostrant Reinhold Company, N.Y- Cincinati, Toronto, London, Melburne.

16

Xeriscape in Landscape Design

Ayten Özyavuz[1] and Murat Özyavuz[2]
[1]Namık Kemal University
[2]Namık Kemal University, Faculty of Agriculture,
Department of Landscape Architecture, Tekirdağ
Turkey

1. Introduction

The term Xeriscape comes from the Greek word xeros, meaning dry. The concept originated in Denver, Colorado, in the early 1980s. Because of severe drought conditions, Denver had rationed water and prohibited irrigation of lawns and yards. A number of terms describe waterconserving landscaping. Among them are "xeriscaping," "low water use," "droughttolerant," waterwise," and "desert" landscaping. Xeriscaping, a widely promoted term the past several years, is a word of Greek origin with *xeros* meaning dry, combined with landscaping. Drought-tolerant indicates the ability of a plant to survive on limited water, although these plants usually look better as water is increased. With improper watering, a drought-resistant plant may become a water guzzler in the landscape. As a result, vegetation in yards withered, and Denver landscapers began promoting what they called Xeriscape, a landscaping approach that uses small amounts of water but maintains a traditional look. Since that time the Xeriscape concept has been adopted in many areas of the country experiencing drought or long term dry conditions, and actual Xeriscape practices have evolved differently in various places (Welsh, 2000). The goal of a xeriscape is to create a visually attractive landscape that uses plants selected for their water efficiency. Properly maintained, a xeriscape can easily use less than one-half the water of a traditional landscape. Once established, a xeriscape should require less maintenance than turf landscape. A Xeriscape-type landscape can reduce outdoor water consumption by as much as 50 percent without sacrificing the quality and beauty of your home environment. It is also an environmentally sound landscape, requiring less fertilizer and fewer chemicals. And a Xeriscape-type landscape is low maintenance — saving you time, effort and money. Any landscape, whether newly installed or well established, can be made more water efficient by implementing one or more of the seven steps. You do not have to totally redesign your landscape to save water. Significant water savings can be realized simply by modifying your watering schedule, learning how and when to water, using the most efficient watering methods and learning about the different water needs of plants in your landscape (Wade et al., 2002). In urban areas, about 25 percent of the water supply is used to water landscapes and gardens. In the summer, as much as 60 percent of the water the average household uses may be for landscape maintenance. Many traditional landscapes require large amounts of water, and much of this water isapplied inefficiently (Texas Agricultural Extension Service, 2003).

Benefits of Xeriscape

Saves Water
For most of North America, over 50% of residential water used is applied to landscape and lawns. Xeriscape can reduce landscape water use by 50 - 75%.

Less Maintenance
Aside from occasional pruning and weeding, maintenance is minimal. Watering requirements are low, and can be met with simple irrigation systems.

No Fertilizers or Pesticides
Using plants native to your area will eliminate the need for chemical supplements. Sufficient nutrients are provided by healthy organic soil.

Improves Property Value
A good Xeriscape can raise property values which more than offset the cost of installation. Protect your landscaping investment by drought-proofing it.

Pollution Free
Fossil fuel consumption from gas mowers is minimized or eliminated with minimal turf areas. Small turf areas can be maintained with a reel mower.

Provides Wildlife Habitat
Use of native plants, shrubs and trees offer a familiar and varied habitat for local wildlife.

2. Xeriscape principles

The seven water-saving principles of Xeriscape landscaping are not new; they have been practiced in the landscape industry for decades. Combining all seven into a comprehensive program of landscape water conservation is what makes Xeriscape landscaping unique. The principles are (Smith and Larson, 2003; Wade et al., 2002; Welsh, 1999; Welsh, 2000);

- Planning and design
- Soil analysis
- Practical turf areas
- Appropriate plant selection
- Efficient irrigation
- Use of mulches
- Appropriate maintenance

2.1 Planning and design

The first step in planning a water-efficient landscape is the process of the site analysis (Kelly, et al., 1991) One of the most important steps is to plan your landscape design. First assess the topography and determine drainage patterns. Examine your site conditions and pinpoint both shady and sunny areas. Decide whether any of the existing vegetation should be preserved. A base map is a plan of the property drawn to scale on graph paper showing the location of the house, its orientation to the sun, other structures on the site, unusual features such as stone outcroppings and existing vegetation (Wade et al., 2002).

To begin your plan, overlay the base map and site analysis sheet with another piece of tracing paper. On this sheet indicate the *public, private* and *service* areas of your landscape. Consider how these areas will be developed based on space requirements for each activity. The *public* area is the highly visible area that most visitors see, such as the entry to the home. In a traditional landscape, this area typically receives the most care, including the most water. Therefore, the careful design of this area is important for water conservation. This area can be designed to require minimal water and maintenance without sacrificing quality or appearance. The *private* area of the landscape, usually the backyard, is where most outdoor activity occurs. It is generally the family gathering area. It may also include a vegetable garden or fruit orchard. The landscape in this area needs to be functional, attractive and durable, but it also should be designed to require less water than the public area of the landscape. The *service* area is the working or utility area of the landscape, an area usually screened from view and containing such items as garbage cans, outdoor equipment, air-conditioning units or a doghouse. In terms of routine maintenance, this area would be designed to require the least care and water of the three areas. In addition to dividing the landscape into use areas, a Xeriscape plan further divides the landscape into three water-use zones: *high* (regular watering), *moderate* (occasional watering) and *low* (natural rainfall) (Wade et al., 2002). To incorporate Xeriscape concepts into your design, some additional thought is needed. The information you generate by drawing a plot plan and doing a site analysis should be integrated to identify microclimates in your yard. Microclimates are created by differing physical and environmental conditions within the landscape. Moisture, sun, shade, air movement, and heat all contribute to create zones that have varying water requirements (Welsh, 2000).

High water-use zones

Very low water zones are of two kinds. Decks and paved areas require no water. These areas help provide recreational and living space and are very practical. However, for paved areas, you should consider using permeable materials such as bricks or paving stones rather than concrete or asphalt to encourage rain to soak into the ground rather than run off. Protected areas where the exposure and shade conditions work together to inhibit evaporation are also very low water-use zones. In these areas, irrigation is needed only to establish new plants. Existing, well-established vegetation in these zones should be retained and new vegetation should be selected on the basis of minimal water use. Because very low water zones require little or no irrigation once they're established, they offer the greatest potential for saving water. Such shaded areas not only reduce water demand, they can also lower indoor temperatures and reduce summer cooling bills.

Low water-use zones

Low water zones are somewhat exposed areas that must be watered to keep plants flourishing but where water can be conserved by mulching and using an efficient low-volume irrigation system or by taking advantage of runoff from downspouts, driveways or patios.

Moderate water-use zones

Moderate water zones are exposed areas with turf or plants with higher water requirements. This zone should be kept small and should be limited to focal points, such as entrance areas, and functional areas, such as lawns. Identifying water-use zones in your yard helps you to group plants with similar water needs together for watering efficiency.

2.2 Soil analysis

A thorough analysis of both the physical and chemical characteristics of the soil is important when developing a water-wise landscape. Since plants with deep roots continue to have access to moisture after surface soil begins to dry out, a primary goal of Xeriscape is to encourage plants to develop deep root systems. In urban areas where the soil may be compacted, it will often be necessary to physically improve your soil before you can grow deep-rooted plants. Physical improvement of soil involves tilling to break up compaction and provide aeration and adding organic matter to keep soil porous. In addition, it may be necessary to chemically improve the soil with nutrients or other materials. Landscape architects emphasize that both kinds of soil improvements are important to developing healthy, deep roots, and that heavy fertilizing will not compensate for insufficient physical soil preparation. Before landscaping, take a sample of your soil to your local county Extension office for testing. Your county Extension agent will provide you with a recommendation for lime and fertilizer based on the analysis. The soil test report will give you information on pH, nutrients, volume weight, and humic matter as well as recommendations for correcting any deficiency the analysis reveals. Your goal in soil analysis is to create an ideal soil environment for the expanding root system. An ideal soil has good aeration and drainage, yet holds adequate moisture and nutrients for optimum root growth. If your soil is deficient in phosphorus, potassium, calcium, or magnesium, recommendations will be made for improvement. However, the lab analysis is not useful for sulfur, nitrogen, and boron. You may want to add a commercial fertilizer such as sulfate of ammonia or composted manure to supply both nitrogen and sulfur (Wade et al., 2002; Welsh, 2000).

2.3 Practical turf areas

Turfgrass is one of the most versatile and functional plants in the landscape. It provides one of the best recreational surfaces for outdoor activities. From a water management standpoint, turf is recognized as one of the most effective plant covers to reduce runoff and erosion while recharging the ground water, which results in more efficient use of rainfall (Wade et al., 2002). Along with minimizing turf perimeter, an important factor in conserving water in lawn areas is selecting a water-conserving, warm-season turfgrass species and cultivar. Warm-season species recommended for North Carolina are centipedegrass, zoysiagrass, and bermudagrass. Within each species are a number of cultivars with slightly different characteristics, including the transpiration rate or rate at which the grass gives up moisture to the air.Turf can help control erosion; it can contribute to temperature modification; it can reduce urban glare; and it can help control dust and mud. Turf is also useful for slowing runoff from landscape areas and can be of practical benefit in areas like swales. Grass is also functional in open recreational areas and can be maintained without heavy use of chemicals that have recently caused health concerns (Welsh, 2000). Use turf where it aesthetically highlights the house or buildings and where it has practical function, such as in play or recreation areas. Grouping turf areas can increase watering efficiency and significantly reduce evaporative and runoff losses. Select a type of grass that can withstand drought periods and become dormant during hot, dry seasons. Reducing or eliminating turf areas altogether further reduces water use (United States Environmental Protection Agency, 2002). Also consider the ease of watering turf areas. Areas that are long and narrow, small, or oddly shaped are difficult to water efficiently. Confine grass to blocky, squarish areas that are easier to maintain.

2.4 Appropriate plant selection

Appropriate plant selection means selecting plants that not only are compatible with the design but also are well suited to the planting site and local environment. It involves selecting plants according to the soil type and light level of the site. Ideally, the plants you select should be adaptable to local fluctuations in temperature and soil moisture. Most plants have a place in Xeriscape. It is important to use healthy plants adapted to our area (that is, plants that can take hot, humid weather as well as hot, dry weather), plant them in the right place, and give careful attention to getting them well established (Figure 1). Encouraging the growth of deep roots by preparing the soil and using appropriate irrigation practices is crucial to helping plants establish themselves. Select trees, shrubs and groundcovers that are adapted to your region's soil and climate (Wade et al., 2002).

Fig. 1. Spartium junceum L. (Deep roots) (Ganos Mountains, Tekirdağ, Turkey)

Native plants are not necessarily the most drought tolerant. Even though a plant may be native to the area, it may not adapt to an adverse new environment (microclimate). When forced to grow in a harsh new environment, native plants can become a high-maintenance nightmare. In addition to the adaptability of a plant to the site, other important criteria to consider include (Florida's Water Manegement Districts, 2004; Wade et al., 2002)

Mature size and form (height and width)

Will the plant remain in scale with the rest of the landscape as it matures, or will it likely overgrow the site and compete with other plants for space, nutrients and water?

Growth rate (Sun and shade requirements, soil needs, water needs, sat and cold tolerances)
Slow-growing dwarf shrubs and ground covers used around the base of the home require little routine pruning.

Texture
Is the leaf texture fine, medium or coarse, and does it combine well with the adjacent plants?

Color

Is the flower or foliage color compatible with other plants or the background color of the building? (Figure 2)

Functional use

Is the plant suitable for the location and intended purpose; i.e. under low windows, along the perimeter of the property as screening hedge, or as a ground cover?

Choose plants that can survive on normal rainfall in your area or that require minimal irrigation. Existing native-plant communities are an example of the "right plant in the right place." Match these factors with your soil and climatic conditions.

Fig. 2. Alkanna tinctoria TAUSCH. (example of many slope area) (Ganos Mountains, Tekirdağ,Turkey)

2.5 Efficient irrigation

A water-wise landscape requires a minimal amount of supplemental water from irrigation. When irrigation is used, water is applied efficiently and effectively to make every drop count (Wade et al., 2002). Irrigating lawns, gardens, and landscapes can be accomplished either manually or with an automatic irrigation system. Manual watering with a hand-held hose tends to be the most water-efficient method. Using irrigation water efficiently also requires us to select the appropriate type of irrigation for the plants and for each area of the landscape. Trees and shrubs in the low water-use zone would need supplemental water only during establishment (first 8 to 10 weeks after transplanting); plants in moderate water-use zones require water only during periods of limited rainfall when they show signs of stress. For these plants, a temporary system such as a soaker hose or hand watering may be all that is required. On the other hand, high water-use zones require frequent watering and may warrant a permanent system with automatic controls. Whenever possible, use highly efficient watering techniques, such as drip irrigation (Wade et al., 2002).

2.6 Use mulches

Mulching is one of the most beneficial landscape practices. Mulches conserve moisture by preventing evaporative water loss from the soil surface and reducing the need for supplemental irrigation during periods of limited rainfall. By maintaining an even moisture supply in the soil, mulches prevent fluctuations in soil moisture that can damage roots. Placing a layer of mulch directlyaround shrubs and trees and on flower beds helps to conserve water. In fact, mulch

- Helps retain moisture in the soil
- Decomposes slowly, adding nutrients to the soil
- Provides habitat or cover for beneficial soil organisms
- Shades soil from the baking sun, reducing the need for water
- Protects against soil erosion and compaction caused by rain
- Reduces weed growth
- Reduces maintenance chores; keeps lawn mowers and weed trimmers from damaging trees and other plants
- Looks good in the landscape

2.7 Appropriate maintenance

The objective of Xeriscape maintenance is to discourage water-demanding new growth on plants. In other words, keep plants healthy, but do not encourage growth at all times. Depending on your current level of maintenance, this may require you to fertilize less often with less fertilizer, to prune only when necessary and lightly when essential and, of course, to irrigate less. Remember, a Xeriscape-type landscape is a low-maintenance landscape. By working smarter, not harder, in the landscape, you'll save time, energy and water without sacrificing the beauty of the environment. Proper watering, weeding and pruning, mowing, and limited fertilization and pest control will keep your Xeriscape healthy and beautiful. Mow your turf grass high (maximum height of one inch for Bermudagrass and two inches for others) and often and leave the short clippings to decompose and replace nitrogen in the soil. Every time you cut your grass, you weaken the root system to some degree, and the more you cut the top growth, the more you restrict root system development. When you remove more than 40 percent of the top growth, the roots stop growing. By mowing high you encourage the development of a deep root system, which is a key to drought tolerance and weed resistance. Higher grass also shades the soil more, acting as a living mulch.

3. References

Florida's Water Manegement Districts, 2004. Water Efficient Landscaping: PreventingPollution&Using Resources Wisely, EPA Water Resources Center, 15 p.

Kelly, J., Haque, Mary., Shuping, D., Zahner, J. 1991. Xeriscape: Landscape Water Conservation in the Southeast, Clemson University, 37 p.

Smith, C.R and Larson, R. 2003. Xeriscape Plant Selections and Ideas, North Dakota University, USA.

Texas Agricultural Extension Service, 2003. Xersicape: Landscape Water Conservaiton, The Texas A&M University System, 16p., USA.

United States Environmental Protection Agency, 2002. Water-Efficient Landscaping: Preventing Pollution&:Using Resources Wisely, 15p.,sUSA.

Wade, L., James, T., Coder K. D., Landry G. and Tyson, A. W. 2002. A guide to developing a water-wise landscape, University of Georgia Environmental Landscape Design Department, Georgia.
Welsh, D. 2000. Xeriscape: North Carolina, National Zeriscape Council, 28p, USA.